AF527894

Conservation of Natural Resources for Food and Environmental Security

Conservation of Natural Resources for Food and Environmental Security

Editors :

S.K. Dubey, R.K. Dubey
A.K. Singh, P.K. Panda
S. Kala & V.N. Sharda

SATISH SERIAL PUBLISHING HOUSE
403, Express Tower, Commercial Complex, Azadpur, Delhi-110033 (India)
Phone : 011-27672852, Fax : 91-11-27672046
E-mail :info@satishserial.com, hkjain1975@yahoo.com
Website : www.satishserial.com

Published by :

SATISH SERIAL PUBLISHING HOUSE
403, Express Tower, Commercial Complex, Azadpur, Delhi-110033 (INDIA)
Phone : 011-27672852 Fax : 91-11-27672046
E-mail : info@satishserial.com, hkjain1975@yahoo.com

ISBN 978-93-81226-15-5

Typeset at: Harminder Singh Kharb *for* Laxmi Art Creation
Printed at: Salasar Imaging Systems, Delhi-110035

Foreword

Over exploitation of natural resources in India to meet the rising needs of burgeoning human and animal population has resulted in severe land degradation problems, diminishing water resources, declining forest cover, loss of biodiversity, pollution of water bodies and changes in climatic patterns leading to concerns on the sustainability of agro-ecosystems. It is now time to adopt strategies like maximization of *in situ* conservation of rainwater and protecting our land resources from menace of water erosion and *in situ* soil degradation like acidity, salinity etc. in order to save and augment the underground water and maintain our rich heritage of biodiversity and soil health through enhancing soil carbon sink through recycling and reuse of organic and industrial wastes, crop diversifications, agro-forestry and use of *in situ* raised and outside available mulch materials and bio-fertilizers.

The States of Uttar Pradesh, Madhya Pradesh, Bihar, Jharkhand and Chhattisgarh constitute the heartland of India's food basket and endowed with vast biodiversity, rich state of natural resources mainly the fertile and productive soils, good annual rainfall, perennial rivers, underground water and crop diversity. Indiscriminate exploitation of soils, underground water and use of agro-chemicals have resulted in gradual but wide spread stagnation and decline in agricultural productivity and depleting water table imposing a great threat to the future sustainable agriculture and food security.

Indian Association of Soil and Water Conservationists (IASWC), Dehradun organized a three day conference, "Food and Environmental Security through Resource Conservation in Central India: Challenges and Opportunities (FESCO)" in September, 2009 in association with CSWCRTI Research Centre, Agra and the Department of Land Development and

Water Resources, Government of Uttar Pradesh and came out with salient useful recommendations. It is hoped that the compilation will prove useful to the researchers, teachers, students, development agencies and all stakeholders for understanding the critical natural resource management issues.

I congratulate the editors for bringing out vital information in form of a book.

(S. AYYAPPAN)
Secretary (DARE) & Director General (ICAR),
Govt. of India, Ministry of Agriculture,
Krishi Bhawan, Dr. Rajendra Pd. Road,
New Delhi - 110 114

Preface

India has 146.02 million ha of degraded land with severe menace of water erosion on 93.7 million ha area alone. About 44% of total degraded lands have erosion rate more than the permissible limit of 10 t ha^{-1} $year^{-1}$. Soil erosion may decline agricultural productivity from 5 to 50% depending upon severity that ultimately depends on intensity of rains; soil erodibility and depth; length and steepness of slope; ground cover; crop water erosion resistance potential; crop management practices and use of soil and water conservation practices. Crop yield loss due to soil erosion has been estimated to be in the range of 1.7 to 4.1% in India with average annual loss of 7.2 million tons of food grains and loss of 5.8-8.4 million tonnes of nutrients every year (nitrogen: 0.8 MT, phosphorus: 1.8 MT and potassium: 2.63 MT). Land degradation leads to loss or deterioration of good soil, decline in soil water holding potential, stagnation or slow agricultural growth, decline in nutrient and water use efficiencies and ground water recharge, change of vegetation to less preferred tree and grass species, reduction in potential of land to grow plants and threats to sustainable future agriculture and food security.

Over exploitation of natural resources due to intensive agriculture; improper crop rotations; shifting cultivation; unscientific, arbitrary and non sustainable land management practices; excessive irrigation and salinity in unlined canal command areas; imbalances in fertilizer use and soil nutrient depletions; *in situ* land degradation due to bio-physical and chemical factors; fragmentation and reduction in size of land holdings; excessive pumping of underground water; over grazing and grazing of forests; deforestation beyond permitted limits; forest fires; open cast mining and extension of agriculture to marginal, sloping and steep lands for meeting out ever increasing food, fodder, fuel, fibre, timber and domestic demand in the country. This has resulted in to large scale land degradation, loss of biodiversity, shrinking forests and ground cover and pollution of water bodies in the Central India comprising of Uttar

Pradesh, Bihar, Madhya Pradesh, Chhattisgarh and Jharkhand (85.88 million ha area). The socio-economic constraints mainly the poverty and increasing land less labours; rapid rate of development and industrialization and declining per capita land and water availability as well as per capita income (Rs 11880 against national average of Rs 25825) has occurred mainly due to high human (35.73 million or 34.73%) and livestock (150 million or 31%) population which acted as a direct cause of land degradation. Besides, natural resource rich Central India has poor forest cover (in Uttar Pradesh and Bihar about 6% only), dominance of crop sector in total agricultural economy (crops: 72.3%, livestock: 22.2%, forests: 3.4% and fishing: 2.13%); distinct variation in total annual rainfall (842.3 mm in western Uttar Pradesh to 1394 mm in Chhattisgarh); diversified soils considerably varying in their erodibility and nature of degradation (inceptisols: 49.25%, entisols: 20.82%, alfisols:17.6% and vertisols: 10.7%). The government has launched many projects to reclaim degraded lands and enhance ground cover. Famous among them are Planning Commission's Policy to Promote Agro-forestry and Uttar Pradesh Land Reclamation Project. The situation recalls for development of technologically sound, practically feasible and viable options to control water erosion, improve the soil health and maximize *in situ* and surface rain water conservation.

Central India has good utilizable flow of perennial rivers, Ganga and Narmada (284.5 Km^2/ year) and is beset with three distinct physiographic regions namely the Vindhyan and Satpura Ranges and Narmada Valley; Indo-gangetic Plains and Malwa and Chhotanagpur Plateaus. Vindhyan and Satpura Ranges and Narmada Valley spread over a total length of 1050 Km with the highest elevation of 3000 m and major rivers like Narmada and Ganga and their tributaries, Kali, Sindh, Parbati, Betwa and Ken. This region is characterized by the arid and inhospitable climate; high biotic pressure; heavy grazing; heavy runoff and soil loss; ravine erosion and formation of gullies; nutrient imbalances and frequent deficiencies of N, P and Zn; acute droughts; cracking and swelling of black soils; degradation of tropical deciduous forests on sloping lands; soil depth limitations and coal mine spoils. Indo-Gangetic plain has big rivers, Ganga, Yamuna, Gomti, Ghaghra and Chambal, but is beset with several production constraints mainly large area under ravines (2.51 million ha), salinity and water logging; depleting water table; nutrient imbalances; cereal-cereal crop rotations; decline or stagnation in crop productivity; saline ground water in semi arid tracts; socio-economic variations and poverty and rainfall aberrations. Malwa and Chhotanagpur Plateaus has average elevation of 500 m with generally north facing orientation of topography draining into river Chambal and its tributaries.

Major production constraint in the region comprise of salinity and alkalinity, imperfect drainage in low lying areas and intermittent dry spells.

Central Soil and Water Conservation Research and Training Institute, Dehradun with eight regional centres has developed technological solutions for sustained higher productivity on sloping agricultural lands; rehabilitation of ravines and mine spoils; alternate options to the shifting cultivation; region specific eco-restoration models, rain water management and wasteland development through participatory watershed development projects. The Research Centre, Agra has developed many watershed models in semi arid India, famous among which are Sheetalpur Watershed, Hamirpur; Budhiya ka Tal and Agra watershed. Besides, CSWCRTI, Research Centre, Agra and Datia, Central India has 19 institutes of ICAR, 11 State Agricultural Universities and many NGOs in addition to line departments engaged in efficient utilization and management of natural resources. For in depth discussion and analysis of region specific Natural Resource Management (NRM) issues and problems, the Indian Association of Soil and Water Conservationist decided to hold a conference, Food and Environmental Security through Resource Conservation in Central India: Challenges and Opportunities (FESCO)" at Agra to address the current NRM issues in order to sustain and enhance agricultural productivity for ensuring food security in the states of the Central India.

The conference brought together more than 168 delegates comprising of policy makers, planners, administrators, scientists, researchers, developmental agencies, NGOs and stake holders on a common platform in order to share their rich experiences in solution of major NRM related issues. In all 163 abstracts of papers including 10 key note addresses from renowned experts and poster presentations made were presented in the conference under seven major theme areas.

1. Status of natural resource.
2. Management and efficient utilization of water resources.
3. Conservation technologies for food and environmental security.
4. Livelihood security and watershed management.
5. Rehabilitation of problem soils.
6. Governance and policy issues for community empowerment and
7. Regional experiences and future thrust.

All papers were thoroughly screened for their value, content and relevance to current NRM issues. The 43 chapters of this book were selected following critical appraisal of content, through review, editing and compilation of the selected full length papers in order to address key

NRM issues for sustainable agriculture, food security and protection of the environment in the states of the Central India. We hope that the book will be of great value to the scientists, researcher, planners, line departments, NGOs and all other agencies working in the field of natural resource management.

Editors

Acknowledgement

The final compilation and publication of this book is the outcome of the collective efforts of many organizations and individuals. Holding conference, Food and Environmental Security through Resource Conservation in Central India: Challenges and Opportunities (FESCO)" at Agra during September, 16-19, 2009; was an outcome of funds received from the sponsors and cosponsors. We are indebted to the IASWC and other organizations that funded to hold FESCO-2009. IASWC was the main sponsor while major co-sponsors comprised of Department of Land development and water Resources, Government of Uttar Pradesh; Department of Land Development, Ministry of Rural Development, Government of India; National Rain-fed Area Authority and National Bank of Agriculture and Rural Development. We are thankful to all sponsors who provided a platform for in depth and detailed discussions on major natural resource management issues in Central India as such and its different regions for sustainable agriculture and food security. We are grateful to then the Director General, Indian Council of Agricultural Research (ICAR), Dr Mangla Rai for his constant encouragements in organizing this conference. We also express our deep gratitude to Dr S. Ayappan, Director General, ICAR, New Delhi for his encouragements in publication of this book.

We want to express our deep gratitude to Dr A.K. Singh, Deputy Director General (Natural Resource Management), ICAR, New Delhi; Shri Yogesh Kumar, IAS & Principal Secretary, Department of Land and Water Resources, Government of Uttar Pradesh; Dr J.S. Samra, Chief Executive Officer, National Rain-fed Area Authority; Shri S.S. Chauhan, National Bank for Agricultural and Rural Development and Dr A. Raizada, Secretary, IASWC who all the way encouraged and participated in this conference through funding, personal participation and addressing the conference. We feel extremely happy to extend our sincere thanks to the cabinet minister, Honorable Shri Ashok Kumar Dohrey, Department of Land Development and Water Resources, Government of Uttar Pradesh who acted as the chair person and to the Honorable Shri Ramkrishna Kushumaria, Cabinet Minister, Government of Madhya Pradesh who acted as the Chief Guest in FESCO-2009. We are also thankful to the Ministers of the Government of Uttar Pradesh mainly Shri Fateh Bahadur Singh, Shri Narayan Singh and Shri Nakul Dubey; Ministers of the Government of Madhya Pradesh Shri Rajendra Shukla and Shri Gopal Bhargava who send their immemorial message for successful organization of the conference and achieving research targets in big interests of natural resource management in the Central India. We specifically thank the chairman and members of different committees constituted for successful organization of FESCO-2009.

We take the opportunity to thank all the officers and staff of CSWCRTI, Research Centre, Agra who took all the pains in the successful organization of FESCO-2009. Special thanks are due to Shri B.P. Joshi, T5 and Shri J.S. Deshwal, T5 of CSWCRTI, Research Centre, Agra who all the way worked hard to compile and publish abstracts and souvenir in FESCO-2009. We sincerely thank to all the 168 participants of the conference who nicely presented their papers and made a meaningful and memorable interaction. We the editors hope that the selected papers presented in this proceeding comprehensively depict diverse variety of major NRM issues that needs to be attended for sustainable agriculture and food security in states of the Central India.

Editors

Contents

Section-5 : Rehabilitation of Problem Lands, Salins, Alkali, Ravines, Riveine Lands, Diava Land etc.

Section-6 : Governance and Policy Issues for Community Empowerment

Section - 1

Status of Natural Resources

Chapter 1

Status of Natural Resources for Food and Environmental Security in Central India

V. N. SHARDA*

*Member, Agricultural Scientist Recruitment Board, New Delhi-110012

Central India comprising the states of Uttar Pradesh (U.P.), Madhya Pradesh (M.P.), Bihar, Jharkhand and Chhattisgarh (Fig.1) plays a vital role both socio-culturally and economically in many respects. This region is unique and characterized with diverse natural resources (physiographic, soil diversity, bio-diversity, water resources and climatological parameters like rainfall, bright sunshine hours, storms and fog etc.), nature of human resources (majority of tribal population) and the socio-economic aspects (low level of literacy, low per capita income etc.). The region can be classified into two broad ecosystems such as flood prone areas and dry land areas. The flood prone areas covering parts of Bihar, eastern Uttar Pradesh, parts of Madhya Pradesh and Chhattisgarh throw different type of challenges where as the low rainfall and dryland areas of Agra and Bundelkhand region, the hilly tracts of Madhya Pradesh, Jharkhand and

Chhattisgarh pose another set of problems to be addressed in a holistic manner. Hence, it is imperative to formulate befitting strategies to efficiently conserve and utilize natural resources to ensure livelihood security of the rural and tribal people through their economic upliftment.

1. Demographic features

The demographic features which significantly affect the development and efficient utilization of natural resources include human and livestock populations and socio economic status of the people.

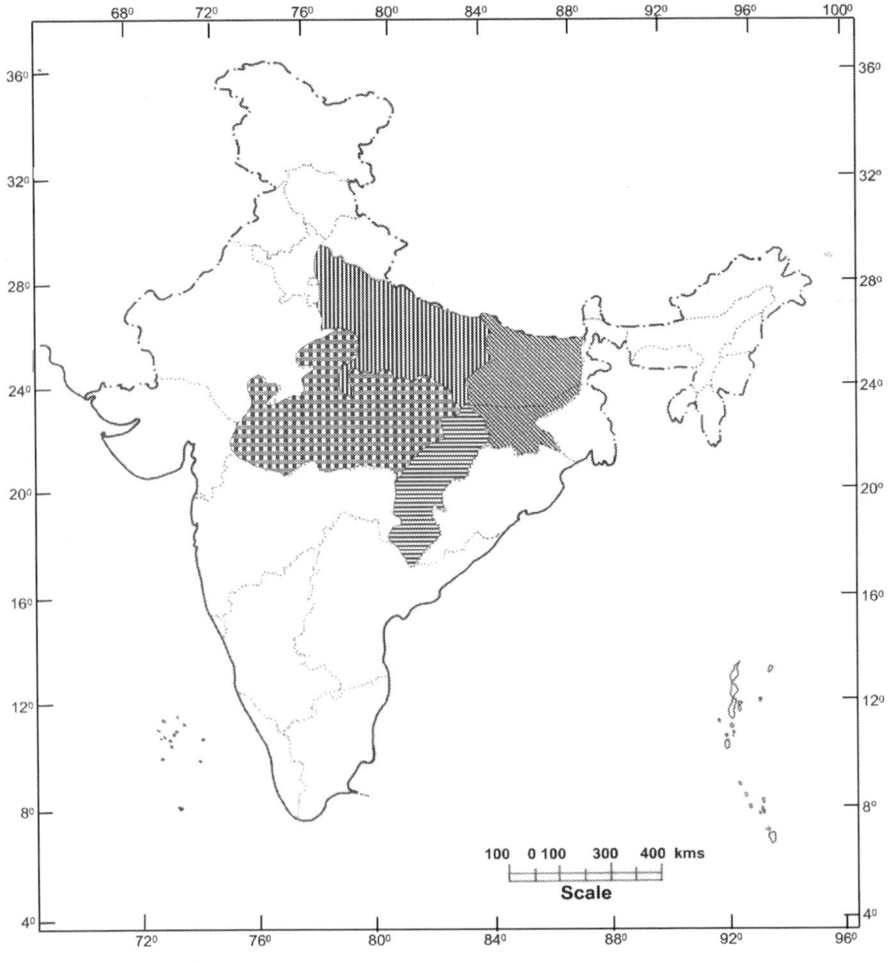

Fig. 1: The map showing five states of Central India

Demography and Agricultural Economy

The central region, spreading over five states, covers an area of 85.88 m ha, equivalent to around 26 per cent of the total geographic area of the country (Table 1). With an average density of 452 people per sq. km, it supports about 1/3rd (358 million) population of the country against a population density of 325 of the country as a whole. It is mainly attributed to about 2 and 3 times higher population density in the states of U.P. and Bihar, respectively, compared to the national average. The state of Chhattisgarh, a newly carved state out of M.P. has the lowest population density of 154 persons per sq km followed by M.P. and Jharkhand. On an average, 80 per cent of population lives in rural areas with highest percentage (90 per cent) in Bihar. Literacy rate in the region is 57 per cent which is lower than the national average of 65 percent with highest rate in Chhattisgarh while the state of Bihar has the lowest literacy rate of 47 per cent.

Table 1: Demographic features of various states in Central region

S No.	States	Area (sq. km.)	Population (no.)	Density (Person per sq. km.)	Rural population (%)	Literacy (%)	Per capita income at current prices (Rs.)
1.	Bihar	94163	82998509	881	89.54	47.00	5772
2.	Chhattisgarh	136034	20833803	154	79.91	64.66	15073
3.	Jharkhand	79714	26945829	338	77.76	53.56	13013
4.	Madhya Pradesh	308000	60348023	196	73.54	63.74	14069
5.	Uttar Pradesh	240928	166197921	690	79.22	56.27	11477
	Total	**858839**	**357324085**	**452**	**79.99**	**57.05**	**11880**
	India	**3287263**	**1028737436**	**324**	**72.17**	**64.84**	**25825**

Source: Census of India, 2001

Wide variations have also been observed in the pattern of population growth among the states (Table 2). It is evident from the table that as per Census year 1971, Uttar Pradesh and Bihar were the most populous states in the region with population density of 324 and 300, respectively. As per Census 2001, though U.P. and Bihar continued to be the most populous states in the region, the population density has increased by almost two and three times, respectively.

The composition of agricultural economy of different States in the region is presented in Table 3. The contribution of crop sub-sector to the agricultural economy in the region has decreased from 74per cent in 1981 to 72per cent in 2002 whereas for livestock sub-sector, it has increased from 19per cent to 22per cent and that of fisheries from 0.7per cent to 2.1per cent. Significant differences have been noted over the years amongst the states regarding magnitudes of contributions of various sub-sectors. The contribution of crops is highest in all states followed by livestock, forestry and fishery sectors. The contribution of fisheries sub-sector in the state of Bihar has increased significantly over the last two decades. These trends must be internalised while allocating resources and setting priorities for research and technology development programmes in different states of the region.

Table 2 : Population ('000) dynamic of the region

State	1971	1981	1991	2001
Bihar (Undivided)	56353 (324)	69915 (402)	86374 (497)	82878 (880)
Madhya Pradesh (Undivided)	41654 (94)	52179 (118)	44977 (149)	60385 (196)
Uttar Pradesh	88341 (300)	110862 (377)	139112 (473)	166052 (689)
India	548 (173)	683 (216)	846 (274)	1027 (324)

Figures in parenthesis are population density (persons per sq km).
Source: Census of India, 2001

Table 3 : Composition of agricultural economy in the region (per cent)

State	Crop sector			Livestock			Forestry			Fishing		
	1981	1991	2002	1981	1991	2002	1981	1991	2002	1981	1991	2002
Bihar	73.2	63.8	68.8	21.3	31.1	24.2	3.8	3.3	2.4	1.7	1.7	4.7
M.P.	70.3	72.5	75.2	18.3	24.9	19.8	11.3	2.3	4.6	0.1	0.2	0.4
U.P.	78.9	75.8	73.1	18.3	23.3	22.5	2.5	0.5	3.2	0.3	0.4	1.3
	74.13	**70.7**	**72.3**	**19.3**	**26.4**	**22.2**	**5.8**	**2.03**	**3.4**	**0.7**	**0.76**	**2.13**

Socio-economic status

Both naturally and culturally, Central India is one of the richest region of the country with a very strong heritage and cultural value system. However, in spite of being blessed with vast natural resources, the states in the central region remained socio-economically poor and could not attain the desired degree of growth rate mainly due to inefficient and inequitable utilization and development of available resources. The per capita income is only Rs. 11,880, which is about 46 per cent of the national average thus indicating poor economic status of the region. All the five states in the region have per capita income lower than the national average with state of Bihar having the least per capita income of Rs.5, 772 which is just one fifth of the national average.

Livestock

The central region has over 150 million livestock constituting 31 per cent of total livestock of the country (Table 4). Uttar Pradesh dominates in the livestock population followed by M.P., Bihar, Jharkhand and Chhattisgarh. Maximum population is of cattle (64.7 million) followed by buffaloes (39 million), goat (38 million) and sheep (3 million). Milk production (27 million tonne) in the region is around 30 per cent of total production in the country of which U.P. alone contributes 18 per cent (16.5 million tonnes). Similarly, 12 per cent of poultry in the region produces only 9.23 per cent of eggs. Hence, the region has enormous potential to increase milk production through scientific techniques, improved breeds and management of pastures and common property resources for quality fodder production.

2. Resource Base

The Central India is endowed with rich and diverse natural resource base in terms of climate, soil, water resources, vegetation etc. However, the resources need to be judiciously managed to meet the growing demands of human and animal populations and bring about economic prosperity in the region with equitable growth and development.

Natural vegetation

In Central India, tropical and sub-tropical forests comprising of multipurpose tree species, fruit plants, shrubs, medicinal herbs, bamboos and canes, wild life and other important flora and fauna is a source of livelihood for a large number of people especially in the tribal

Table 4: Livestock and poultry population (000 no.) and milk (000 tonne) and egg (crore no.) production

State	Cattle	Buffaloes	Sheep	Goat	Horse and Ponies	Other livestock	Total livestock	Milk production*	Poultry	Egg production*
Bihar	10729	5743	382	9490	117	701	27162	2974	13911	78.9
Chhattisgarh	8882	1598	121	2336	4	553	13493	831	8181	88.0
Jharkhand	7659	1343	680	5031	5	1108	15826	1330	14429	69.8
Madhya Pradesh	18913	7575	546	8142	32	409	35617	5506	11705	90.2
Uttar Pradesh	18551	22914	1437	12941	154	2535	58531	16512	11718	90.2
Total	**64734**	**39173**	**3166**	**37940**	**312**	**5306**	**150631**	**27153**	**59944**	**417.1**
India	**185181**	**97922**	**61469**	**124358**	**751**	**15320**	**484321**	**90733**	**489012**	**4520.1**

Source: www.indiastat.com
Livestock Census, 2003, Dept. of Animal Husbandary, Dairying and Fisheries, Min. of Agri., Govt. of India.
* 2004-05

areas. Besides generating revenue and employment, the forests have a critical bearing on water resources in the region. Approximately 17.5 m ha area in the region is under forest constituting about 25 per cent forest area of the country (Table 5). The total forest area in the region is 20.33 per cent of the total geographical area which is almost at par with the national average of 20.64 per cent (Table 5). The state of Chhattisgarh has maximum percentage of forest area and alone contributes around 32 per cent of the total forest area in the region. The state of Chhattisgarh, Jharkhand and M.P. have high percentage of forest cover in the region, more than the national average. However, the states of U.P. and Bihar have only about 6 percent of the total geographic area of these states under forest cover which is mainly attributed to high population density leading to over-exploitation of vegetal cover. Hence, vigorous efforts are needed to enhance the forest cover to be at least on par with the national average.

Table 5 : Area under forest (sq km) in different states

States	Geo-graphical area	Total forest	Dense forest	Open forest	Mangrove	Scrub	Non forest
Bihar	94163	5558	3027	2531	-	150	88455
Chhattishgarh	135191	55998	38980	17018	-	88	79105
Jharkhand	79714	22716	11681	11035	-	807	56191
Madhya Pradesh	308245	76429	41843	34586	-	2378	229438
Uttar Pradesh	240928	14118	5996	8122	-	749	226061
Total	**858241**	**174819**	**101527**	**73292**		**4172**	**679250**
India	**3287263**	**678333**	**390564**	**287769**	**4461**	**40269**	**2568661**

Source: Statistical Abstracts India (2005-06)

Soil resources

Traditionally, soil resources of the region are divided into four broad types; viz; alluvial, black, red and laterite. Taxonomically, these soils are classified into five soil orders (Table 6). Majority of the soils of Central India belong to orders of Inceptisols (49.25per cent) and Entisols (20.82per cent) followed by Alfisols (17.6per cent) and Vertisols (10.7per cent). In U.P., nearly 70.1 per cent area is under Inceptisols followed by Entisols (18.9 per cent), Alfisols (4.9per cent), Vertisols (1.6per cent) and Mollisols (0.2per cent). In Bihar, Inceptisols cover 34.9 per cent of the total geographical area followed by Alfisols (34.5), Entisols (27.3) and Vertisols (0.6). In M.P., the Inceptisols are the dominant soils, covering 40.5 per cent; Entisols and Alfisols 19.1 per cent each, Vertisols 20.7 per cent and Mollisols occupy only 0.6 per cent of the total geographical area. Bihar and eastern U.P. have large areas under *Tal, Diara and Chaurs* lands. The soils of *Diara* land occurring in alluvial and flood plains of U.P. and Bihar are young and stratified with AC profiles and have very faint and poor pedogenic manifestations, which are classified as Typic Ustifluvents. A sizable population lives in this area and is fully dependent on agriculture. The *Tal* land in Bihar extends from Fatuah to Lakhisarai over an area of about 1034 sq. km with a length of about 105 km. The topography of *Tal* land is such that rain water from the area gets accumulated up to a depth of about 5 metres. Since the area is low lying and doesn't have drainage facilities, it remains inundated for a period of 3 to 4 months in a year, covering almost the entire *kharif* season. The *Diara* lands in Bihar constitute an area of 11.59 lakh hectare. The topography of this land is undulating with confused patterns of upland and low land. The *Chaurs* is a large tract of land in North Bihar which remains water logged for a variable period and represents a fragile and unstable eco-system. About 4 lakh hectares is *chaur* land with varying depths of water during different periods of the year. The red, black and yellow coloured soils of Rajmahal traps of Bihar have been classified under red soils as Alfisols, black soils as Inceptisols, Vertisols and Entisols and yellow soils as Inceptisols. The land surface being uneven, particularly in Jharkhand, U.P. and M.P., is subjected to rill and gully erosion causing loss of soil and plant nutrients. High rainfall and longer fallow season results in high runoff and soil loss leading to severe land degradation problems as well as groundwater table depletion every year in Bundelkhand region of U.P. and M.P. Most of alluvial plains present in Bihar and eastern U.P., fall under very good lands while most of the areas in Jharkhand and Chattisgarh have moderately good lands. Lands with severe limitations for crop growth are found in western parts of Jharkhand, eastern parts of Chhattisgarh and Bundelkhand region. Besides, intensive agriculture practiced with high use of selective agro-chemicals has become unsustainable due to imbalances in soil fertility, multi-nutrient deficiencies and deterioration of land quality and soil health. Declining trends in productivity as recently been observed in many existing cropping systems have resulted in declining fertilizer use efficiency and organic matter levels in the soils. The most critical region specific constraints which are responsible for deceleration in productivity are presented in Table 7.

Table 6 : Soil Resource of the Region ('000 ha)

Soil order	Bihar	M.P.	U.P.	Total	%	Soil related constraints
Alfisol	5992	8500	1557	16049	17.6	Weak soil structure, crusting, compaction and erosion by water
Entisol	4741.3	8461.1	5778.8	18987.2	20.8	Erosion, nutrient depletion, low organic matter
Inceptisol	6060	17915.3	20908.3	44883.6	49.2	Erosion, nutrient imbalance and low organic matter
Vertisol	106.8	9189.6	479.1	9775.5	10.7	Massive structure, poor tilth, moisture stress and water erosion
Mollisol	-	260.2	103.3	363.5	0.4	Nutrient imbalance and depletion of soil organic matter
Rockout	18.7	12.7	412.9	444.3	0.5	---
Misc.	469.2	-	202.2	671.4	0.7	---
TGA	**17388**	**44344.9**	**29441.6**	**91124.5**		

Table 7 : Regions Specific Factors for deceleration in productivity

Region/State	Principal Crops	Region Specific Constraints
Bihar and Eastern U.P.	Rice, wheat, maize, jute, gram, peas, sugarcane	Flood/water logging, improper drainage, salinity/alkalinity, arsenic contamination
Western U.P.	Sugarcane, maize, wheat, rice, potato	Ground water depletion, decreasing total factor productivity, micronutrient deficiency

Jharkhand and Chattisgarh	Rice, wheat, soybean, maize, ragi, pigeonpea, potato	Moisture stress, drought and soil acidity, iron toxicity
Bundelkhand (U.P. and MP)	Jowar, wheat, gram, rice	Drought, moisture stress, soil crusting and cracking, soil salinity/alkalinity

Principal crops of the region

Due to high rainfall and waterlogged conditions, rice is the predominant *kharif* crop of the entire region occupying nearly 70-90per cent area under both irrigation and rainfed situations. At some places, pulses and oilseeds are also grown during *kharif* season. Wheat, potato, sugarcane, pulses and oilseeds are the major *rabi* crops while jute is also grown as pre-monsoon crop. The details of principal crops in different states are presented in Table 7. Major crops grown in black Vertisol region are soybean, sorghum, maize, rice, pigeonpea, wheat and chickpea whereas in Inceptisol soils of western U.P., sugarcane, maize, wheat, potato and oilseeds are mainly grown. The area and productivity trends of major food grain crops of the region are exhibited in Table 8. The analysis of data revealed that in spite of increase in irrigated area, there is a declining trend in productivity of major food grain crops in major part of the area (27.42 M ha) covering Bihar and U.P. The poor yield of crops in Jharkhand and Chhattisgarh is attributed to adoption of traditional crops and varieties under rainfed conditions. Agricultural development is much lower than its potential in the region, thus limiting employment in the agriculture sector and causing a large section of the population to remain below the poverty line and suffer from malnutrition. For example, Bihar occupied about 3per cent of the total cultivated area and 8per cent population of the country to produce about 6.9per cent (14.56 M tonnes) of the total food grains in 1999-2000. The average yield of food grains in the state was 1620 kg/ha as against the national average of 1697 kg/ha. The yield of rice in Bihar was 15.40 q/ha as against national average of 19.90 q/ha. Similarly, the yield of wheat was 20.61 q/ha as against the national average of 27.55 q/ha. Enhancing productivity of these areas is possible by adopting appropriate IPNM and intercropping systems. In spite of low cropping intensity (120 to 130per cent), high rainfall and longer fallow season result in high runoff and soil losses leading to severe land degradation problems. Integrated farming systems in rice-based rainfed ecosystem in the region through introduction of fish, duck and pig culture have tremendous potential in the region by taking into consideration the socio-economic background of the farming community.

Table 8 : Area and productivity of food grain crops of the region during 2000-2007

State	2000-2001			2006-07		
	Area (M ha)	Productivity (Kg/ha)	% coverage of irrigation	Area (M/ha)	Productivity (Kg/ha)	% coverage of irrigation
Bihar	7.12	1694	49.5	6.70	1656	59.3
Jharkhand	1.84	1095	-	2.38	1550	4.3
U.P.	20.30	2105	64.9	20.04	2057	71.6
M.P.	10.78	945	32.8	11.78	1167	39.7
Chattisgarh	4.93	589	-	5.06	1148	24.3
Total	**44.97**			**45.96**		

Water resources

The Central region of the country is covered by five major river systems, viz; Ganga, Godavari, Cauvery, Narmada and Mahanadi. Surface water potential and utilisable surface water resources of these river systems are given in Table 9. The Ganga river system alone contributes a maximum utilisable flow of 250 km^3/year in the region out of 690 km^3/year available in the country. The Godavari, Cauvery, Narmada and Mahanadi put together have an utilisable surface water availability of around 180 km^3/year. The region has a vast potential of ground water resources. The annual replenishable ground water resource is around 163 billion cubic meter per year out of which around 152 billion cubic meter is the net ground water availability per year (Table 10). Yet, the ground water development in the region is very poor except in the state of Uttar Pradesh (70per cent). In other states, ground water development is even less than 50per cent of the net available quantity.

The Central region receives good amount of rainfall. Long term average shows that Bihar, Chhattisgarh, Jharkhand and eastern Madhya Pradesh receive more than 1000 mm of annual rainfall whereas in U.P. and western M.P., the annual rainfall is less than

1000 mm (Table 11). Despite good average annual rainfall, the region experiences frequent droughts due to erratic and uneven distribution of rainfall.

Table 9 : River basin wise surface water resources potential and utilisable surface water resource in km³/year

River Basin	Average annual flow	Utilisable flow
Ganga	525.02	250
Godavari	110.54	76.3
Cauvery	21.36	19
Mahanadi	66.88	49.99
Narmada	45.64	34.5

Source: Integrated water resources development – A plan for action, Report of The National Commission for Integrated Water Resources Development, Ministry of Water Resources, New Delhi, 1999.

Table 10: Ground Water Scenario

S. No.	States	Dynamic Resources			
		Annual replenishable ground water resource (BCM)	Net annual ground water availability (BCM)	Annual ground water draft (BCM)	Stage of ground water development (%)
1.	Bihar	29.19	27.42	10.77	39
2.	Chhattisgarh	14.93	13.68	2.80	20
3.	Jharkhand	5.58	5.25	1.06	20
4.	Madhya Pradesh	37.19	35.33	17.12	48
5.	Uttar Pradesh	76.35	70.18	48.78	70
	Total	**163.24**	**151.86**	**80.53**	**39.4**

Source: Central Ground Water Board, 2006
BCM: Billion Cubic Metre

The available abundant water resources need to be efficiently and effectively utilised to improve overall productivity and prosperity of the region. It is essentially required as in all the states of the region, the irrigation potential created and utilized is lower than the national average (Table 12).

Table 11: Seasonal and total rainfall (mm) in different states

State	June	July	August	September	Monsoon	Total
U.P. East	105.3	309.2	301.3	197.8	913.6	981.4
U.P. West	68.6	268.4	286.2	149.6	772.8	842.3
M.P. East	143.6	371.0	382.0	200.7	1097.3	1190.1
M.P. West	108.1	304.5	315.0	176.7	904.3	927.1
Bihar	175.2	348.1	297.8	227.1	1048.2	1219.3
Jharkhand	193.2	338.1	329.3	244.0	1104.6	1352.1
Chhattisgarh	188.6	394.4	391.2	231.6	1205.8	1394.0

Source: Statistical Abstracts of India (2005-06)

Table 12: Statewise gap between irrigation potential created and utilized in Central India

S. No.	States	Irrigation potential up to 2007 (in '000 ha)			
		Created	Utilised	Gap	% utilised
1.	Bihar	8609.46	6085.97	2523.49	70.69
2.	Chhattisgarh	1770.2	1325.85	444.35	74.90
3.	Jharkhand	1258.34	906.29	352.05	72.02
4.	Madhya Pradesh	4033.33	3252.31	781.02	80.64
5.	Uttar Pradesh	34127.05	27157.55	6969.5	79.58
	Total	**49798.38**	**38727.97**	**11070.41**	**75.56**
	Overall India	**109100.98**	**90419.00**	**18681.98**	**82.88**

Source: Central Water Commission, 2007

A major part of the cultivated area covering 12.4 and 5.6 m ha in U.P. and M.P., respectively is under irrigation through tubewells and other wells (Table 13). In spite of availability of good surface water resources, about three fourth area in U.P., two third in M.P. and Bihar and half of area in Jharkhand is being irrigated by tube wells and other wells by exploiting the ground water. Only in Chhattisgarh, about 75 percent of the area is dependent on canals and tanks for irrigation purposes.

Table 13: Area ('000ha) irrigated from different sources in Central states

State	Canal	Tanks	Tube wells and other wells	Other sources	Total
Bihar	966	111	2251	134	**3462**
Chhattisgarh	768	50	165	77	**1090**
Jharkhand	743	147	1165	328	**2384**
Madhya Pradesh	976	127	3734	820	**5631**
Uttar Pradesh	2747	67	9371	206	**12391**

Source: Statistical Abstracts of India (2005-06)

Biodiversity

The conservation and sustainable use of natural ecosystems and biodiversity is essential to support sustainable development with biological resources providing raw materials for livelihoods, sustenance, trade, medicine and industrial development. Protecting forests and other natural ecosystems can provide a host of services that reduce human vulnerability to natural hazards along with the benefits of global value such as carbon sequestration, hydrological and geochemical cycling of essential elements etc. Protected areas, which form the corner stone of biodiversity and the natural habitats within them, can protect watersheds and regulate the water flow, prevent soil erosion, influence the rainfall regime and local climate, conserve genetic reservoirs etc. Richness of biodiversity is fundamental to agricultural production and food security besides being a valuable ingredient of environmental conservation. India has the distinction of being as one of the 12 Mega centres of plant diversity in the world with 1, 26, 188 species.

Central region of the country is rich in biodiversity because of its diverse physiographic and climatic conditions. Central, eastern and southern parts of the region are rich, whereas northern and western parts are deficient in forest cover. Variability in climatic and edaphic conditions brings about significant difference in the forest types in the region. There are four important forest types, viz; Tropical Moist, Tropical Dry, Tropical Thorn and Sub-tropical broadleaved Hill forests. The forest area can also be classified based on the composition of forest and terrain of the area. Based on composition, there are three important forest formations, namely Teak forest, Sal forest and Miscellaneous forests. Bamboo bearing areas are also widely distributed in the region. The region is also a treasure house of wild economic plants, particularly the edible and medicinal plants, which are largely utilized locally or in several Ayurvedic medicines but are commercially underutilized. To obviate pressure on the natural forests, plantations have been undertaken in forest and non forest areas to supplement the availability of fuel wood, small timber, fodder etc. Positive impact of the introduction of the exotic tree species like *Acacia auriculiformis* and Eucalyptus in some of the degraded forest tracts in the region has increased the availability of fuel wood in the region.

Predominant pattern of agricultural growth, ever increasing industrial development and growth of urban centres has caused erosion of biodiversity in the region. The greater demand of water for all these enterprises has led to the destruction of natural aquatic habitat and erosion of aquatic biodiversity. In sixties, green revolution resulted in intensive use of canal irrigation water and chemicals. The eighties witnessed extensive digging of bore-wells to exploit ground water resources for agriculture. This changed the cropping sequence in the region to mainly paddy -wheat rotation. There is an urgent need for meaningful crop diversification which is congenial to environment and meets the needs of the people.

The conflict between agriculture and biodiversity is inevitable which can be overcome by evolving strong institutional mechanisms. India is a signatory to conservation of biological diversity which stipulates safety of genetic resources through a regulatory control mechanism. Several *in situ* conservative initiatives have been taken in the Central region of the country involving local community.

3. Deterioration of Natural Resource Base

The degradation of natural resources is posing a serious threat to the sustainability of production systems in the country. The burgeoning population pressures coupled with unscientific land use practices and shrinking per capita land and water resources are largely responsible for over-exploitation and reduced productivity of natural resources. The scenario of natural resources in the Central region is briefly presented in the following sections:

Land degradation

The degradation of land is a multi-faceted phenomenon, which can be manifested in a variety of forms. In Central region of the country, about 45 per cent area suffers from degradation due to water erosion, physical deterioration and other complex chemical processes. As evident from Fig. 2, about 15, 35, 49, 45 and 62 per cent of the total geographical area in the states of Bihar, Chhattisgarh, Jharkhand, Madhya Pradesh and Uttar Pradesh, respectively suffers from various forms of land degradation compared to 37 percent in the country. It is also observed that around 38 per cent of the total geographical area in Central region is degraded exclusively by water induced soil erosion which in conjunction with salt-affected and acid soils works out to be 41.6per cent. Among the states, Uttar Pradesh has maximum (56.4 percent) TGA under water erosion (including salt-affected and acid soils) followed by Madhya Pradesh (44.4 percent), Jharkhand (44.0 percent), Chhattisgarh (28.0 percent) and Bihar (11.4 percent). In the five states, water erosion alone contributes 83 percent to the total degraded area of 39.04 M ha which in conjunction with salinity/ alkalinity and acidity goes up to 91.4 percent. Thus water induced soil erosion poses a serious problem for sustainable production in the region. Besides, the Central region is also affected by special land degradation problems such as formation of gullies and ravines in around 2.5 million hectare of land which accounts for 63 per cent of the total ravine affected area in the country (Table 14).

Table 14 : State wise distribution of gullies and ravinous area

State	Area (lakh ha)
Uttar Pradesh	12.30
Bihar+Jharkhand	6.0
Madhya Pradesh + Chattisgarh	6.83
Total	**25.13**
India	**39.75**

The Central Soil and Water Conservation Research and Training Institute, Dehradun in collaboration with National Bureau of Soil Survey and Land use Planning, Nagpur computed potential erosion rates for different states of the country considering six erosion categories, viz; very slight, slight, moderate, moderately severe, severe and very severe with potential erosion rates of < 5, 5-10, 10-15, 15-20, 20-40 and > 40 t/ha/yr. Table 15 presents the potential erosion rates of more than 10 t/ha/yr categories. As evident, about 52 percent area suffers from potential erosion rate of more than the permissible limit of 10 t/ha/yr against national average of about 39 percent. Similarly, about 16 percent area falls in very severe category with erosion of more than 40 t/ha/yr against national average of about 11 percent. Among the states, Bihar has the least (13 percent) area under more the permissible rate category while Chhattisgarh has the highest percentage (65 percent) well comparable to Jharkhand, Madhya Pradesh and Uttar Pradesh with an average of 52 percent. A similar trend is exhibited under very severe erosion category.

Table 15: Area affected (per cent) under various potential soil erosion rates in Central India

States	Moderate (10-15) t/ha/yr	Moderately severe (15-20) t/ha/yr	Severe (20-40) t/ha/yr	Very severe (> 40) t/ha/yr	Total (>10) t/ha/yr
Bihar	6.23	3.43	2.73	0.58	12.97
Chhattisgarh	7.99	6.45	18.22	32.64	65.3
Jharkhand	15.55	11.44	20.95	16.83	64.77
Madhya Pradesh	12.92	9.53	18.93	17.87	59.25
Uttar Pradesh	27.58	9.95	8.26	13.44	59.23
Mean	14.05	8.16	13.82	16.27	52.3
Overall country	11.22	6.46	9.92	11.13	38.73

Recently, the Institute estimated production and monetary losses due to water erosion in different states of the country by integrating the potential erosion rates with experimentally derived productivity loss factors under major soil groups of the country in different agro-climatic regions. It is observed that Central India suffers a loss of 4.27 million tonnes in production of major rainfed cereal, oilseed and pulse crops equivalent to INR 34950 million in monetary terms due to water erosion. Madhya Pradesh state suffers the most in both production (2.03 million tonnes) and monetary (INR 18799 million) terms and the minimum losses are observed in Bihar (0.123 and 866.7, respectively). However, Chhatisgarh has maximum percent loss of 26.7 percent in its production of cereal, oilseed and pulse crops followed by Madhya Pradesh (19), Jharkhand (15), Uttar Pradesh (9.8) and Bihar (4.6), respectively.

Bihar

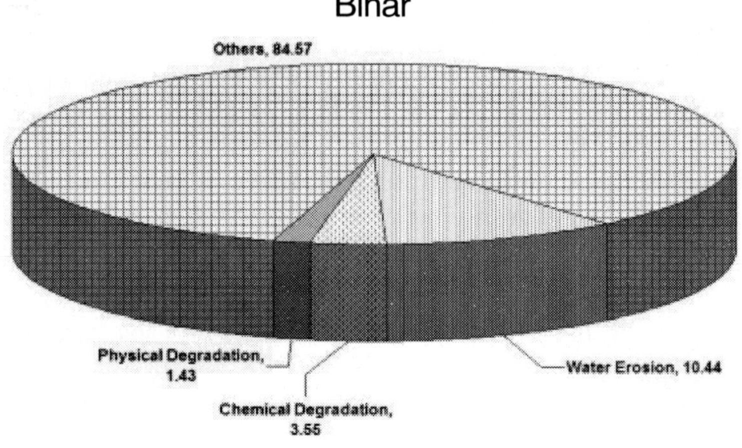

Chhattisgarh

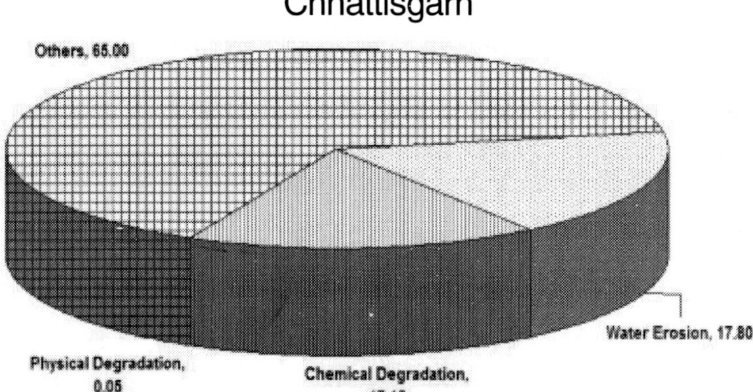

Jharkhand

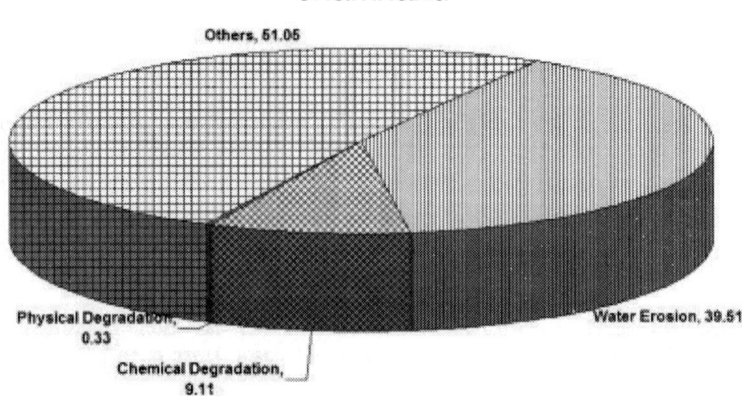

fig. 2 contd...

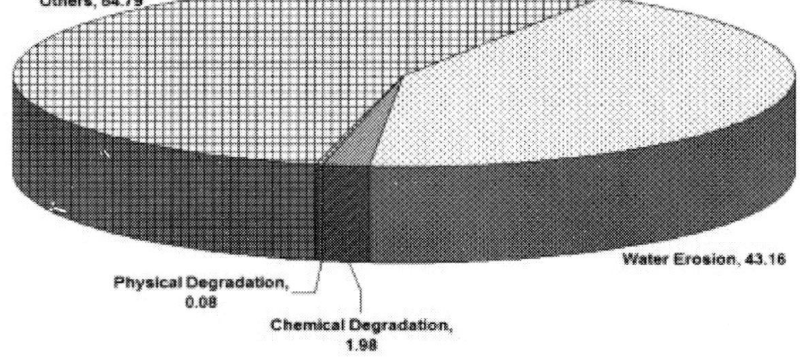

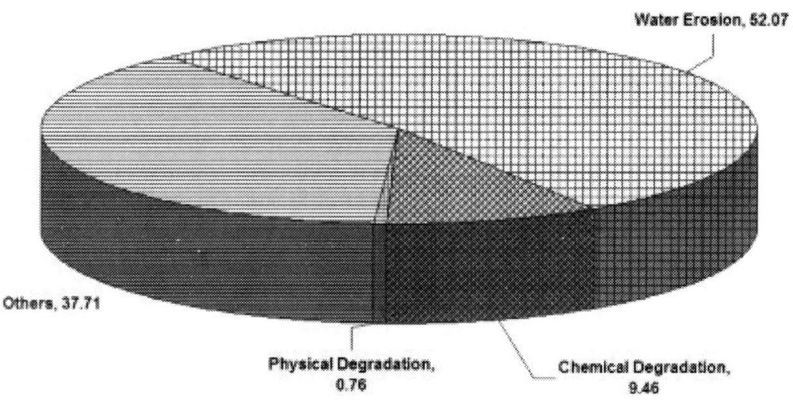

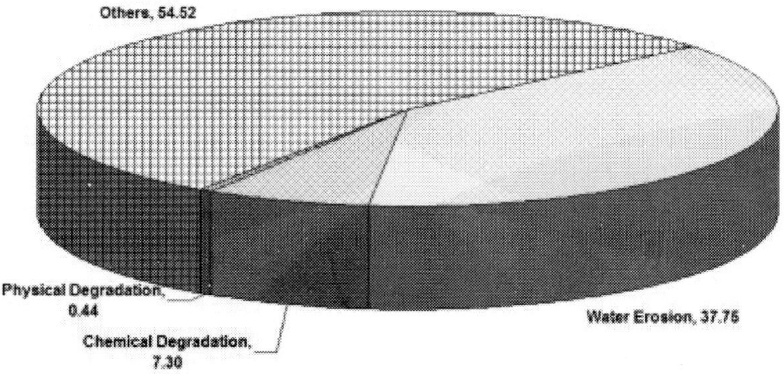

Fig. 2: Land degradation in Central India due to various processes (NBSSLUP, 2009)

Depletion of forest cover

The increasing population demands, intensive agricultural practices, growing urbanisation and industrialisation, frequent natural calamities, initiation of power projects near forests, construction of roads, building of dams on rivers etc are the major factors responsible for depletion of forest cover in the region. A comparative study on depletion of forest cover during 2003-2005 (Table 16) reveals that 233 sq. km of forest cover has been lost in the Central region which works out to be about 32 per cent of the total depletion in the country. Maximum loss of forest cover is registered in the state of Madhya Pradesh (132 sq km) followed by Chhattisgarh (129 sq km). However, the forest area has marginally increased in the states of Jharkhand (22 sq km) and Bihar (6 sq km) while no change in forest cover has occurred in the state of Uttar Pradesh during this period.

Table 16: Comparative situation of forest cover in India *(sq. km)*

S. No.	States	2005 Assess-ment	2003 Assess-ment	2001 Assess-ment	1999 Assess-ment	Change in 2001	Change in 2003	Change in 2005
1.	Bihar	5579	5573	5375	4830	545	198	6
2.	Chhattisgarh	55863	55992	57730	56693	1037	-1738	-129
3.	Jharkhand	22591	22569	22531	21644	887	38	22
4.	Madhya Pradesh	76013	76145	75282	75137	145	863	-132
5.	Uttar Pradesh	14127	14127	10778	10756	22	3349	0
	Total	**174173**	**174406**	**171696**	**169060**	**2636**	**2710**	**-233**
	Overall Country	**677088**	**677816**	**653898**	**637293**	**16605**	**23918**	**-728**

Source: Compiled from State of Forest Report, 1999-2005, Forest Survey of India, Min. of Environment and Forests, Dehradun

Declining water table

About 75 per cent irrigated area in U.P., 66 per cent in M.P., 65 per cent in Bihar and 49 per cent in Chhattisgarh is being irrigated by tube wells and dug wells. Maximum development of tube wells and dug wells for groundwater utilization has been observed in the Gangetic region mainly following paddy-wheat rotation. Due to increasing demand of irrigation, the ground water is being extracted at a rate more than the permissible limits in good quality aquifers.

Central ground water board has been monitoring the fluctuations in ground water table through a network of observation wells. Maximum percentage of observation wells showing decline in ground water is in the state of M.P. (78.6per cent) followed by Chhattisgarh (76.4per cent), Jharkhand (68.7per cent), U.P. (80.0per cent) and Bihar (49.3per cent) (Table 17). The maximum percentage of observation wells showing decline of more than 4 m is observed in M.P. (21.1per cent) followed by Chhattisgarh (14.9per cent), U.P. (13.6per cent), Jharkhand (7.5per cent) and Bihar (4.5per cent).The rate of decline is less intensive in the state of Bihar where the fall is 38.8, 5.97 and 4.45 per cent in categories of < 2, 2-4 and > 4 meters, respectively. Hence it is evident that in M.P., Chhattisgarh and U.P., the water table depletion is an acute problem both in terms of area and rate of decline. It calls for immediate remedial measures and suitable regulatory mechanism to prevent sharp decline in water table.

Table 17 : State wise number of observation wells showing decline in water level (from May 2000 to May, 2003)

State	No. of observation wells	No. of observation wells/ Percentage showing decline						Percentage of observation wells showing decline
		Decline 0-2m		Decline 2-4m		Decline >4m		
		No.	%	No.	%	No.	%	
Bihar	201	78	38.8	12	5.97	9	4.5	49.3
Chhattisgarh	268	132	49.2	33	12.3	40	14.9	76.4
Jharkhand	93	46	49.4	11	11.8	7	7.5	68.7
Madhya Pradesh	748	285	38.1	145	19.4	158	21.1	78.6
Uttar Pradesh	1028	573	55.7	110	10.7	140	13.6	80.0

Source: www.indiastat.com

4. Soil and Water Conservation Programmes in Central India

The Central Government has launched various schemes, mostly based on participatory watershed approach, like Desert Development Programme (DDP), Drought Prone Area Programme (DPAP), National Watershed Development Programme for Rainfed Areas (NWDPRA) and Integrated Watershed Development Programme (IWDP) in the states of Central region to reclaim degraded areas and enhance their production potentials. These schemes were initiated during the period 1973-74 to 1990-91. Prior to this, River Valley Projects (RVPs) and Flood Prone Rivers (FPRs) programmes were undertaken during 1961-

62 and 1980-81, respectively to treat catchment areas of rivers across the region and prevention of floods. With the launch of massive watershed development programmes in the country during 1990's, all the previous programmes were converged to develop different areas by adopting participatory watershed management approach.

A total of 123.17 lakh hectares have been treated at a cost of Rs. 396871.8 lakhs till the end of X Five Year Plan (Tables 18 & 19). Maximum areas has been treated under DPAP in the region followed by IWDP. State wise trend shows that maximum area has been treated under various schemes in M.P. though budget utilized was highest in U.P.

Table 18: Area (lakh ha) treated in different states under various schemes up to March, 2007

S. No.	States	DDP/ NAP (NAEB)	DPAP	IWDP	RVP &FPR	NWDPRA	SWC /EAP	Total
1.	Bihar	0.12	2.72	3.21	1.64	0.44	5.64	13.77
2.	Chhattisgarh	0.44	5.80	3.84	0.31	1.34	-	11.73
3.	Jharkhand	0.39	7.98	1.36	-	0.71	-	10.44
4.	Madhya Pradesh	0.79	16.33	7.64	11.56	11.79	0.81	48.92
5.	Uttar Pradesh	0.64	8.89	8.59	7.66	7.14	5.39	38.31
	Total	2.38	41.72	24.64	21.17	21.42	11.84	123.17

Table 19: Funds (Rs. in lakhs) released for soil and water conservation works under different schemes up to March, 2007

S. No.	States	DDP/ NAP (NAEB)	DPAP	IWDP	RVP &FPR	NWD PRA	SWC /EAP	Total
1.	Bihar	2193	3250	2885.3	4136.1	1384.4	-	13848.8
2.	Chhattisgarh	10355	8605	8460.3	1464.4	5639.5	-	34524.2
3.	Jharkhand	8192	6435	1245.3	-	3992.7	-	19865.0
4.	Madhya Pradesh	13927	39824	20935.4	24608.3	27568.7	12635.4	139498.8
5.	Uttar Pradesh	13280	19045	20142.3	25769.8	24046.9	86851.0	189135.1
	Total	47947	77159	53668.6	55978.6	62632.2	99486.4	396871.8

Source: Watershed Development in India, Status and Perspective (Sharda *et al*, 2008)

5. Critical NRM issues in Central India

Central India may be broadly divided into three physiographic zones from NRM point of view i.e. Vindhya and Satpura ranges and Narmada valley, Indo-Gangatic plains and Malwa region and Chhota Nagpur plateaus. Vindhyan range has a total length of 1050 km with the highest elevation of 3000 m dating back to 1.6 to 1.7 billion years and having arid and inhospitable climate. The southern slope of Vindhyan region drains into Narmada while northern slope drains into Ganges and its tributaries such as Kali, Sindh, Parbati, Betwa and Ken. Vindhyan tableland is a plateau and lies north of central part of range inhabiting cities of Bhopal and Indore while Vindhyan scraplands are sloping. Degradation of tropical deciduous forests except Sal and Lendia in Vindhyan range in the north to G B Pant Sagar (dam) is notable due to high biotic pressure caused by thermal power plants, coal mining, heavy grazing of cattle etc (Pant et al, 1992). Triangular Satpura range runs parallel to Vindhya range and west flowing rivers Tapti and Narmada on two sides and has a length of 900 Km with highest elevation of 1000 m. Indo-Gangetic plains having main rivers Ganges, Yamuna, Gomati and Chambal comprise of Terai, Banger (old alluvium) and Khader (new alluvium) belts. Malwa plateau has an average elevation of about 500 m and generally slopes towards north and drains into Chambal and its tributaries. Western part of Malwa plateau drains into Mahi river. Chhota Nagpur plateau covers most of Jharkhand and adjacent parts of Bihar and Chhattisgarh with approximate area of 65000 km^2 and consists of three smaller plateaus, viz; the largest Ranchi plateau (average elevation: 700 m and mostly covered by dry deciduous forests), Hazaribagh plateau and Koderma plateau. Major critical NRM issues in Vindhyan and Satpura ranges and Narmada valley include heavy runoff and soil loss, soil nutrients imbalances due to the deficiencies of N, P and Zn and risk of acute drought during *kharif* season. Black soils, having problem of cracking are interspersed with patches of red soils in the region and mainly rice, sorghum, pigeon pea. soybean, gram, wheat, cotton and vegetables are grown in *kharif* (rainfed) and *rabi* (partly irrigated) seasons. Ravine erosion, formation of gullies of variable size is also a major problem in the area. Indo-Gangetic plains in Central India comprise of three agro-climatic zones (upper, medium and lower) which have further been divided into many sub agro-climatic zones like Rohilkhand, Avadh and South Bihar, North Bihar etc. Though these regions differ widely in critical NRM issues but broadly they comprise of sodic phase inclusion in soils, injudicious use of irrigation water (flooding etc) leading to water logging and soil salinity, nutrient imbalances due to the deficiencies of N, P and Zn, over exploitation of underground water etc. Critical NRM issues in Malwa and Chhotanagpur zone broadly comprise of salinity and alkalinity hazards in irrigated agriculture, imperfect drainage in low lying areas, intermittent dry spells etc. As such, these regions have wide variations in climate, edaphic conditions, cropping systems and major crops raised and socio-economic status of people. Hence, critical NRM issues are discussed hereunder in each zone separately.

A. Vindhya and Satpura ranges and Narmada Valley

I. *Land Use Policy and Livelihood Issues*: Basic principles of land use i.e. land use as per the steepness of slope (agriculture up to 30per cent slope on terraces or following the principles of sloping agricultural land technology (SALT); horticulture and forestry with appropriate grasses or pastoral systems at 30-50per cent slope and beyond 50per cent slope use of land in integration with intensive soil and water conservation measures following participatory watershed approach) need to be adopted for sustaining and efficiently managing the natural resources of soil, water and bio-diversity. Rain water/runoff harvesting on micro-catchment/ farm/village/watershed basis for sustaining livestock as well as agricultural crops/fruit plants/other cash crops through supplemental irrigation utilizing water saving irrigation methods like drip and sprinkler for improved efficiency of scarce and limited harvested water during occasional or acute droughts is desirable in the zone. The zone provides tremendous scope for alternate land use and agro-forestry systems. Soil nutrient imbalances may be suitably managed in the zone through use of optimum dose of fertilizers and promoting bio-fertilizers, practices like green manuring, legume intercropping, *in situ* and brought up mulches etc. specifically in red soils.

II. *Sustainability of Water Resources*: Rain water received in the zone needs to be conserved either *in situ* or harvested at farm/village/ micro catchment/watershed level in order to control heavy runoff and soil losses. The traditional water harvesting structures like village ponds need to be rejuvenated and properly maintained for future prosperity and posterity. The black soils having swelling type montmorillonite clay poses the problem of internal drainage during rainy season and it cracks during dry spells. Perenniality of streams in the hilly eco-system needs to be revived through appropriate afforestation programs and joint forest management. Watershed hydrology in the region is being modified at different scales through watershed development programmes, construction of multi-purpose dams of variable capacity and interlinking of rivers affecting water regime not only in the catchment areas but also in the command areas. Such upstream-downstream linkages need to be established for sustainable development of water resources. The inter-state conflicts on sharing of water resources should be resolved through scientific evaluation following river basin approach.

III. *Catchment Area Treatment and Ravine Reclamation*: Catchment area treatment is necessary to check high erosion rates from agriculture as well as other land uses. Heavy sedimentation has environmental implications like pollution of rivers and water courses and sedimentation in hydro-electric dams exceeding the designed rates which reduces their life span apart from damage to machineries of the power station. Different catchments or sub-basins need to be prioritized on the basis of runoff and sediment yields, degradation status etc and treated accordingly. Catchment area treatment with appropriate vegetative and engineering measures covering the agricultural lands, stabilization of stream banks of rivers, use of gully erosion control measures like drop structures, vegetative barriers, brush wood check dams, gully plugs, chute structures; plantation of trees and grasses for stabilization of steep slopes etc. may be adopted to bring the runoff and erosion losses within the permissible limits. Leveling and terracing in small gullies may be promoted for conserving soil and water as well as sustaining agriculture in the zone. Agro-forestry systems need to be adopted not only for resource conservation but also minimize risks during crop failures.

IV. *Management of Common Property Resources (CPRs):* Community based natural resource management is the key for equitable, sustainable and environmental resource use (Kull, 2002). CPRs include common grass lands, pastures, panchayat and community lands, water resources etc. These resources are spread over an area of 12.6 M ha in Central India with maximum area in Madhya Pradesh (undivided) (7.9 M ha) followed by Uttar Pradesh (undivided) (3.10 M ha) and Bihar (undivided) (1.26 M ha) and heavily suffer from severe degradation problems. The poorly managed CPRs may be put to productive use through participatory community management programs with special emphasis on conflict resolution and equitable distribution of benefits. The practice of joint forest management (JFM) has succeeded in halting forest degradation in Jharkhand but its poverty reduction objective has yet to be realized (Sanjay Kumar, 2000). Sakurai and Palanisami (2001) found that "Tank Irrigation" management as a local common property resource was more successful than the open well system.

B. Indo-Gangetic Plains

I. *Sustainability of Agriculture Production*: Consequent upon green revolution in India, cereal-cereal cropping sequences like rice-wheat, maize-wheat, sorghum-wheat, bajra-wheat etc. with high yielding crop varieties were continuously practiced with imbalanced dose of fertilizers in the Indo-Gangetic plains. This resulted in over-exploitation of nutrients and substantial reduction in soil organic matter in the most productive upper horizon (0-30 cm) of soil leading to wide spread deficiency of N, P and Zn. Farmers are frequently complaining of either stagnated or declining crop yields; in spite of higher production cost thus reducing profitability of agriculture gradually. Besides reduction in profitability of agriculture, non-adoption of resource conserving technologies in conjunction with crop production techniques is further causing productivity fatigue of soil resources with negative impact on soil and environmental quality. The crop production needs to be integrated with crop diversification, inclusion of legumes in crop rotations, use of organic and green manures, bio-fertilization, residue management, *in situ* legume or weed mulching in wide spaced rainy season crops, agro-forestry etc in Indo-Gangetic plains to restore the soil health and fertility. Traditional organic manures like FYM or compost need to be applied in integration with inorganic fertilizers to replenish nutrients deficiency in soil-plant system. Besides, judicial use of canal and underground water by adopting efficient irrigation methods at critical stages of plant growth is necessary for sustainable agriculture.

II. *Depleting Ground Water*: Ground water exploitation has aggravated in Indo-Gangetic plains with the advent of green revolution in India for irrigating *rabi* crops (mainly wheat), raising upland paddy and cultivation of sugarcane and potato crops in a major part. The recharging of ground water has been badly affected due to conversion and leveling of village or community ponds for other purposes, high rates of runoff and soil loss along the river network, abandonment of lands due to low crop productivity and lack of public awareness about soil and water conservation. Ground water level in many parts of the zone has reached a critical limit. *In situ* rain water conservation through field bunding and/or land leveling, grassing field bunds etc apart from storage of runoff water in the farm/village ponds and artificial recharge of ground water through percolation tanks and roof top water harvesting in the watersheds are the viable options to augment the ground water recharge.

III. *Water logging, Salinization and Alkalization*: Soil salinity in patches due to rise of salts through capillary action on evaporation or soil drying owing to presence of hard calcium carbonate (kankar) layer in sub-soils or due to irrigation with poor quality water and along the unlined canals due to seepage in nearby areas invariably prevails in the entire Indo-Gangetic alluvium belt. Alkali soils formed through pedogenic processes are also found in the zone. To check soil salinity and alkalinity, recommended measures need to be adopted.

C. Malwa and Chhotanagpur Plateaus

Soil salinity and alkalinity problems in the zone may be tackled by adopting the measures as discussed in the previous section. The irrigation water may be treated with recommended quantity of gypsum or gypsum may be directly applied in the field before irrigation of crops. To handle frequent and variable droughts in the zone, *in situ* soil moisture conservation through use of bioengineering measures such as *in situ* grown legume or weed mulches, legume intercropping, vegetative barriers etc. as per prevailing rainfall and topographic conditions is needed.

6. Climate Change Impact in the Region

Climate-related disasters have brought widespread misery and huge economic losses to India, adversely affecting public health, food security, agriculture, water resources and biodiversity. The situation is likely to worsen if human beings continue to pump 'greenhouse gases' (GHGs) like carbon dioxide into the atmosphere. These gases trap heat from the sun and thus lead to 'global warming'. As the Earth's temperature rises, a series of reactions take place such as rise in sea levels and inundation of land, changes in weather patterns and impact on agricultural productivity, faster precious fresh water evaporation, increase in disease carrying vectors leading to epidemics etc. Some impacts of climate change on agriculture and water resources in the region are briefly discussed in the following sections:

6.1 Impact on Agriculture

Most of the states in the country largely depend upon rain water for irrigation which is highly variable both spatially and temporally. Scientists predict that because of global warming, already fickle weather system could become even more undependable. Semi-arid regions of western India are expected to receive higher than normal rainfall as temperatures soar, while central India will experience a decrease of about 10 to 20 percent in winter rainfall by 2050. Agriculture will be adversely affected not only by an increase or decrease in the overall amounts of rainfall, but also by shifts in the timing of the rainfall. For instance, over the last few years, the Chhattisgarh region has received less than its share of pre-monsoon showers in May and June. These showers are important to ensure adequate moisture in fields being prepared for rice crops. The state of Madhya Pradesh, where soybean is grown on 77 percent of total agricultural land, could dubiously benefit from an increase in carbon dioxide in the atmosphere. According to some studies, soybean yields could go up by as much as 50 per cent if the concentration of carbon dioxide in the atmosphere doubles. However, if this increase in carbon dioxide is accompanied by an increase in temperature, as expected, then soybean yields could actually decrease. If the maximum and minimum temperatures go up by 1°C and 1.5°C, respectively, the gain in yield comes down to 35 per cent. If maximum and minimum temperatures rise by 3°C and 3.5°C respectively, then soybean yields will decrease by five per cent compared to 1998. Changes in the soil, pests and weeds brought by climate change will also affect agriculture in India. For instance, the amount of moisture in the soil will be affected by changes in factors such as precipitation, runoff and evaporation.

6.2 Impact on water resources

The enhanced surface warming over the Indian sub-continent by the end of the next century would result in an increase in pre-monsoonal and monsoonal rainfall and no substantial change in winter rainfall over the central plains is predicted. This would result in an increase in the monsoonal and annual runoff in the central plains with no substantial change in winter runoff. Studies also indicate that there will be an increase in evaporation and soil wetness during monsoon and on an annual basis (Lal and Chander, 1993).

The hydrologic sensitivity of the Kosi basin to projected land use, and potential climate change scenarios has been analysed by Sharma *et al.* (2000 a, b). It was found that runoff increase was higher than precipitation increase in all potential climate change scenarios related to contemporary temperature. The scenario of contemporary precipitation and a rise in temperature of 4^0C caused a decrease in runoff by 2-8per cent. It is also projected that soil moisture may increase marginally by 15-20per cent over parts of Central India. This increase is confined to the monsoon months of June through September. During the rest of the year, there is either no change in soil moisture, or a marginal decline possibly due to the increase in temperature leading to enhanced evapotranspiration (Lal and Singh, 2001).

Gosain et al. (2006) and Gosain and Rao(2003) projected that the quantity of surface runoff due to climate change would vary across river basins as well as sub-basins in India. However, there is general reduction in the quantity of available runoff. An increase in precipitation in the Mahanadi, Ganga, Godavari and Cauvery is projected under climate change scenario; however the corresponding total runoff for all these basins may not increase. This may be due to increase in evapotranspiration because of increased temperature or variation in the distribution of rainfall.

7. Flood Problems in Central India

Most parts of Central India are subjected to severe flood problem. In the Ganga basin, the flood problem is mostly in the areas on the northern bank of the Ganga river caused by the northern tributaries of Ganga. Uttar Pradesh, Bihar, and West Bengal are the worst affected states in the Ganga basin. In eastern Uttar Pradesh, the rivers that cause flooding include the Sarada, the Ghagra, the Rapti, and the Gandak besides the main Ganga river. North Bihar suffers from floods almost every year due to spillage of rivers. The rivers, such as the Burhi Gandak, the Bagmati, the Kamla, the Kosi in the lower reaches, and the Mahananda, spill over their banks causing considerable damage to crops, property and life. Large scale devastation takes place in Uttar Pradesh and Bihar due to high floods in the Ghaghra, the Rapti, the

Gandak, the Kosi, the Mahananda and the Bagmati and Adhwara leading to loss of lives, dwellings, properties, installations, communication, and infrastructure facilities. For example, high flood in the Kosi river in 2008 was the worst in recent flood history of India which affected over 2.3 million population, destroyed 2.5 lakh houses and caused damage to around 1 lakh hectare crops in northern Bihar. In Madhya Pradesh, the problem is not serious but it has also experienced some isolated incidents of heavy floods in recent years. In the central and southern parts of the country, floods are observed in the Narmada, the Tapi, the Godavari and the Mahanadi. There have been instances of floods in these rivers due to very heavy rainfall in their catchments, e.g., flooding of Hoshangabad (Narmada) in 1999. The Tapi and the Narmada occasionally carry high flows which affect areas in the lower reaches in Gujarat.

8. Interlinking of Rivers (ILR) for Mitigation of Floods and Droughts

In view of the large variations in rainfall over space and time, the country experiences frequent floods in some parts and severe droughts in others. Floods are a recurring feature particularly in the Brahmaputra and Ganga river systems which carry 60 per cent of the water resources of our country. On the other hand, large areas in Rajasthan, Gujarat, Andhra Pradesh, Karnataka and Tamil Nadu experience frequent droughts. Interlinking of rivers is considered to be a viable alternative to tackle such problems. Interlinking of rivers in India is expected to greatly reduce the regional imbalances in the availability of water in different river basins.

Central region of the country particular Uttar Pradesh, Bihar and Madhya Pradesh would benefit significantly through interlinking of rivers. Around 4 M ha land can be brought under irrigation with link canal length of 1827 km having 56785 ha of land requirement (Table 20).

Epilogue

Central region of covers around 26 percent of total geographical area of India and supports 35 and 31 percent of human and livestock population, respectively. Amidst plenty of natural resources, the region has not made substantial progress as compared to other regions of the country due to several impending factors including inefficient management of natural resources. Low literacy rate and poor economic status of the people are some of the indicators of the backwardness of the region. From natural resources point of view, the region suffers from various serious problems such as floods in Bihar, eastern Uttar Pradesh, parts of Madhya Pradesh and Chhattisgarh and droughts in Agra and Bundelkhand region. Vindhyan and Satpura ranges and Narmada valley experience heavy runoff and soil losses and soil nutrients imbalances due to deficiencies of N, P and Zn. Indo-Gangetic plains of Central India suffer from salinity, alkalinity and waterlogging due to injudicious use of irrigation water. To adequately address the issues related to NRM and low productivity in the region coupled with poor socio-economic status, appropriate strategies need to be evolved which may include integrated land use planning for enhancing resource conservation and employment generation, catchment area treatment through suitable mechanical and bio-engineering measures for reducing soil erosion and sedimentation of reservoirs, inter-linking of rivers to mitigate regional imbalances of water availability thereby reducing floods and droughts, and development of optimal cropping pattern in different regions taking into account the constraints prevailing in the region such as soil salinity, alkalinity, waterlogging etc.

Table 20: The characteristic of link canals in Central India

Link	Canal Length (km)	Approximate land requirement for the link canal (ha)	Irrigated Area, (lakh ha)
Gandak-Ganga	639 (tailing in Rae Bareli, U.P.)	23170	17.49 (0.44 in Nepal)
Ghagra- Yamuna	417 (to tail in Etawah district in U.P.)	5796	11.7 (2.54 in Nepal)
Chunar- Sone	149.1 (98 km in Bihar and 51.1 km in U.P.)	1572	0.668
Teesta- Ganga	206 (Bihar)	2280	6.536
Mahanadi-Godavari	210.45	22267	3.52
Par – Tapi - Narmada	205.34	1700	3.04
Total	1826.89	56785	42.954 (Including 2.98 lakh ha in Nepal)

Source: Dams, Rivers and People (Vol 4, Issue 11-12, Dec 2006-Jan 2007)

References

Anonymous (2005). Statistical Abstracts of India (2005-06). Central Statistical Organisation, Ministry of Statistics and Programme Implementation, Govt. Of India, New Delhi.

Census of India (2001). Directorate of Census, Ministry of Home Affairs, Government of India. Central Ground Water Board, Hydrology Project, Ministry of Water Resources, Government of India.

Central Ground Water Board. Hydrology Project, Ministry of Water Resources, Govt. of India. (2006-07).

Gosain, A.K. and Rao, S. (2003). Impacts of climate on water sector. In: Climate change and and India Vulnerability Assessment and Adaptation (Eds.P.R. Shukla, S.K. Sharma, N.H. Ravindranath, A. Garg and S, Bhattacharya), Universities Press (India) Pvt.Ltd., Hyderabad.

Gosain, A.K., Rao, S. And Basuray, D. (2006) Climate change impact assessment on hydrology of Indian river basins. *Current Science*, **90** (3): 346-353.

Kull, C. A. (2002). 'Empowering pyromaniacs in Madagascar: ideology and legitimacy in community-based natural resource management', *Development and Change* **33** (1): 57-78.

Kumar, S. (2000). 'Does 'participation' in common pool resource management help the poor? A social cost-benefit analysis of joint forest management in Jharkhand, India', *World Development* **30**(5): 763-782.

Lal, M. and Chander, S. (1993). Potential impacts of green house warming on the water resources of the Indian subcontinent.JEH, 1(3): 3-13.

Lal, M. and Singh, S.K. (2001). Global warming and monsoon climate. *Mausam*, **52**: 245-262.

NBSSLUP. (2009). Soil Resource Management Reports. National Bureau of Soil Survey and Land Use Planning, Nagpur.

Pant, D. N.; Das, K. K. and Roy, P. S. (1992). Mapping of tropical deciduous forest and land use in part of vindhyan range using satellite remote sensing. *J. Ind. Soc. Remote Sens.* **20** (1): 9-20.

Sakurai, T. and Palanisami K. (2001). 'Tank irrigation management as a local common property: the case of Tamil Nadu, India', *Agricultural Economics* **25** (2-3): 273-283.

Sharda, V.N., Juyal, G.P. and Naik. B.S. (2008). Watershed Development in India: Status and Perspective. Allied Printers, Dehradun, 219 p.

Sharma, K.P., Moore, B. and Vorosmarty, C.J. (2000a). Anthropogenic, climatic and hydrologic trends in the Kosi Basin,Himalaya. *Climate Change*, **47** (1-2): 141-165.

Sharma, K.P., Vorosmarty, C.J and Moore, B. (2000b). Sensitivity of the Himalayan hydrology to land use and climatic changes. *Climate Change*, **52**: 287-313.

□□□

Chapter 2

Soil Health is the Health and Wealth of the Nation and Its Productivity is Nation's Wealth Reserve

R.N. GUPTA*

* Ex. Professor of Soil Conservation (Kanpur). Corresponding address: Green House, B-17 Sector H, Aliganj Extension, Lucknow-226024 Tel. M: 9919026620

History is full of examples of man and soil working together for their mutual benefit and evolution. Where such an evolution is in the negative direction, desertification and environmental degradation may lead to horrible situation like the European soils during the dark ages. Not to quote the extreme, when the production dwindled to nearly one tonne per ha or looking at the wastelands in India which, in the words of Dr. Swaminathan, are wasted lands due to human interference. However, if the evolution is in the positive direction, even the poor soils are rebuilt like the same European made productive soils as of today.

Food, cloth, feed, fodder, fuel, timber and a host of other necessaries come from the soil. There is an intimate relationship between the quality of soil, its nutritive value of food produced and health of human and animal who feed on it. The physical and mental lethargy of certain sections of our population is certainly due to nutrient deficiency of food produced on poor soils. A poor soil is the habitat of poor people who in turn make the soil poorer. A vicious circle of land-people-poverty syndrome is thus built around impoverished farms. On the other hand, a productive soil has on it people who are healthy and wealthy and who in turn keep the soil productive. Looking at the level of human endeavor, Punjab productivity is 2.43 times of Bihar soils and fertilizer use efficiency 1.25 times higher.

Criteria for soil health

Health of the soil is measured from the level of available nutrients (primary, secondary and micro nutrients), its physical and biological status, when the soil can hold moisture, breathe, retain and release nutrients to the growing plants.

Nutrient availability

Till recently the major nutrient balance world wide was said to be reasonably in equilibrium where annual removal of plant nutrients was balanced by global fertilizer consumption with additional biological N fixation. Vlek (1997) estimations are that 230 Mt of plant nutrients are removed annually from agricultural land which is balanced by 130 Mt of fertilizer consumption plus 90 Mt of biological nitrogen fixation, notably inorganic: BNF ratio being 100: 69. However, Indian agriculture is said to be running in a deficient nutrient balance of nearly 9 million tonnes of NPK. Tandon (2004) has estimated that nutrient addition due to fertilizers, organics and BNF amounted to 24 Mt when removal by crops along with erosion total to be 33 Mt. of NPK. In addition there is nutrient imbalance addition ratio of 2.8: 1.04 against removal ratio of 2.2: 1:3.9 and unspecified negative balance of secondary and microelements. More strikingly, bulk of the fertilizer supply is diverted to irrigated agriculture while the rainfed areas are most vulnerable to soil degradation that suffer from fertility losses and humus decomposition leading to carbon emission to atmosphere. Coming to U.P., way back in 1987-88, it was a estimated that phosphorus removal from the gross cropped area of 244.28 lakh ha was of the order of 4.52 lakh tonnes P_2O_5 against the application of 3.30 lakh tonnes of fertilizers, leaving annual drain of soil fertility to the tune of 1.22 lakh tonnes of P_2O_5. The bulk of this drain was thought to come from 70.91 lakh ha of rainfed areas. Low and decreasing soil fertility is seen in the State where more and more blocks are added annually from low to very low levels of available N and P. There are just a few blocks left out of 820 blocks which have even moderate levels of N & P. (Director of Agriculture, U.P. 2005-07). In addition the soils show increasing deficiency of Zn and S every year.

Soil erosion and nutrient availability

Soil productively has been found to be closely related to soil degradation, be it erosion, salinization, water logging or of any other form, which, in the larger sense, undermines the production capacity of the ecosystem, disrupting the C, N and S cycles (Lal and Stewart, 1990). Runoff from 141.3 m ha of degraded lands in India is reported to wash away nutrients equivalent to 5.37 million tonnes NPK annually (Bali and Kanwar, 1997) with greater additionality of soil humus. As a mean of several runoff plots, Goel *et. al.* (1968) estimated annual nutrient losses equivalent to 4.17, 7.19 and 32.91 kg/ha of N, P_2O_5 and K_2O respectively. More startling was that Gupta and Singh (1967) found preferential removal of available P_2O_5 in soil erosion. The deteriorating soil fertility is reflected in the miserably low levels of N, P and humus in the eroded soils of the State (Table 1). The eroded soils have been found to respond heavily to the application of fertilizers (Gupta and Khanna, 1964). The data in Table 2 show that N response is poor unless phosphorous fertility is built up. Very few soil conservation projects have been found in India where engineering erosion control measures have been supplemented with productivity building measures with the result that the efficiency of conservation measures has been low. Universal soil loss equation (USLE) has been found an effective tool in assessing soil erosion from a farm or watershed and its applicability to alluvial soils has been simplified to measure soil

Table 1: Fertility status of ravenous soils (Chappar Chata Dist. Etawah)

Depth cm	$pH_{2.5}$	EC dSm^{-1}	O.C.%	Total N%	Av.P_2O_5*	Av.K_2O*
0-23	7.6	0.41	0.15	0.014	20/Low	150/M
23-53	8.0	0.43	0.13	0.012	16/ Low	200/M
53-83	8.0	0.82	0.12	0.012	12/ Low	250/H
83-113	8.2	0.90	0.10	0.010	20/ Low	250/H

Source: Sachan , Das and Gupta (1976) *= Kg per ha.

loss (Gupta 2009) and suggest effective measures to control erosion and build soil productivity.

Salinization and soil productivity

Soil salinization is a heinous form of soil degradation, akin to human leprosy, destroying productivity and environment quality. The soils are high in pH, low in humus and available nutrients (Table 3), impermeable to water and drainage non-existent. Extensive research during sixties and seventies of the last century (Agarwal and Gupta 1968, Agarwal *et. al.* 1979) demonstrated the usefulness of organics and BNF in combination with gypsum, fertilizers and leaching in the reclamation of sodic soils and building their productivity. However, the World Bank added sodic land reclamation project run for 15 years

Table 2: Response of fertilizers on eroded soils of Rahman Khera Farm (q/ha)

P/N	N_0	N_{30}	N_{60}
P0	9.42	10.81	11.98
P30	11.63	14.83	16.28
P60	14.10	15.49	16.69

Source:Gupta and Khanna (1964)

Table 3 : Fertility status of alkali soils

	Depth (cm)					
Soil property	0-15	15-30	30-45	45-60	60-90	90-120
pH_2	10.2	10.4	10.3	10.2	10.1	10.0
ESP	89	90	84	80	78	76
$CaCO_3$ %	3.4	3.5	3.8	4.5	9.8	10.6
Organic C %	0.12	0.10	0.08	0.06	0.05	0.04
Available N kg/ha	45	41	36	32	32	30
Available P kg/ha	38	34	30	28	25	25
Available K kg/ha	478	475	460	450	455	452
DTPA ext Zn mg/kg	0.42	0.40	0.38	0.35	0.30	0.25
DTPA ext Mn mg/kg	2.6	2.8	2.0	1.2	1.0	1.0

Source: Anand Swarup (2004) Chemistry of sodic soils, Advances in sodic land reclamation 27-47 International Conf. on Sustainable Management of Sodic Lands Feb. 9-14, 2004

could not adequately appreciate the benefits of organics and BNF in the reclamation project with the result that there was only marginal improvement in the physical and chemical properties of the soil and productivity started declining soon after reclamation (Table 4) Reclamation requirements are that excess sodicity is kicked out of the root zone, loss of soil productivity is built up and conditions are created that chances of reversal become remote. In fact we are able to meet only half way so far as reclamation is concerned and crop productivity has been achieved.

Table 4 : Trend of wheat yields on reclamation of sodic soils

Reclamation Year	2000-01	01-02	02-03	03-04	04-05	05-06
1999-2000	32.31	32.17	28.16	25.09	25.92	23.32
2000-01		30.54	28.71	26.28	25.96	23.32
2001-02			26.51	25.30	26.03	24.64
2002-03				24.56	25.38	23.30
2003-04					23.81	24.42
2004-05						22.82
Average	32.31	31.35	27.79	25.30	25.42	23.79

Source : Crop cutting trails of SLRP II

Soil Physical Attributes

Physical soil conditions affecting plant growth are relatively less understood. Productivity is affected by soil texture, structure, aeration, permeability and temperature provided nutrient and moisture availability are not limiting factors. Quite early it was found that a soil pore volume of nearly 50% equivalent to a bulk density of 1.35 gm/cc, a quarter of which was made-up of aerated pores was optimum for the growth of wheat crop, when fertilizer response was also highest (Singh and Gupta, 1971). The oxygen diffusion rate of 26 x 10^{-8} gm cm^{-2} min^{-1} is the requirement of wheat as compared to rice requirement of 0.36×10^{-8} gm cm^{-2} min^{-1} (Table 5). It can be easily understood how continuous rice-wheat rotation being practiced must have damaged the physical soil health for wheat production unless soil structure is built up through BNF and organic manuring.

Table 5: Oxygen diffusion rate (ODR) requirements of different crops

Crop	ODR X 10^{-8} gm cm^{-2} min^{-1}
Maize	20.0
Soybean	28.4
Wheat	26.3
Peas	32.5
Toria	22.9
Rice	0.36

Source: Sinha and Ghildyal (1982)

Soil Organic Matter

The benefits of soil organic matter in improving soil quality are well known. It is one of our most important natural resources affecting the level of crop production. Humus imparts life to the soil. In every form of soil degradation, organic matter suffers heavy losses. In several runoff experiments SOC losses ranged between 18-150 with a mean of 61.86 kg/ha/yr when the soil erosion losses ranged between 1.1-16.7 t /ha/yr and the relationship being practically found to be linear. SOC being light and located near the surface is preferentially lost over all other soil constituents either mechanical or chemical, ER ratio being 4.61 (Goel *et al.* 1968). In the process of alkalization humus being soluble in alkali is lost /washed out easily from the soil, leaving it into black alkali, very poor in organic matter. The place of organic matter in land conservation policy needs to be adequately appreciated specially when agriculture is being practiced in India on carbon related negative nutrient balance, large parts of forest are degraded and over exploitation of land is a common phenomenon. All these factors lead to greater mineralization and dwindling of carbon stock in our soils. It is reported that carbon emission due to water erosion alone amounts to 6 million tonnes C/yr (Lal, 2004). It is suggested that our land use planning should be reviewed against carbon emission. All agricultural and animal waste should be returned to the soil to improve its C stock. In all spheres of agricultural management, rehabilitation of degraded land and forest conservation, immense potential exists in India for carbon sequestration to the tune of 39-49 Tg C/ year (Lal, 2004). Increasing organic matter in the soil will increase its productivity and Nation's wealth reserve.

References

Agarwal, R.R and Gupta, R.N. (1968). Saline and Alkali Soils in India. ICAR, New Delhi

Agarwal, R. R, Yadav, J, S. P., and Gupta, R. N. (1979) Saline and Alkali Soils in India ICAR, New Dehli.

Bali Y. P. and Kanwar J, S. (1977) Soil degradation in India FAO Soils Bull. 34: F.A.O, Rome

Director of Agriculture UP (2005 – 07). Intensive crop management practices for rabi season Extension & Training Bureau Press, Lucknow

Goel, K. N, Khanna, M. L and Gupta R. N. (1968). Effectof degree and length of slope and soil type on plant nutrient losses by water erosion in alluvial tracts of U.P. *J. Soil and Water Conservation*, India **16**: 1-6

Gupta, R.N. and Singh, N. (1967). Selectivity of erosion processes with respect to soil P in alluvial tracts of Uttar Pradesh. *Journal of Indian Soc. Soil Sci.* **15**: 262-268.

Gupta, R. N. and Khanna, M.L. (1964). Phosphorus fertility status and fertilizer needs of eroded alluvial soils of Utter Pradesh. *Journal of Indian Soc. Soil Sci.* **12** (2): 113-119.

Gupta, R.N. (2009). Reconsideration of soil textural parameters required to compute the USLE K factor in the Gangetic plain of India *Indian Journal of Soil Coservation* (under publication).

Lal, R. (2004). Soil Carbon sequestration in India Climate Change **63**(3): 277-296.

Lal, R. and Stewart, B.A. (1990) Soil degradation and global threat, Advances in Soil Science Vol. 11 Soil degradation XIII- XVIII, Springer Verlag, London.

Sachan, S.S., Das, S.K. and Gupta, R.N. (1976). Fertility status of ravines in Uttar Pradesh.

Sinha, A.K. and Ghildyal, B.P. (1982) Soil aeration Review of Soil Research in India vol I ; 77-80. Indian Society of Soil Science, New Delhi.

Singh, Y. P. and Gupta, R.N. (1971) Fertilizer response to the physical effects of soil compaction *J. Indian Soc. Soil Sci.*, **19**(4): 345-52.

Tandon, H.L.S. (2004). Fertilizers in Indian Agriculture from 20th to 21st century FDCO. New Delhi pp 240.

Vlek, P.L.G., Kuhne, R.F. and Denich, M. (1997). Nutrient resources for crop production in the tropics Phil. Trans. Soc., London **B352**: 975-985.

❑❑❑

Chapter 3

Assessment of Wastelands / Degraded Lands in Central India Using Remote Sensing and G.I.S. Techniques with Special Reference to Uttar Pradesh

ALOK MATHUR
Remote Sensing Applications Centre, U.P., Sector-G, Janakipuram, Lucknow, U.P.

Land degradation has been a major cause of concern and therefore, many workers have attempted to determine the area under various land degradation classes adopting different approaches. As a result, there are as many estimates for area under land degradation in India. With the advent of remote sensing technology, the first attempt was made using 1980-82 Landsat data on 1:1 million scale to prepare a wasteland map of the country. The map for the first time showed the spatial extent and distribution of wastelands which included some degraded lands also in all the states of country. However, due to the small scale and low spatial resolution of 80 m. isolated patches of land less than 100 ha. could not be mapped. Since then the remote sensing techniques have been operationally utilized for mapping degraded lands and wastelands with continuous improvements in spatial resolution and mapping scale.

Database with spatial distribution of various soil & land attributes is a pre requisite for planning of any land development program. With remote sensing techniques coupled with geographical information system (G.I.S.) it is now possible to generate not only these kinds of maps for reclamation/development of degraded lands but also for monitoring purpose. The remote sensing and G.I.S. techniques can be utilized for updating the database by virtue of tremendous potential of remote sensing data namely large synoptic coverage, multi temporal & multi spectral data acquisition, repeativity / revisit capability and computer compatibility.

The physical processes which contribute to land degradation are mainly water and wind erosion, compaction, crusting and water logging. The chemical processes include salinisation, alkalization, acidification, pollution and nutrient depletion. The biological processes on the other hand are related to the reduction of organic matter content in the soil, degradation of vegetation and impairment of activities of micro flora and fauna. Wastelands have been defined as uncultivated non forested land. In remote sensing, vegetative cover, soil characteristics and spatial patterns are used for interpreting the various categories. The wasteland maps can not be considered as such a base data for degraded lands. Some of the categories of wastelands, such as, the steep sloping lands, scrub and without scrub lands, snow covered lands etc. have been recognized as wastelands, which are not necessarily, the degraded lands.

Extent of Degraded Lands and Wastelands in Central India

Maximum area in land degradation suffers from the problem of water erosion, followed by soil acidity, water logging and soil salinity / alkalinity in five states of central India viz. Bihar, Chattisgarh, Jharkhand, Madhya Pradesh and Uttar Pradesh. According to land degradation assessment on 1:250,000 scale by NBSS & LUP (2005), Madhya Pradesh (including Chattisgarh) has 59% of its geographical area facing one or other kind of land degradation process. Area under wastelands is much less compared to degraded lands which is indicative of large areas under cultivation or forest cover is under land degradation and if conservation measures are not adopted, they will gradually convert to wastelands. The wasteland mapping on 1:50,000 scales by NRSC (2005) also shows that 18.5% of geographical area of Madhya Pradesh is covered by various kinds of wastelands mainly gullied/ravines, lands with scrub and degraded forest. Jharkhand

(14%) has large areas under degraded forest wasteland class with some other prominent wasteland like land with scrub, gullies/ravines and barren rocky/stony wasteland. Uttar Pradesh having 7% area under wastelands also predominantly under similar classes as in Jharkhand but U.P. in addition has the saline/alkali wasteland too. Bihar & Chattisgarh has only 5.8% and 5.6% area respectively under wastelands.

Harmonised data (Anonymous 2008) shows 14.58 m ha & 14.0 m ha wasteland/degraded lands in Uttar Pradesh & Madhya Pradesh respectively, while Bihar has least (1.46 m ha) area (Table-1).

Table 1 : Extent of Wastelands/Degraded Lands

State	Wasteland/ Degraded Land[1]		Land Degradation[2]		Wastelands[3]	
	Area	% of T.G.A.	Area	% of T.G.A.	Area	% of T.G.A.
Bihar	1.46	15.5	6.28	36.1	0.54	5.88
Jharkhand	3.83	48.0	6.28	36.1	1.11	14.0
Madhya Pradesh	14.00	45.4	26.21	59.1	5.71	18.5
Chattisgarh	4.71	34.8	26.21	59.1	0.76	5.6
Uttar Pradesh	14.58	60.5	15.32*	52.0	1.69	7.0

T.G.A.- Total Geographical Area
*Including Uttarakhand
1. After harmonization of wasteland/degraded lands datasets of India (2008).
2. As per NBSS & LUP-ICAR-2005 on scale of 2,50,000
3. As per wasteland atlas of India, NRSC, Hyderabad 2005

Remote Sensing Capability

Aerial photographs were initially used for mapping degraded lands. The application of remote sensing data from space borne sensors started with the launch of first earth resources satellite Landsat-1 in 1972 which had a spatial resolution of 80 m. However, many other satellites viz. Landsat TM, SPOT, Indian Remote Sensing satellites (IRS) and IKONOS, with better spatial and spectral resolution enabled to map & monitor the degraded and wastelands more accurately. Soil texture, soil color, soil organic matter content, moisture content and iron oxide content affect the spectral reflectance from a bare soil surface. Degraded lands and wastelands are well manifested on the false color composite (FCC) and therefore can be easily interpreted for various categories. In 1982, the first wasteland map of country was prepared by NRSC, Hyderabad on 1:1 m scale using Landsat data of 80 m resolution, which was followed by mapping of wastelands on 1:50,000 scale with Landsat TM (30 m) and IRS LISS-II (36 m) datasets of February/March (NRSA, 2000). It was for the first time that whole country had a spatial data base showing distribution of various kinds of wastelands. Village boundaries with names were also added to this data base which facilitated the identification of villages where reclamation/development programs can be taken up. Subsequently, the exercise was repeated with a IRS-LISS-III data set of Feb./March 2003/04 (NRSA 2005) and in addition to village boundaries & other base line information, watershed boundaries with codes were also added. The other biggest advantage was that all data set was digitally created and all information generated is amenable to computer analysis. Here, the GIS has been extensively utilized and data set is available in the from different thematic maps viz. wasteland class map, village boundary map, base layer map showing major road, railway line, drainages etc., watershed layer & administrative layer having district, tahsil boundaries. State-wise wasteland area is given in Table 2 (NRSA, 2005). So far only one season data (Rabi) was utilized for mapping of wastelands but under the updation mapping for complete country, three season data (rabi, kharif & zaid) of 2006 are being utilized in order to study the spatial and temporal changes in wastelands.

For the first time in the country, NRSC, Hyderabad has started a nation-wide project on mapping of degraded lands on 1:50,000 scale using the same three season data set (rabi, kharif & zaid) of 2006 in order to avoid the variations in spatial extents of wastelands because of different data set. By using the same data sets it is hoped that the area statistics of common classes will remain uniform/comparable to a great extent.

Table 2 : State wise - category wise wastelands area in km²

S. No.	Wasteland class	Uttar Pradesh	Bihar	Jharkhand	Chattisgarh	Madhya Pradesh
1	Gullied and/or ravinous land	3252.78	135.53	411.09	50.79	5274.32
2	Land with /Without Scrub	3618.94	783.97	2308.78	4004.89	29053.88
3	Waterlogged and marshy land	1848.57	1403.21	35.88	0.00	50.55
4	Saline/Alkaline	4842.49	76.20	0.00	0.00	0.00
5	Shifting Cultivation	0.00	0.00	0.00	69.39	0.00
6	Degraded Forest	2166.37	2858.79	7764.09	2949.54	21737.52
7	Sands	626.39	30.83	2.31	3.79	0.00
8	Mining wastelands	9.12	10.84	88.03	3.25	121.48
9	Industrial wastelands	31.94	0.18	63.07	4.91	0.00
10	Barren Rocky/Stony waste /Sheet Rock	379.75	103.97	416.11	394.98	779.19
11	Steep sloping area	11.36	32.67	35.99	12.39	80.51
	Total wasteland area	**16984.16**	**5443.68**	**11165.26**	**7584.15**	**57134.03**
	% To T.G.A.	**7.0**	**5.8**	**14.0**	**5.3**	**18.5**

(Based on 2003 IRS-LISS-III Feb/March Data)

Utilization of Remote sensing and G.I.S. based information

Wastelands maps showing district and village boundaries, watershed boundaries with codes have found many end users who are directly associated with the implementation of reclamation/ development programs for wastelands. When the Uttar Pradesh Bhumi Sudhar Nigam, Lucknow (U.P.) took up the large scale sodicland reclamation program with the financial assistance from World Bank, only wastelands maps were available to locate and select the potential areas. Based on these maps, initially 12 districts of Uttar Pradesh were identified with select group of villages showing high density of sodic areas. Subsequently these villages were mapped showing sodiclands at plot level initially using aerial photographs. In the second phase of project the IRS data of two seasons (rabi & kharif) were utilized to assess the status of sodiclands in 17 project districts in the year 1999-2000 and LISS-III+PAN merged satellite data (5.8 m) was used instead of aerial photographs for mapping of sodiclands at plot level in selected villages of 21 districts. It was for first time in the country that remote sensing data was extensively utilized for mapping sodicland area at cadastral level for an operational large scale reclamation project. As a result, a total of 4154 villages were mapped in 21 districts covering 4.07 lakh ha. of sodic area during two phases of the project. Out of total area mapped at cadastral level, 72% was totally barren, 18% supported only paddy crop and remaining 10% was supporting poor wheat crop in rabi season besides paddy in kharif.

For identification of suitable salt-affected areas for the third phase of the World Bank aided reclamation project, wasteland maps prepared from 2003 satellite data have been used and an area of more than 4.0 lakh ha. was earmarked in 28 badly affected afflicted districts of Uttar Pradesh. With the help of G.I.S. data base, and groundwater level data it was possible to identify the suitable blocks in each selected district. An example of Sultanpur district U.P. is given in Table-3

Table 3 : Block wise Area of Sodic Wasteland of District Sultanpur, U.P.

S. No.	Block	Slight Salinity/ Alkalinity (ha)	Moderate Salinity/ Alkalinity (ha)	Strong Salinity/ Alkalinity (ha)	Total area (ha)	Water table post-monsoon 2006
1	Bhetua	-	107.49	6875.1	6982.59	> 2 M
2	Bhater	-	701.1	6094.4	6795.5	> 2 M
3	Akhandnagar	-	212.7	3567.11	3779.81	> 2 M
4	Amethi	-	125	224.5	349.5	> 2 M
5	Gauriganj	4.74	151	5309.6	5465.34	> 2 M
6	Jaisingnagar	-	30.54	1673.2	1703.74	> 2 M

Remote sensing data has proved a very effective tool in monitoring the spatial extent of salt-affected lands in particular and other degraded/wastelands in general. The wasteland statistics show an overall decrease in wastelands between the first estimation carried out between 1986-2000 and 2003 using remote sensing data of matching spatial resolution and one season (rabi) only. (Table-4). The spatial extent of saline/alkali wastelands was monitored in parts of 14 districts of Uttar Pradesh between 1998 and 2007, where reclamation was extensively supported in selected villages by the project. The IRS satellite data with spatial resolution of 23 m showed overall conversion of 40% single cropped sodic area and 32% barren sodic area into double cropped areas after project intervention in selected villages falling inside the study area.

Table 4 : Change in Wasteland Area in km^2

S. No.	State	1986-2000		2003	
		Wasteland Area	% to TGA	Wasteland Area	% to TGA
1	Bihar	5880.0	6.2	5443.68	5.8
2	Chattisgarh	10185.98	7.5	7584.15	5.3
3	Jharkhand	15884.00	19.9	11165.26	14.0
4	Madhya Pradesh	60732.53	19.7	57134.03	18.5
5	Uttar Pradesh	22971.60	13.2	16984.16	7.0

With the launch of new sensors of better spatial resolution it has now become feasible to monitor the changes in land use at plot level within a village. Geographical Information System (G.I.S.) and Global Positioning System (G.P.S.) are now integral parts of remote sensing technology for generating accurate and timely assessment of wastelands and degraded lands at national/ regional/ state / block/ plot level/ watershed level.

References

Anonymous (2008). Harmonization of wastelands/degraded lands datasets of India. Published by NRAA, Ministry of Agriculture, Govt. of India, NASC Complex, New Delhi.

NBSS & LUP (2005). Land Degradation Scenario of India and Programme/ Schemes for development of degraded lands.

NRSA (2005). Wastelands Atlas of India. Ministry of Rural Development, Department of Land Resources, Govt. of India.

□□□

Chapter 4

Diatomological Analysis of Water Bodies of Semi Arid Region of Central India

MEENAKSHI SINGH[1], PAYAL LODHA[2], G.P. SINGH[2] AND RAJESH SINGH[3]

[1] Research Scholar, [2] Associate Professor, Algal Biotech Lab, Department of Botany, University of Rajasthan, Jaipur [3] Sr. Scientific Officer, Biology Division, Forensic Science Laboratory, Jaipur (Rajasthan)

Increased emphasis is being laid on modern trends of aquatic biology i.e., on the use of biological organisms as indicators of water pollution. The intimate association which algae particularly diatoms belonging to the Bacillariophyceae make them an interesting tool of study in this respect (Desikachary, 1954; Zahamensky, 1973; Tarar & Bodhke, 1998 and Gandhi, 1999). Recent studies on the diatom diversity in Indian lakes in relation to their environmental conditions (Murugesan & Sivasubramanian, 2008 and Mishra *et al,* 2009).

The present study deals with diatomological and water analysis of two selected lakes i.e., Mansagar and Mawatha water bodies. Although both comes under the same Amer tehsil of Jaipur district but represents distinct environmental characteristics. Mansagar Lake (popularly known as Jal Mahal) is an eighteenth century pleasure palace nestled amongst the Nahargarh hills are noted for its sophisticated design and grand architecture. This lake is an east westerly elongated shallow lake with irregular margins and has dense growth of aquatic macrophytes and algae. Its location is 26° 92′ N latitude 75° 82′ E longitude and has water spread area of 24.6 hectare of maximum depth 5.5 meters having water holding capacity of 54.12×10^4 cubic meters.

Mawatha Lake is situated adjacent to the Amer Fort surrounding famous "Kesar kyaari" garden and is having high religious value due to high sulphur content. It is a natural lake and completely depends on rainfall and often dries during summer months. It is located at Latitude 27° 00′ N and Longitude 75° 51″ E. The water spread area of this lake is 14.2 hectares with maximum depth of 7 meters and water carrying capacity of 39.76×10^4 cubic meters. The aim of proposed investigation is to generate an extensive reference database of diatoms of Mansagar and Mawatha water bodies which are presumably helpful in diagnosis of suspected drowning cases.

Material and Methods

Sample collections were made from April 2009 to March 2010 in different areas of Mansagar and Mawatha lakes. Collections were made with the help of planktonic meshnet (pore size 40 ìm) and were preserved in Lugol's solution. Detailed diatom study were done by following Hot HCl and KMnO4 method (recommended technique of acid digestion) as mentioned in Methods manual for collection of diatom samples, version 1.0, 2006. Qualitative analysis of diatoms were done by studying 5 to 10 permanent slides with DPX mounting medium under Nikon's light microscope at 100 X oil immersion lens whereas quantitative estimation were based on count of 50 or above diatoms per slide by using Haematocytometer (Pachar, J.V. and J.M. Cameron, 1993). Morphological characterstics of diatoms were examined under the Nikon Labophot-II microscope with H-III photomicrographic attachment. To estimate the diatom diversity these resources were used: H.P. Gandhi, 1955; Husted, 1959; Patrick and Reimer, 1966; Prescott, 1975; Desikachary, 1989 and Anand, 1998.

Monthly samples were collected and analyzed as per standard methods of APHA (2000) and Trivedi and Goel (1986) for the examination of selected physico-chemical parameters of water like pH, atmospheric and water temperature, total dissolved solids, total suspended solids, total hardness, electrical conductivity, dissolved oxygen, biological oxygen demand and chemical oxygen demand. All chemicals, reagents and solvents used for the analysis were of analytical grade.

During the course of the present investigation, soil samples from selected lakes bed were also analyzed simultaneously for basic soil properties like pH, percentage of sand, silt and clay, electrical conductivity, organic carbon, available nitrogen, potassium, phosphorus and their oxidized states to understand the co-relation between their abundance and distribution of diatom communities.

Results and Discussion

A number of diatom species belonging to 13 genera of diatoms were reported. Particularly, centric diatoms like *Coscinodiscus, Stephanodiscus* and *Cyclotella* were most dominant during the pre-monsoon period. Mawatha Lake showed distinct number of dominant genera of pennate diatoms like *Nitzschia, Pinnularia, Tabellaria, Cymbella, Diatoma* etc. Results revealed a great diversity in Mansagar water body as compared to Mawatha water body. Results also depicted that few diatoms like *Melosira, Synedra, Cyclotella, Coscinodiscus, Amphora, Cocconeis, Stephanodiscus, Gomphonema,, Fragillaria, Diatoma, Tabellaria* were restricted to particular water body whereas *Nitzschia, Cymbella, Pinnularia, Navicula* were reported from both the water bodies. Seasonal variations of diatom diversity of Mansagar Lake are well depicted in Fig. 2-5 and Mawatha Lake is represented by Fig. 6-8.

The physico-chemical analysis of Mansagar lake water showed great variation in values range, in accordance with the change in season i.e., BOD 17-48 mg/l, COD 67-216 mg/l, DO 1.3-4.48 mg/l, TDS 1390- 4382 mg/l, EC 280-2820 μmho/cm^2, pH 7.8-9.1, Total Hardness 270-474 mg/l, Water Temperature 18.5-31.0°C and Atmospheric Temperature 22.8 - 35.0°C (Table 1).

The water quality analysis of Mawatha Lake showed differences in values of various physico-chemical parameters like BOD 12 -377 mg/l, COD 38-1488 mg/l, DO 0.05- 4.0 mg/l, TDS 384- 1184 mg/l, EC 245-1271 μmho/cm^2, pH 7.8-8.7, Total Hardness 284-474 mg/l, Water Temperature 20.8-29.8 °C and Atmospheric Temperature 22.1- 34.6°C (Table 2).

Table 1 : Physico-chemical parameters of water sample of Mansagar Lake, Jaipur, Rajasthan (India).

Sampling months	Water Temp. °C	Air Temp °C	pH	EC µmho/ cm^2	DO %	BOD	COD	Hardness	TDS
Apr. 09	22.4	30.5	9.1	280	2.4	22	192	342	1406
May 09	31.0	35.0	8.3	438	2.2	25	96	350	4382
June 09	28.0	33.0	7.80	348	4.48	28	78	362	1754
July 09	27.2	34.6	728	350	4.02	25	67	360	1648
Aug 09	24.0	26.7	8.2	390	3.4	48	113	270	1006
Sep 09	23.5	28.0	8.63	471	2.8	31	128	374	1710
Oct 09	22.0	24.6	9.06	604	1.3	27	117	370	1800
Nov 09	21.0	23.0	8.11	970	4.14	17	124	474	2228
Dec 09	24.0	28.1	8.90	2140	2.35	21	216	436	2384
Jan 2010	21.0	26.0	9.01	2060	3.8	26	196	396	2060
Feb 2010	21.2	23.2	9.1	2820	3.6	26	120	290	1976
Mar 2010	18.5	22.8	8.43	1990	4.4	26	140	274	1390

All values in mg/l otherwise given.
Abbreviations: DO: Dissolved Oxygen, BOD: Biological Oxygen Demand, COD: Chemical Oxygen Demand, EC: Electrical Conductivity, TH: Total Hardness & TDS: Total Dissolved Solids.

Table 2 : Physico-chemical parameters of water sample of Mawatha Lake, Jaipur, Rajasthan (India).

Sampling months	Water Temp. °C	Air Temp °C	pH	EC µmho/ cm^2	DO %	BOD	COD	Hardness	TDS
Apr. 09	23.0	31.2	8.7	245	1.8	123	480	284	1008
May 09	29.8	34.2	7.9	260	0.15	194	1041	364	964
June 09	27.2	34.6	7.8	350	0.05	125	67	360	648
July 09	-	-	-	-	-	-	-	-	-
Aug 09	23.0	25.8	9.06	290	4.0	412	38	332	182
Sep 09	23.2	27.6	7.27	1045	2.8	377	1488	448	1184
Oct 09	22.8	24.6	8.5	1271	1.3	280	1382	426	1084
Nov 09	21.2	23.8	8.11	970	0.34	117	124	474	728
Dec 09	20.8	22.1	8.05	340	0.07	21	86	436	384
Jan 2010	-	-	-	-	-	-	-	-	-
Feb 2010	-	-	-	-	-	-	-	-	-
Mar 2010	-	-	-	-	-	-	-	-	-

Note: '-'represents dried water body due to no rainfall.
All values in mg/l otherwise given. Abbreviations: DO: Dissolved Oxygen, BOD: Biological Oxygen Demand, COD: Chemical Oxygen Demand, EC: Electrical Conductivity, TH: Total Hardness & TDS: Total Dissolved Solids.

Five soil samples from both water bodies bed were taken, air dried at room temperature and analyzed for various physico-chemical properties like percentage of sand, silt & clay, pH, EC, OC, Nitrogen, available Phosphorus and Potassium and their oxidized states (Fig. 1). The average readings obtained were compared with the normal soil range and the results showed that Mawatha lake bed have highly good quality nutrient rich soil as compared to the Mansagar Lake (Table 3).

Table 3: Showing comparative soil analysis of Mansagar & Mawatha water bodies.

SOIL ANALYSIS	UNITS	MANSAGAR	MAWATHA	NORMAL SOIL
Sand	(%)	66	60	81
Silt	(%)	16	15	11
Clay	(%)	18	22	8
pH	(1:2)	8.15	7.55	6.50 -7.50
EC	*milli mhos/cm*	0.30	0.32	0.5 - 0.75
OC	(%)	1.5	3.0	1.25 - 3.00
Avail. 'N'	*kg/ha*	1390.19	1302.64	200 - 500 to 224 - 560
Avail. 'K'	*kg/ha*	1021.44	1908.48	-
Avail. 'K_2O'	*kg/ha*	1235.94	2309.26	125 - 250 to 169.40 - 38.80
Avail. 'P'	*kg/ha*	202.16	96.32	-
Avail. 'P_2O_5'	*kg/ha*	462.95	220.57	20 - 50 to 27.10 - 67.76

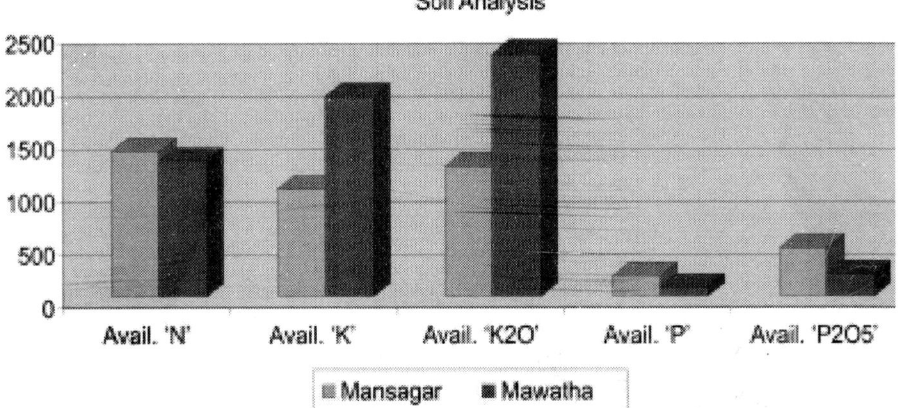

Fig. 1 : Comparative soil analysis of selected water bodies.

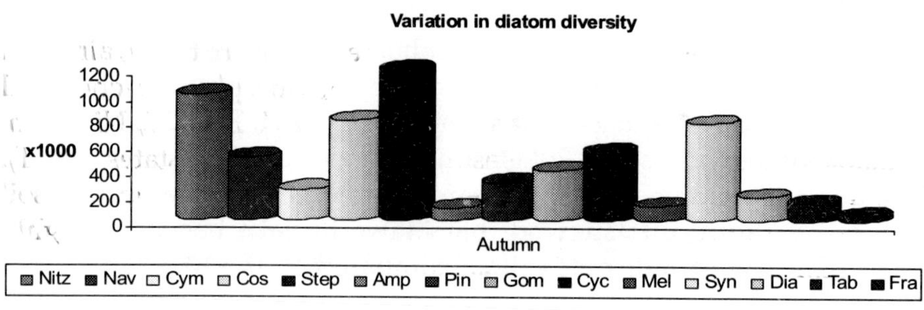

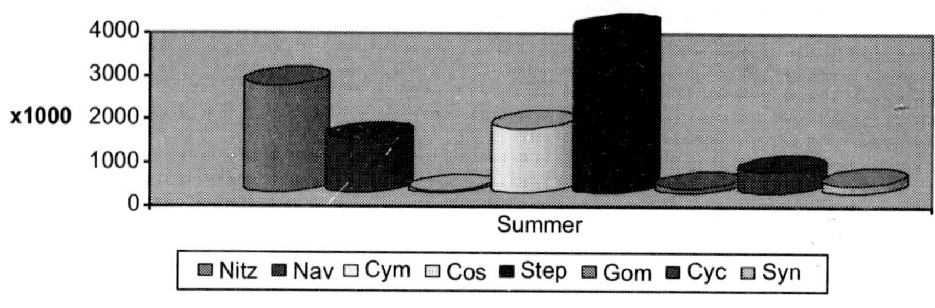

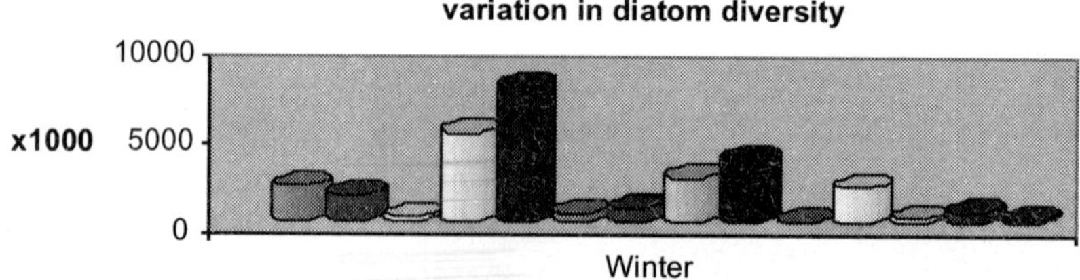

Fig. 2-5 : Showing seasonal variation on generic level of Mansagar Lake, Jaipur Generic Abbreviations Used: *Nitz = Nitzschia, Nav = Navicula, Cym = Cymbella, Cos = Coscinodiscus, Step = Stephanodiscus, Amp = Amphora, Gom = Gomphonema, Mel = Melosira, Syn = Synedra, Dia = Diatoma, Tab = Tabellaria, Fra = Fragillaria, Pin = Pinnularia & Cyc = Cyclotella*

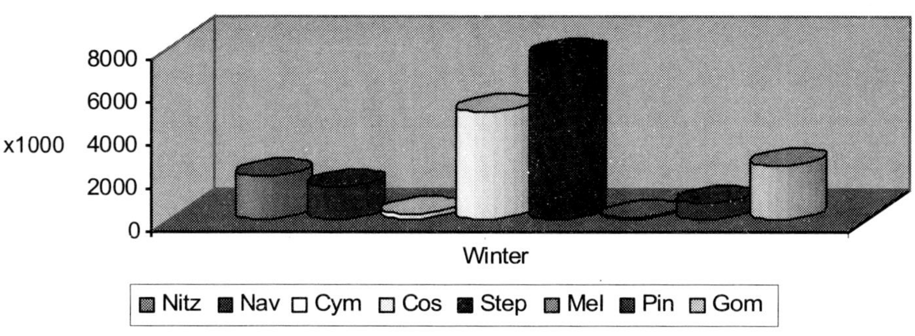

variation in diatom diversty

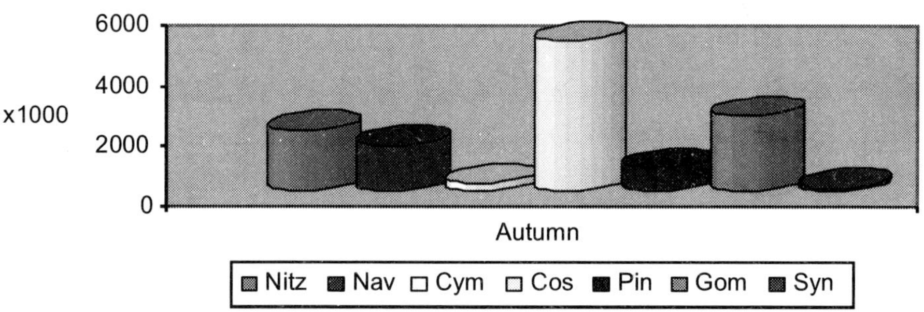

variation in diatom diversity

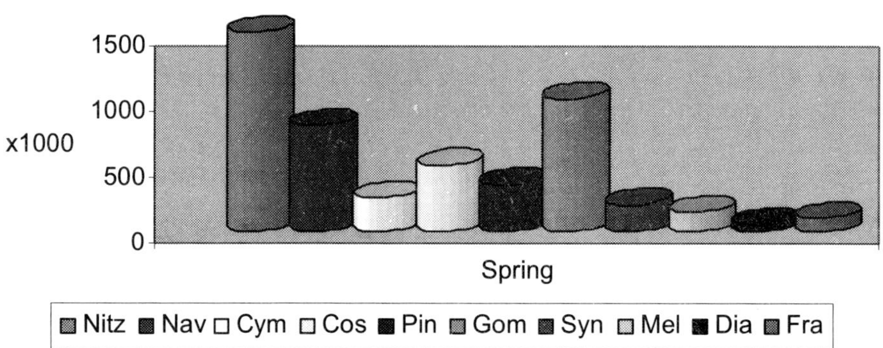

Fig 6-8 : Seasonal variation on generic level of Mawatha Lake, Jaipur. Generic Abbreviations Used: *Nitz = Nitzschia, Nav = Navicula, Cym = Cymbella, Cos = Coscinodiscus, Step = Stephanodiscus, Gom = Gomphonema, Mel = Melosira, Syn = Synedra, Dia = Diatoma, Fra = Fragillari & Pin = Pinnularia*

Conclusion

The present studies lead us to conclude that spatial and temporal diatom variations were evident in all the physico-chemical parameters examined. In general, water quality characterstics studied showed alarming sustainable management of these ecosystems. In case of Mansagar Lake, species level dominance of genera *Nitzschia, Navicula, Coscinodiscus, Gomphonema, Cyclotella, Stephanodiscus* and *Synedra* showed tolerance to high limits of EC, TDS, COD, pH and hardness. High BOD values during spring and winter season were due to rise in input of surface run off and raw domestic sewage of Jaipur city. The observed low DO values may be due to decomposition of organic matter and decay of vegetation.

In case of Mawatha Lake, reductions in the generic variety of diatoms were observed in comparison to Mansagar Lake. Maximum dominance of diatoms like *Nitzschia, Pinnularia, Synedra, Melosira, Coscinodiscus* and *Navicula* were found in winter season in accordance with medium values of BOD, DO, EC, pH, TDS and hardness. Sensitive Diatoms like *Pinnularia, Melosira, Fragillaria, Stephanodiscus* and *Cymbella* showed reduced species level diversity in spring season due to increase terrigenous organic matter (nutrient rich lake bed) as well as anthropogenic activities. Thus, selected diatom taxa can serve as good indicators of water pollution.

References

Anand, N. (1998). Indian fresh water microalgae. Pub. Bishen Singh Mahendra Pal Singh, Dehradun, India. pp. 1-94.

APHA, American Public Health Association, (2000). Standard methods for examination of water and waste water. Washington DC, USA, 21$^{st\ Edition}$.

Desikachary, T.V. (1989). Atlas of Diatoms (Marine diatoms of the Indian Ocean region). Madras Science Foundation, Madras. Fasc. VI: 1-13 pl. 622-809.

Gandhi, H.P. (1955). A contribution to our knowledge of the Fresh-water diatoms of Pratabgarh, Rajasthan. *The Journal of the Indian Botanical Society*. **34** (4): 307-338.

Gandhi, H.P. (1999). Fresh water diatom of Central Gujarat. Pub. B. Singh and M.P. Singh, Dehradun, India pp. 324.

Huested, F. (1959). Die Kieselalgen Deutschlands, Osterreichs Und der Schweiz, Vol. 2. Koeltz Scientific Books, USA. pp. 845.

Methods manual for the collection, preparation and analysis of diatom samples. Version 1.0 February 2006.

Murugesan, S. and V. Sivasubramanian, (2008). Freshwater diatoms from Porur Lake, Chennai. *Indian Hydrobiology.* **11** (1): 149-154.

Pachar, J.V. and J.M. Cameron, (1993). The diagnosis of drowning by quantitative and qualitative diatom analysis. *Med. Sci. Law.* **33** (4): 291-299.

Patrick, R. and C.W. Reimer, (1966). The diatoms of the United States, exclusive of Alaska and Hawaii. I. Monograph of the *Acad. Nat. Sci. Philad.* **13** (1): pp 668.

Prescott, G.W. (1975). Algae of the Western Great Lakes Area. Michigan State University, USA. Pp. 998.

Tarar, J.L. and S. Bodhke, (1998). Studies on diatoms of Nagpur, *Phykos,* **37** (1&2): 104-108.

Trivedi, R.K. and P.K. Goel, (1986). Chemical and Biological Methods for water pollution studies. Enviornmental publications, Karad, Maharashtra, India. pp. 1-248.

Zahamensky, L. (1973). Contribution to the knowledge of algal flora of fish pond Bratislara zelenza. *Studinica Zbslov. Nar. Muz. Prippcedy* **19** (2): 113-128.

□□□

Chapter 5

Assessment of Permissible Soil Erosion Limits for Different Landscapes of Chhattisgarh in India

BRIJ LAL LAKARIA[1], H. BISWAS AND D. MANDAL[2]
[1] Indian Institute of Soil Science, Bhopal-462038 [2] Sr. Scientist, Central Soil and Water Conservation Research and Training Institute, Uttarakhand (India), Central Soil and Water Conservation Research and Training Institute, Research Centre, Datia-475 661 (M.P), India

Problems of land degradation coupled with loss of soil productivity and declining soil quality continue to be the subject of major concern for environmentalists and is gaining widespread attention of scientific community. Though the loss of any amount of soil by erosion is generally not considered beneficial but years of field experience as well as scientific research indicate that some loss can be tolerated without affecting the crop production significantly (Schertz, 1983). Land restoration strategies in India, aimed at the conservation of depleting natural resources on a sustainable basis, are mostly based on a single generalized tolerance limit of 11.2 Mg ha^{-1} yr^{-1} which is considered on the basis that this level of soil erosion equals to that of soil formation (Mannering, 1981; McCormack *et al.*, 1982) but this limit varies with climate, soil, topography and biotic factors. Also, this limit simply implies that soil erosion at a rate below 11.2 Mg ha^{-1} yr^{-1}

does not require immediate conservation measures. For instance, according to the soil erosion status report (Tamgadge *et al.*, 2003) about 26.2, 15.0, 9.0, 10.1, 17.7, 10.5 and 11.5% area under very slight (<5 Mg ha^{-1} yr^{-1}), slight (5-10 Mg ha^{-1} yr^{-1}) moderate (10-15 Mg ha^{-1} yr^{-1}), moderately severe (15-20 Mg ha^{-1} yr^{-1}), severe (20-40 Mg ha^{-1} yr^{-1}), very severe (40-80 Mg ha^{-1} yr^{-1}) and extremely severe (>80 Mg ha^{-1} yr^{-1}), respectively. It indicates that 24% area in the region has only slight erosion level and thus do not require any conservation measure which could be misleading as one is not sure how much erosion a particular land is able to permit without affecting sustainable production. Further, this limit does not take into account soil characteristics affecting erosion so as to decide whether this quantum of erosion may be permitted or there is need to arrest even this small amount of erosion. The permissible erosion rate of annual soil loss is a threshold limit based on different criteria that will permit a high level of crop productivity (McCormack *et al.*, 1982). The aim of any conservation planning is to maintain soil erosion at a rate either equal to or less than the permissible level to prevent further soil degradation. On shallow soils with an acceptable limit of 2 Mg ha^{-1} yr^{-1}, erosion at a rate 11.2 Mg ha^{-1} yr^{-1} could lead to relatively rapid productivity loss in comparison to some deep soils with erosion permissible limit of 10 Mg ha^{-1} yr^{-1}. A typical range of 'T' values from 2.5 to 12.5 Mg ha^{-1} yr^{-1} was earlier proposed by USDA-NRCS (1999).

There has been little research on quantitative assessment of permissible erosion limits and the parameters used for assessment have been different for many studies (Lal, 1998; Renschlor and Harbor, 2002). However, for practical purposes, permissible erosion limits may be estimated from rooting depth (USDA-SCS, 1973; DLWC, 1997) and soil groups based on state of soil (USDA-NRCS, 1999). In this paper, we attempted to work out the erosion permissible limits for different landscapes in the Chhattisgarh region of India by integrating the effect of sensitive parameters *viz.*, bulk density, infiltration, erodibility, soil carbon content and fertility parameters.

Materials and Methods

Study area

The study region, rice bowl of India, is located between 17° 41' to 24° 10' N latitude and 80° 14' to 84° 24' E longitude and covers an area of 14.8 M ha and accounts for 4.5 per cent of total geographical area of India. The region has geological formation of granite, sandstone, laterite, shale, limestone and alluvium. The geological era is Archaean, Dharwar, Cuddapah, Vindhyan and Permian. The region exhibits a wide range of macro and micro climate, geology, landscapes and vegetation, which are reflected in a large variety of soils. The soils in the study area belong to five orders *viz;* Inceptisols (33.6%), Vertisols (19.9%), Alfisols (35.2%), Entisols (10.4%), Mollisols (0.8%) and rock out crops (0.07%); six sub-orders, ten great groups and twenty two sub-groups. The climate is mainly tropical, hot sub-humid with a mean annual rainfall ranging from 750 mm in northern part to 2224 mm in south-east resulting in slight to extremely severe degree of erosion.

Land assessment and quantifying permissible erosion limits

The state of soil was assessed on the basis of sensitive indicators representing a particular soil function. Potential indicators (Table 1) based on soil functions related to erosion and selected on the basis of sensitivity analysis (Nearing *et al.*, 1990) were used for developing an integrated index for assessing permissible erosion limits. Potential indicators were assigned weightage to express their relative importance. Since the primary function of the soil with respect to erodibility is to permit infiltration (Karlen and Stott, 1994), the highest weight of 0.35 was assigned to this soil function. Water transportation and absorption was assigned a weight of 0.10 based on the assumption that these are complementing the infiltration (Mandal, 2005). Resistance to physical degradation was assigned a weight of 0.25. Functions responsible for resistance to biochemical degradation and the sustainability of plant growth were each assigned a weight of 0.15. The function representing the ability of soil to sustain plant growth was assumed to be less important than those contributing to infiltration, transport or erodibility.

Table 1 : Soil function and potential indicators with respective weight.

S. No	Soil function	Potential indicator	Weight
1	Accommodate water entry	Infiltration rate	0.35
2	Facilitate water transport and retention	Bulk density	0.10
3	Resist physical degradation	Erodibility factor K	0.25
4	Resist biochemical degradation	Total organic carbon	0.15
5	Sustain plant growth	Soil pH	0.15

The indicator values were transformed into dimensionless scores ranging from 0 to 1 through fuzzy modeling (Wymore, 1993) using a scoring algorithm. Different attribute values were converted to common membership grades varying from 0 to 1 (McBratney and Odeh, 1997). The basic model in study is MF $(x_i) = [1/(1+\{(x_i-b)/d\}^2)]$. Where, MF$(x_i)$ represents individual membership function values for i[th] soil property x. This model was used as an asymmetric model with two variants where only the lower and upper boundaries of a class have practical importance. These two variants are:

1) Asymmetric left (more is better)

$$MF(x_i) = [1/(1+\{(x_i-b_1-d_1)/d_1\}^2)] \text{ if } x_i < (b_1 + d_1) \quad (1)$$

2) Asymmetric right (less is better)

$$MF(x_i) = [1/(1+\{(x_i-b_{2+}d_2)/d_2\}^2)] \quad \text{if } x_i > (b_2 - d_2) \quad (2)$$

The Model parameters include lower crossover point, central concept (b), upper crossover point and width of transition zone (d). The lower and upper crossover points represent the situation where a land attribute is at a marginal level for a given purpose, while 'b' is for an ideal level (Burrough *et al.*, 1992). For example, for infiltration an ideal value was set at 5 cm/h following the critical level concept developed by Lal (1996), while crossover point (marginal) was set between 1 and 2 cm hr^{-1}. Similarly the value of K (erodibility), presented in an ordinal form consisted of five classes (Table 2) and has an

Table 2 : Range of soil properties, their ranking and model used to convert soil attributes into 0-1 scale.

Ranking	Soil attributes									
	Infiltration (cm hr^{-1})		Bulk density (Mg m^{-3})		Erodibility Factor (K)		Total organic carbon (%)		pH	
	Range	Score	Range	Score	Range	Score	Range	Score	Range	Score
1	0.5–1.0	0.2	< 1.40	1.0	< 0.10	1.0	<0.50	0.2	< 5.0 > 9.0	0.2
2	1.0–2.0	0.3	1.40-1.47	0.8	0.10– 0.29	0.8	0.50 - 0.75	0.3	5.0 - 5.5 8.5 – 9.0	0.3
3	2.0–3.5	0.5	1.48-1.55	0.5	0.30–0.49	0.5	0.75 - 1.00	0.5	5.5 – 6.0 8.0 – 8.5	0.5
4	3.5–5.0	0.8	1.56-1.63	0.3	0.50– 0.69	0.3	1.00 - 1.50	0.8	6.0 - 6.5 7.5 – 8.0	0.8
5	> 5.0	1.0	> 1.63	0.2	> 0.70	0.2	> 1.50	1.0	6.5 - 7.5	1.0
Model	Model 3		Model 2		Model 2		Model 3		Model 3 Model 2	

asymmetric right function (model 2), that is less is better, because as the K value increases resistance to erosion decreases (Harris *et. al.*, 1996). The associated score of 1 represents the highest potential function for that system, *i.e.* the indicator is non-limiting to particular soil functions and processes. An increasing logistic curve (model 1), more is better was adopted for sensitive indicators such as infiltration, organic carbon and fertility and a lower asymptotic curve for bulk density (Grossman *et. al.*, 2001).

Ratings obtained for different soil functions when converted to 0 to 1 scale were multiplied by their respective weights. The aggregate of all the weighted parameters was then used to quantify the state of soil (Q) for each soil mapping unit:

$$Q = q_{we}\, w_{we} + q_{wt}\, w_{wt} + q_{rpd}\, w_{rpd} + q_{rbd}\, w_{rbd} + q_{spg}\, w_{spg} \qquad (3)$$

Where 'q' is the individual ratings for a particular soil function such as q_{we} for infiltration rate; q_{wt} the water transportation; q_{rpd} the rate of physical degradation; q_{rbd} the rate of biological degradation; q_{spg} the ability of soil to sustain plant growth, and w is the weight assigned to each function. Soils in each landscape were grouped into three soil groups: I ($Q < 0.33$), II ($Q = 0.33 - 0.66$) or III ($Q > 0.66$) based on the aggregated score (Q) as obtained in equation 3. Thus, soils under group III perform all functions at optimal levels and may erode at higher rates than those under groups I or II without affecting long term productivity. Permissible erosion limit was arrived at based on the soil group (Q) of the landscape unit and soil depth (Table 3) according to guidelines developed at the Iowa State University Statistical Laboratory (USDA-NRCS, 1999).

Table 3 : Permissible erosion limits (Mg $ha^{-1}yr^{-1}$) to soil mapping units based on soil depth and aggregated score.

Soil depth (cm)	Group I (Q < 0.33)†	Group II (Q = 0.33 – 0.66)	Group III (Q > 0.66)
Less than 25	2.5	2.5	7.5
25-50	2.5	5.0	7.5
50-100	5.0	7.5	10.0
100-150	7.5	10.0	10.0
More than 150	10.0	12.5	12.5

Information of soil properties with respect to different soil function was collected under different landscapes (Tamgadge *et. al.*, 1996). Soil erodibility factor, 'K' of USLE, was estimated from texture and organic matter content (Wischmeier *et. al.* 1971). In all, 277 soil mapping units were there under different landscapes, *viz*; hilly region (62 nos. spread over 27804 km^2), undulating and rolling lands (28 nos. spread over 216607 km^2), plateau region (100 nos. spread over 53671km^2), plain lands (58 nos. spread over 34721km^2) and valleys (29 nos. spread over 15323 km^2).

Results and Discussion

The range and mean values of soil properties are presented in Table 4. Infiltration rate varied from very slow to excessive in almost all the landscapes. The plain lands and valleys have lower infiltration rate owing to fine textured soil containing higher proportion of clay fraction as compared to soils in hilly, undulating and rolling lands. Bulk density, an indicator to represent water movement and transport in soil, varied depending upon variations in soil texture and organic carbon content. Mean value of bulk density was lowest in plateau region. The erodibility factor, varied from 0.19 to 0.35. Lowest mean erodibility factor was observed in valley regions. Total organic carbon

Table 4 : Soil properties under different landscapes in Chhatisgarh region

Landscape		Infiltration (cm hr^{-1})	Bulk density (Mg m^{-3})	Erodibility Factor (K)	Total organic carbon (%)	pH
Hilly region	Mean	4.61±1.15	1.63±0.06	0.24±0.04	1.55±0.85	6.2±0.7
	Range	1.0 - 5.0	1.54 – 1.72	0.19 – 0.35	0.40 – 3.12	4.9 – 7.5
Undulating and rolling lands	Mean	2.96±1.00	1.63±0.03	0.25±0.01	1.29±0.43	5.9±0.6
	Range	1.0 - 5.0	1.56 – 1.69	0.22 – 0.27	0.65 – 1.87	5.5 – 7.3
Plateau region	Mean	3.171±1.04	1.59±0.04	0.25±0.01	1.23±0.34	6.0±0.7
	Range	1.0 - 5.0	1.54 – 1.69	0.23 – 0.27	0.77 – 2.17	4.9 – 7.3
Plain lands	Mean	2.09±1.13	1.67±0.04	0.25±0.01	1.10±0.15	6.7±0.6
	Range	1.0 - 5.0	1.55 – 1.72	0.23 – 0.27	0.60 – 1.50	5.6 – 7.0
Valleys	Mean	1.07±0.37	1.70±0.03	0.24±0.01	1.24±0.18	6.3±0.7
	Range	1.0 - 3.0	1.62 – 1.73	0.22 – 0.26	1.05 – 1.65	5.4 – 7.0

† Q is soil state (the total aggregated score)

content ranged between 0.40 to 3.12 per cent with highest under hilly region and subsequently decreased towards valley region which is largely attributed to the higher C addition in forest through leaf fall (Mandal *et. al.*, 2006). Soil pH varied between 4.9 and 7.5, however, the mean soil pH under all the landscapes was near neutral.

Soil state and permissible erosion limits

The aggregated score (Q) which represent the soil state varied from 0.47 to 0.89 among different landscapes. It indicates a medium to good soil state under all the landscapes (Table 5). The mean aggregated 'Q' values were highest in the hilly region followed by plateau region, undulating and rolling lands, plain lands and valleys. Since none of the landscape had mean 'Q' less than 0.33, hence, the soils of the region were placed into group II and III. It indicates that the limitations are moderate to low and these landscape units can be managed with suitable conservation measures to maintain optimum productivity level. Soil depth in all the landscapes varied from 25 to even above 150 cm. Therefore, a soil mapping unit having a medium soil state *i.e.* soil group II and depth of 25-50 cm could be assigned a 'T' value as low as 5 Mg ha^{-1} yr^{-1}. The state of Soil 'Q' ranked highest in

Table 5 : Soil state (Q) and permissible erosion limits under different landscapes.

Landscape	Mean/Range	Soil state (Q)	Permissible erosion limits (Mg ha^{-1}yr^{-1})
Hilly region	Mean	0.80±0.09	8.06±1.46
	Range	0.47 - 0.89	5.0 - 12.5
Undulating and rolling lands	Mean	0.66±0.09	9.29±1.50
	Range	0.48 - 0.80	5.0 - 12.5
Plateau region	Mean	0.68±0.09	9.88±1.72
	Range	0.51 - 0.88	5.0 - 12.5
Plain lands	Mean	0.61±0.07	10.78±1.83
	Range	0.49 - 0.80	5.0 - 12.5
Valleys	Mean	0.56±0.03	11.64±1.67
	Range	0.53 - 0.68	5.0 - 12.5

hilly region with a mean value ranging between 0.47 and 0.89. However the permissible erosion limits values varied between 5.00 and 12.5 Mg ha^{-1} yr^{-1}. It indicates heterogeneity with respect to soil depth, infiltration, bulk density, organic carbon content and erodibility. The area under permissible erosion limits and their distribution in different landscapes is presented in Table 6 and Fig. 1. Among different landscapes, the permissible erosion limits varied from 5.00 to 12.5 Mg ha^{-1} yr^{-1} in hilly, undulating to rolling land and plateau region, and 7.5 to 12.5 in plain lands and valley regions. None of the landscapes had low erosion limit of 2.5 Mg ha^{-1} yr^{-1} due to their medium to good soil state and better rooting depth. A sizable area of 691.7 km^2 in hilly region, undulating to rolling lands and plateau region is able to permit a soil loss of 5.0 Mg ha^{-1} yr^{-1} only. These areas are more sensitive and require suitable soil conservation measures. Preferably these areas should be brought under permanent vegetation and should not be cultivated (Mandal *et. al.*, 2006). While about 43277 km^2 (29%) area can tolerate erosion more than the default erosion limit of

Table 6 : Area of Chhattisgarh landscapes under different permissible erosion limits (km^2).

Landscape		Permissible erosion limit (Mg ha^{-1} yr^{-1})				Total
		5.0	7.5	10.0	12.5	
Hilly region	Area	69.6	24676.2	1454.4	1603.8	27804.0
	Percent	0.3	88.7	5.2	5.8	
Undulating and rolling land	Area	414.6	4268.6	11256.3	667.4	16606.9
	Percent	2.5	25.7	67.8	4.0	
Plateau region	Area	207.5	9331.0	33964.6	10168.3	53671.4
	Percent	0.4	17.4	63.3	18.9	
Plain lands	Area	0.0	2989.3	11097.7	20633.8	34720.8
	Percent	0.0	8.6	32.0	59.4	
Valleys	Area	0.0	3292.1	1827.3	10203.7	15323.1
	Percent	0.0	21.5	11.9	66.6	
Total	Area	691.7	44557.2	59600.3	43277.0	148126.2
	Percent	0.5	30.1	40.2	29.2	

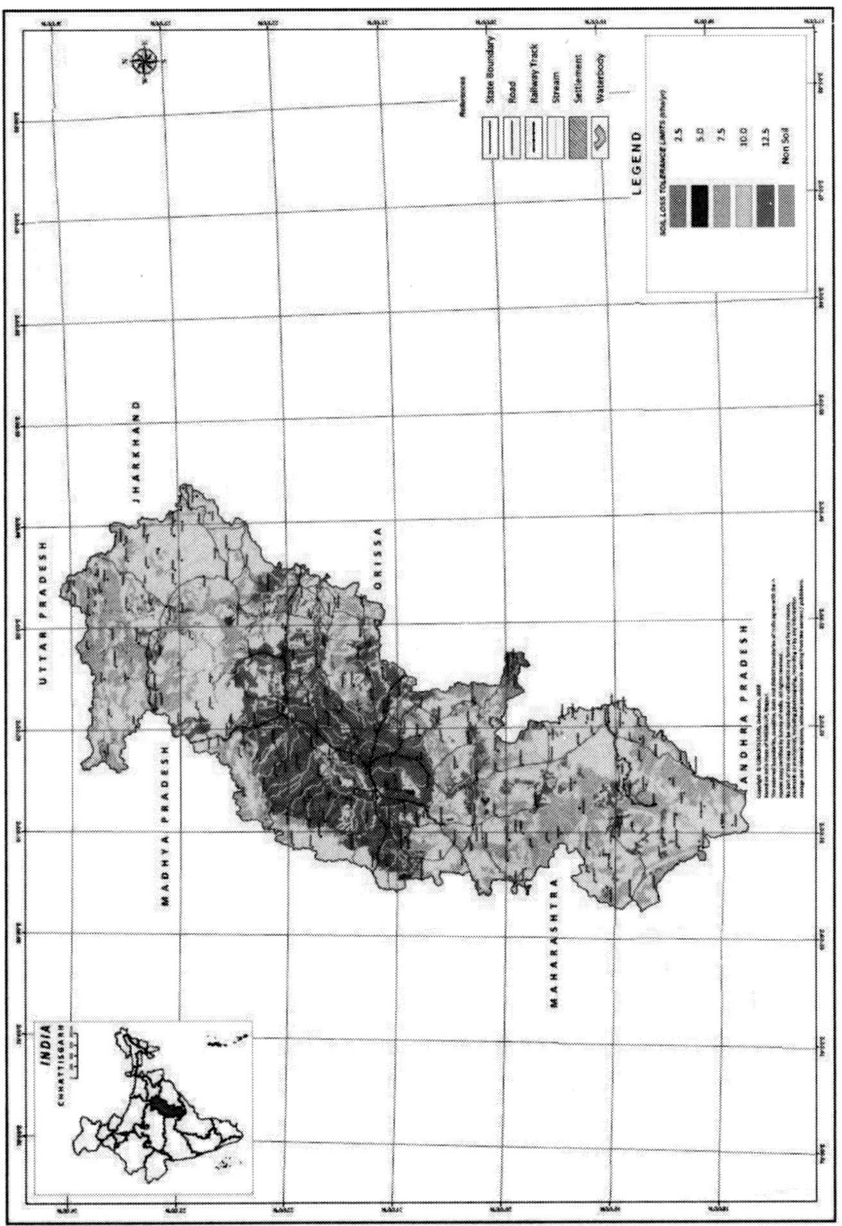

Fig. 1 : Distribution of permissible erosion limits in the region.

11.2 Mg ha^{-1} yr^{-1}. Thus 71% (1048492 km^2) demands necessary suitable soil and water conservation measures. Thus, while preparing conservation plans for different landscapes much emphasis is needed for hilly region where 94.2% (26200 km^2) could permit erosion below 10 Mg ha^{-1} yr^{-1}. Similarly 25.7% of the undulating and rolling lands have erosion tolerance of just 7.5 Mg ha^{-1} yr^{-1} and 67.8% can tolerate a soil loss of 10 Mg ha^{-1} yr^{-1}. Thus 96% area under undulating and rolling landscape is facing severe water erosion problem which suggests that conservation planning is needed for the majority of the landscapes.

Conclusion

The permissible erosion limits were calculated for the various landscape elements of Chhatisgarh region. The values of these limits were greatest for valleys, but declined for plain lands, undulating and rolling lands, plateau region and hilly regions, respectively. This assessment indicated that the permissible soil erosion limit for 71% of the landscape was lower than the default value of 11.2 Mg ha^{-1} yr^{-1} meaning that land management guidelines and conservation measures need to be improved for the majority of the landscape so that the soil resource may be sustained.

References

Burrough, P.A., MacMillan, R.A. & Deursen, W.V. (1992). Fuzzy classification methods for determining land suitability from soil profile observations and topography. *Journal of Soil Science,* **43**: 193-210.

DLWC (Department of Land and Water Conservation) (1997). Soil and Landscape issues in Environmental impact assessment, Technical Report No. **34,** DLWC, Sydney.

Grossman, R.B., Harms, D.S., Seybold, C.A. & Herrick, J.E. (2001). Coupling use dependent and use-invariant data for soil quality evaluation in the United States. *Journal of Soil and Water Conservation,* **56**: 63-68.

Harris, R.E., Karlen, D.I. & Mulla, D.J. (1996). A conceptual framework for assessment and management of soil quality and health. In: *Methods for assessing soil quality.* Soil Science Society of Amerca Special Publication, No. 49. (eds. J.W. Doran & A.J.Jones), pp 61-82. SSSA Madison, Wisconsin.

Karlen, D.L. & Stott, D.E. (1994). A framework for evaluating physical and chemical indicators of soil quality. In*: Defining Soil Quality for a Sustainable Environment* (ed. J.W. Doran *et. al.*), pp 53-72. Soil Science Society of America, Special Publication, No. 35. SSSA, Madison, Wisconsin.

Lal, R. (1996). Methods and guidelines for assessing sustainable use of soil and water resources in the tropics. Scientific Publishers, Jodhpur, India.

Lal, R. (1998). In: *Soil erosion at multiple scales: Principles and methods for assessing causes and impacts*, International Board for Soil Research and Management (IBSRAM), CABI, Wallingford, U.K., pp 149-160.

Mandal D. (2005). Proceedings of Interaction Workshop on Soil Loss Tolerance Limits. CSWCRTI, Dehradun, Uttrakhand, India, 2005.

Mandal, D., Dadhwal, K.S., Khola O.P.S. & Dhayni, B.L. (2006). Adjusted T values for conservation planning in Northwest Himalayas of India. *Journal of soil and Water Conservation*, **61**: 391-397.

Mannering, J.V. (1981). The use of soil loss tolerance as a strategy for soil conservation. In: *Soil Conservation: Problems and Prospects.* (ed. R.P.C.Morgan) John Wiley & Sons, New York.

McBratney, A.B. & Odeh, I.O.A. (1997). Application of fuzzy sets in soil science: Fuzzy logic, fuzzy measurements and fuzzy decisions. *Geoderma*, **77**: 85-113.

McCormack D.E., Young, K.K. & Kimberlin, L.W. (1982). Current criteria for determining soil loss tolerance. In: Determinants of soil loss tolerance. (eds. B.L.Schmidt, R.R.Allmaras, J.V.Mannering & R.I. Papendick.) Agronomy Society of America Special Publication No. **45**, Madison, Wisconsin.

Nearing, M.A., Ascough, D.L. & Latlen, J.M. (1990). Sensitivity analysis of the WEPP hillslope profile erosion model. *Transactions of ASAE*, **33**: 839-849.

Renschlor, C.S. & Harbor, J. (2002). Soil erosion assessment tools from point to regional scales: The role of geomorphologists in land management research and implementation. *Teomorphology*, **47**: 189-209.

Schertz, D.L. (1983). The base for soil loss tolerance. *J. Soil and Water Conserv.*, **38**: 10-14.

Tamgadge, D. B, Tiwari, A. K. and Maji, A.K. (2003). In: *Soil Erosion Chhattisgarh*, Publication no. 103, NBSS&LUP, Nagpur, p. 30.

Tamgadge D.B., Gaikawad., S.T., Nagabhushana, S.R., Gajbhiye, K.S. Deshmukh, S.N. and Sehgal, J. (1996). In: Soils of Madhya Pradesh: their kinds, distribution, characterization and interpretations for optimizing land use, 1996, 59b (Soils of India Series 6), NBSS&LUP, Nagpur, India, p 182.

USDA-SCS (1973). Advisory notice, Soil-6. U.S. Government Printing Office, Washington, D.C.

USDA-NRCS (1999). National soil survey handbook: U. S. Government Printing Office, Washington, D.C..

Wischmeier, W.H., Johnson, C.B. and Cross, B.V. (1971). A sol erodibility nomograph for farmland and construction sites. *Journal of Soil and Water Conservation*, **26**: 189-193.

Wymore, A.W. (1993). Model based systems engineering: An introduction to the mathematical theory of discrete systems and to the tricotyledon theory of system design. CRC Press, Boca Raton, Florida.

□□□

Chapter 6

Mapping Gullied Wastelands Along Gomti River in Uttar Pradesh using IRS LISS-III Data

ALOK MATHUR, S. CHANDRA, M. MISHRA, R.S. RAWAT AND A.R. SIDDIQUI
Remote Sensing Applications Centre, U.P. Sector-G. Jankipuram, Lucknow-2260021

Water erosion is the cause of most wide spread form of land degradation. Splash erosion, sheet erosion, rill & gully erosion are the processes which result into the loss of top soil and terrain deformation. Areas having soil loss more than 10 t ha^{-1} yr^{-1} have been classified as degraded lands by NBSS & LUP. After harmonising the NRSC data set and NBSS & LUP data on wasteland and degraded lands, Maji (2007) reported an area of 73.27 M ha under cultivation in the country affected by the problem of water erosion (>10 t ha^{-1} yr^{-1}). In addition to above 9.3 M ha area under open forest (<40%) is also facing the problem of water erosion in the country. Loss of top soil is continous affecting in turn the agriculture production and causing siltation of reservoirs. Non-availability of correct information on spatial extent and area under various categories has been considered a major constraint for overall treatment of eroded areas. Remote Sensing and G.I.S. techniques have made it possible to generate updated information in a spatial domain at a regular interval.

The problem of gullies and ravines is concentrated in the four states of M.P., Rajasthan, U.P. and Gujrat along the rivers of Chambal, Mahi, Yamuna, Sabarmati & their tributaries. Karale *et al.* (1988) based on remote sensing data computed increment of ravines to the extent of 3-11% over a period of 21 years in Ratlam & Ujjain districts of M.P. Das (1985b) reported that about 8000 ha of good agricultural lands are affected every year by the process of gulliying. Singh *et al.* (1987) monitored the changes in extent of gullied and ravines lands between 1910-11 and 1972-73 and during period 1972-86 in part of Etawah, Agra, Mainpuri & Bhind districts. They reported a decrease by 11 km^2 between 1910-11 and 1972-73, but during 1972-86 periods, the gullied and ravinous area increased by 5 km^2. The average annual increase in eroded areas between 1972-86 was calculated to be 49 ha in the area. Uttar Pradesh is severely affected with the problem of various kinds of wastelands. Salt-affected, gulled/ravinous and waterlogged wastelands are the dominant kinds mapped in U.P. Indian Remote Sensing Satellite data has been successfully utilized in mapping of wastelands at various scales. In the present paper, attempt has been made to study the changes in the extent of gullied wastelands along the river Gomti which is a tributary of river Ganga, between 2003 & 2006 using IRS LISS-III satellite images and Arc GIS.

Study Area

The distribution of gullied or /ravines wastelands is predominant along the rivers like Yamuna, Chambal, Dhasan, Ken & Betwa but they are also found as gullied lands along small river system like that of Gomti in Uttar Pradesh. The Gomti river originates in Pilibhit, India from a pond known as Gomath Tal. It extends 900 km through Uttar Pradesh and meets the Ganga river near Saidpur city in Ghazipur district. Gomti river, excluding the Sai distributary has 8237.9 km long drainage network in 12 districts (Fig. 1). It flows eastwards after Lucknow through Barabanki, Faizabad, Sultanpur, Jaunpur and it confluences with river Ganga in U.P.

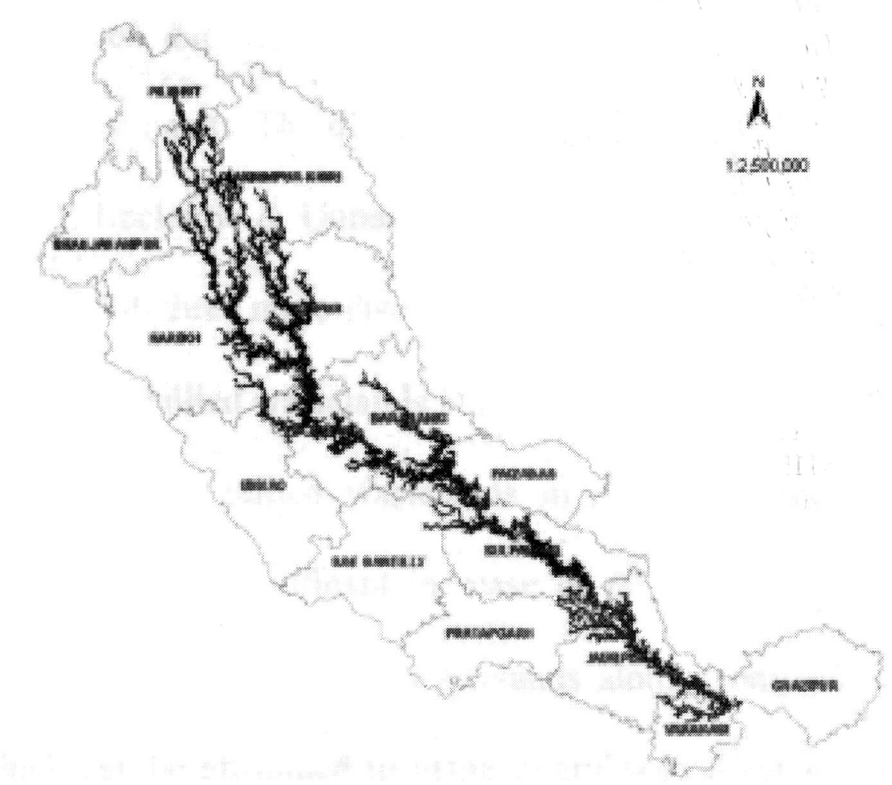

Fig. 1 : Gomti River System.

Materials and Methods

Indian Remote Sensing satellite LISS-III data of 23 m spatial resolution of Feb/March for the year 2003 & Feb/March and Oct. 2006 was utilized for on screen visual interpretation of FCC using Arc GIS 9.1. The gullied wastelands have been interpreted following color, tone, texture, shape, size and association on FCC on a scale of 1:35,000. The Gomti river system excluding Sai river was extracted using Arc info from available digital data base prepared from 1:50,000 scale maps. The change detection in the extent of gullied wastelands was carried out by computation of area statistics of 2003 and 2006 in GIS and further by using 'union' function for analysis of coverages in Arc GIS 9.1. District boundaries were taken from census map of 2001.

Results & Discussion

The IRS LISS-III data with spatial resolution of 23 m. has proven its capability for mapping and monitoring of gullied wastelands along Gomti river system in U.P. The G.I.S. has played a significant role in studying the spatial distribution of gullied wastelands and the spatial changes taken place during the study period. The gullied lands commonly called as broken lands along rivers, without any tree/shrub cover or with tree/shrub density of less than 20% have been mapped as gullied wastelands. These gullied lands at many places covered by thick tree vegetation mainly of Acacia, Eucalyptus etc. have not been mapped as gullied wastelands. The gullied wastelands mapped using 2003 and 2006 satellite data have been shown in Fig. 2 & 3 respectively. The extent of gullied wastelands along Gomti river system increased from 115.62 km^2 to 138.4 km^2, registering an increase of 22.78 km^2 from 2003 to 2006 (Table 1).

Table 1 : Area Under Gullied Wastelands Along Gomti River System in km^2.

S.No.	District	2003	2006	Change in +/- Area
1.	Pilibhit	0	0	0
2.	Lakhimpur Khiri	0.01	0	-0.01
3.	Shahjahanpur	0	0	0
4.	Sitapur	2.83	1.61	-1.22
5.	Hardoi	6.02	3.12	-2.9
6.	Lucknow	0	0	0
7.	Unnao	0	0	0
8.	Barabanki	15.22	18.75	+3.53
9.	Faizabad	7.85	4.44	-3.41
10.	Sultanpur	27.54	87.66	+60.12
11.	Pratapgarh	0.26	0.13	-0.13
12.	Raebareli	2.12	0	-2.12
13.	Varanasi	3.40	1.52	-1.88
14.	Jaunpur	48.56	21.17	-27.39
15.	Ghazipur	1.81	0	-1.81
	Total	115.62	138.4	+22.78

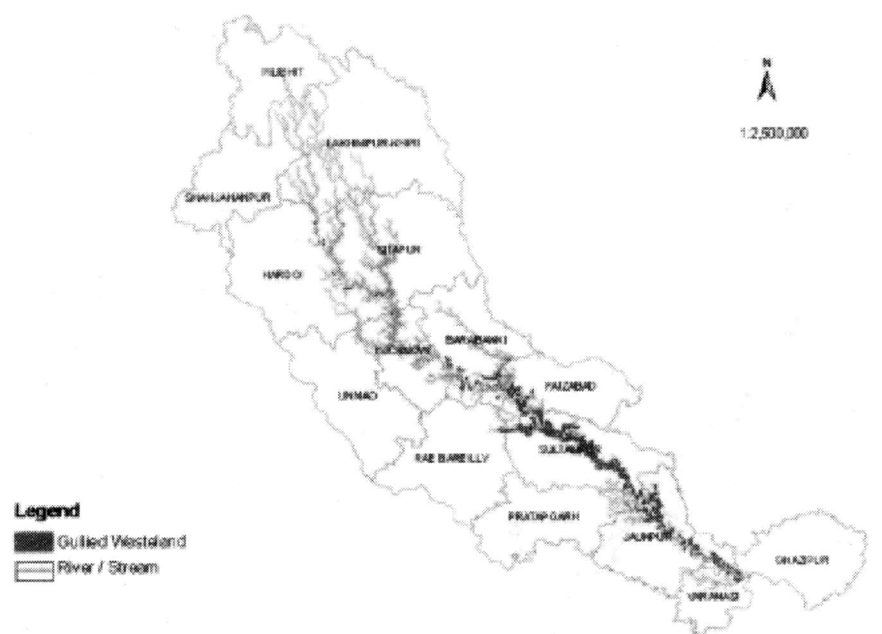

Fig. 2 : Gullied Wasteland Along Gomti River in 2003.

Fig. 3 : Gullied Wasteland Along Gomti River in 2006.

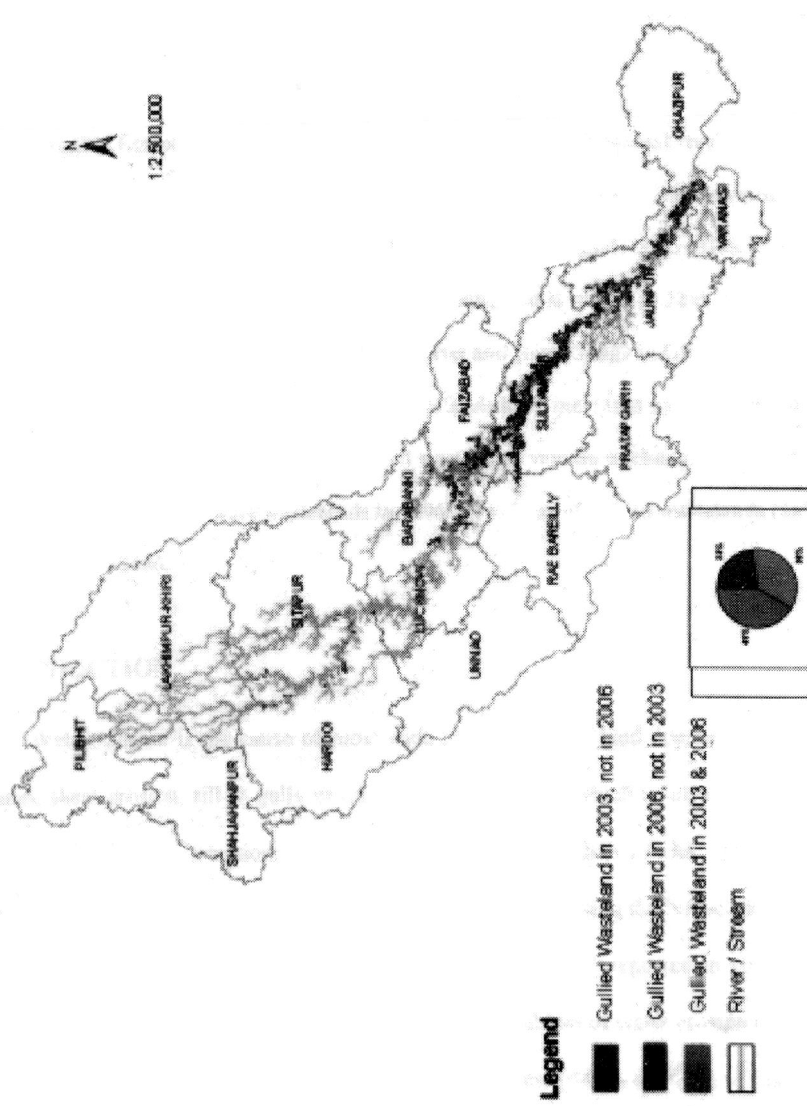

Fig. 4 : Gullied Wasteland Dynamics Along Gomti River in 2003 & 2006.

In 2003 only one season (rabi) satellite data was used while in 2006 two season e.g. rabi & kharif season data has been used, which has probably facilitated more in delineation of gullied wastelands. The distribution of gullied wastelands show that in 2003 Pilibhit, Shahjahanpur, Lucknow & Unnao did not have gullied wastelands along Gomti river system, while in 2006 three more districts namely Lakhimpur Khiri, Raebareilly and Ghazipur also don't show gullied wastelands along Gomti river system. In 2006, there has been decrease in area under gullied wastelands in all remaining districts except for Sultanpur which has shown a significant increase of 60 km^2. Raebareilly & Ghazipur districts showed 2.12 & 1.81 km^2 gullied wastelands along Gomti in 2003 but in 2006 there are none, which can be attributed to using kharif season satellite data additionally and the changes taken place in vegetative cover in 3 year period.

The overlaying of two coverages i.e. gullied wasteland maps of 2003 and 2006 by 'Union' in GIS environment facilitated further analysis of changes taken place in gullied wastelands over the period of study (Fig. 4). The results show that 74.6 km^2 area (41.6%) which was gullied wasteland in 2003 remained as such in 2006 also. In 2006 new areas to the extent of 63.84 km^2 have been added as gullied wastelands, either due to loss of vegetative cover from existing gullies or addition of new gullies. An area of 41.02 km^2 which was lying as gullied wastelands in 2003 is now under good density of vegetation i.e. > 20% in 2006 thereby indicating reduction in soil erosion.

Conclusion

Remote sensing data in association with GIS techniques is a very effective tool in mapping and monitoring the spatial distribution of gullied wastelands over a period of time. The information thus generated indicate not only the changes taken place in the earlier existing wastelands but also the new areas converted into wastelands during the period. The spatial information generated is in digital form and is very useful for soil conservation.

Acknowledgement

Senior author is thankful to Director, RSAC,U.P. for granting permission to present the paper. Secretarial support received from Km. Neeta Srivastava is gratefully acknowledged.

References

Das, D.C. (1985b). Problem of Soil Erosion and Land Degradation in India: Proc. National Seminar Soil Consv. Water Mgmt, New Delhi.

Karale, R.L., Saini, K.M. & Narula, K.K. (1988). *J Soil Wat. Conserv.* India **32**: 75.

Maji, A.K. (2007). Assessment of degraded and wastelands of India *J. of Indian Society of Soil Science* **55**: 427-435.

Singh, A.N.; Pant, B.C. and Siddiqi, I.H. (1987). Monitoring gullied and ravine lands in western Uttar Pradesh using topographical maps and Landsat TM data. Proc. National Symp. Remote Sensing in Land Transformation and Management. Indian Society of Remote Sensing, Dehradun. pp. 191-196.

□□□

Chapter 7

Analysis of Rainfall for Crop Planning in Bundelkhand Zone of Uttar Pradesh

MUKESH CHAND, DUSHYANT KUMAR, DHANANJAI SINGH, NISHI ROY AND D.K. SINGH*

C.S.A. University of Agriculture and Technology, Kanpur - 208002 and
*Dr. Bheem Rao Ambedkar University, Agra, Uttar Pradesh

India's economy is mainly dependent on agriculture, which is based on monsoonal rainfall and their distribution. The growth, development and yield of crops are governed by fluctuation in rainfall. The weekly, monthly and seasonal pattern of rainfall and their probabilities are helpful in crop planning by identifying the period of drought, normal and excess rainfall (Ray *et. al.*, 1980). Rainfall variability is the major factor influencing agricultural productivity and sustainability in the tropics (Virmani, 1994). Panigrahi and Panda (2002) and Kar (2003) also applied Markov Chain model for calculating of initial and conditional probability of dry and wet spells of different duration for various climatic situations and have evaluated its practical importance in crop planning. Bundelkhand zone of Uttar Pradesh is located in south -west corner of Uttar Pradesh extended between 24° 11′ and 26° 27′ N Latitude and 78° 17′ and 81° 34′ E Longitude with an average altitude ranging 250-300 m above msl. The zone accounts for nearly 1/9 of the total area (243286 km^2) of the state. Agriculture is the backbone of rural economy and major area falls under rainfed agro-ecosystem, which experienced moderate to severe drought conditions of prolonged nature once in two to five years. Apart from the prolonged drought, inadequate and erratic distribution of rainfall often exposes the plants to shortage of desired levels of moisture, and thus plant growth takes place under below optimum conditions. As a result, the productivity levels remain low over the region. The occurrence of drought is continuous since 2000 over lasted for more than 7 years, with exception of year 2001, 2003 and 2008. Red and black soils are found across the zone, having poor in organic carbon, nitrogen and phosphorus content. The zone is characterized by semi arid climate with recurrent drought make cultivation an extremely difficult job with uncertain returns. The region faced frequent moisture stress periods during *kharif* season due to breaks in monsoon rains or early withdrawal of monsoon. The drought periods extend to more than two weeks and often occur during mid and later part of the rainy season in August extending to September and merging with the withdrawal of monsoon. The proper understanding and efficient utilization of the natural resources especially rainfall is therefore, of great concern for the improvement and sustainability of agriculture in rainfed areas. Hence , there is a need to quantify regional rainfall variability to assess its effect on agricultural productivity. The present paper deals with annual, seasonal, monthly and weekly variability, initial and conditional probabilities of wet and dry weeks and suitable crop planning strategies for *kharif* and *rabi* seson in Jhansi district of Bundelkhand agoclimatic zone of Uttar Pradesh.

Materials and Methods

Daily rainfall data for 34 years (1975-2008) was collected from agometeorological observatory situated at Regional Research Station, C.S.A. University of Agriculture and Technology, Bharari campus, Jhansi. The weekly, monthly, seasonal, annual and decadal rainfall pattern were critically examined and analyzed. The standard deviation and coefficient of variation were also worked out. The weekly initial and conditional probabilities of rainfall were estimated by using Markov Chain method for getting 5, 15, 20 and 25 mm rainfall in a given week (Robertson, 1982). Monthly probability of rainfall an amount expected at 90, 75, 50, 25 and 10 per cent confidence level was computed by Incomplete Gamma Distribution (Tom, 1958). The analysis of droughts were calculated as per India Meteorological Department criteria as "mild", "moderate" and "severe" based on departure of actual rainfall from the normal or average rainfall of the available data. If the rainfall departure up to 25 per cent from the normal, it termed as mild drought; if the departure range between 25 to 50 per cent from the normal rainfall, it is called moderate drought and if the departure range between 50 to 75 per cent from the normal rainfall, it is called severe drought.

Results and Discussion

Analysis of annual rainfall variability

The mean annual rainfall of Jhansi in bundelkhand zone of Uttar Pradesh for the past 34 years (1975-2008) was found to be 878.3 mm with standard deviation 254.7 mm and coefficient of variation 29.0 per cent. The annual rainfall amount ranged from 385.3 mm (the lowest during the year 2006) to 1307.5mm (the highest during the year 2008) over the years. Out of 34 years of annual rainfall, 18 years recorded excess rainfall over average or normal (878.3 mm) while 16 years recorded below the normal (Fig. 1). A decrease trend in decadal annual mean rainfall has been observed during 34 years. The decade (1975-84) recorded the highest mean annual rainfall (951.4 mm) while the lowest rainfall (848.2 mm) received during the decade 1985-95 (Table 1). The annual rainfall pattern also showed that the frequency of droughts have increased since the year 2000 onwards and zone

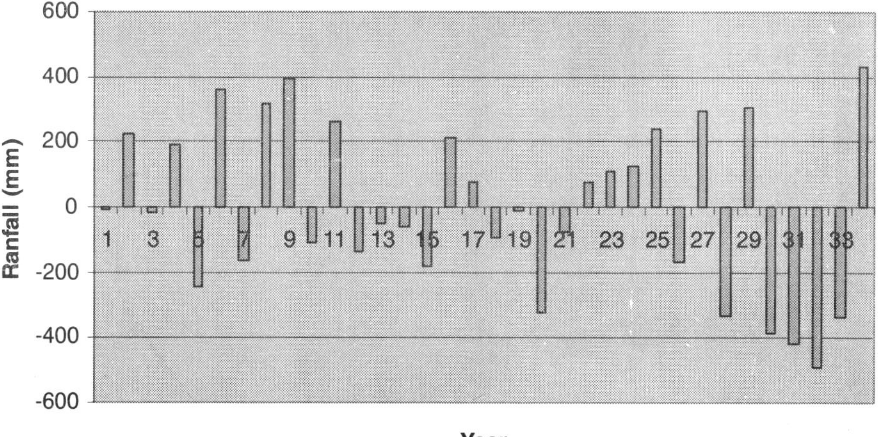

Fig. 1 : Annual variability of rainfall over normal of Jhansi in bundelkhand zone of U.P. (1975-2008)

has been facing continuous drought except the years 2001, 2003 and 2008 when rainfall was adequate, this continuous failure of rainfall has meant that agricultural land can neither provide work nor livelihood to the farmers who are dependent on it (Table 2).

Table 1 : Decadal shift in rainfall pattern of Jhansi in Bundelkhand zone of U.P. (1975-2008).

Month/ Year	1975-84	1985-94	1995-2004	2005-08
Jan	12.2	10.6	8.3	1.15
Feb	14.9	19.3	7.46	7.025
March	3.6	6.0	4.81	16.85
April	3.3	3.8	3.19	0.3
May	16.4	11.9	14.68	9.85
June	90.5	59.1	92.92	212.8
July	277.2	240.0	235.12	179.175
Aug	365.5	263.3	285.35	145.375
Sep	164.3	173.1	195.79	86.075
Oct	13.5	47.9	27.78	1.025
Nov	7.9	1.6	3.11	7
Dec	2.7	11.8	18.77	5.425
Annual	**972.0**	**848.2**	**897.28**	**672.05**

Table 2 : Annual drought analysis of Jhansi in Bundelkhand zone of U.P. (1975- 08).

Year	Rainfall (mm)	Category	Year	Rainfall (mm)	Category	Year	Rainfall (mm)	Category
1975	875.4	Normal	1987	828.2	Normal	1999	1118.8	Normal
1976	1104.3	Normal	1988	815.9	Normal	2000	711.6	Mild
1977	859.5	Normal	1989	700.5	Mild	2001	1173.7	Normal
1978	1066.9	Normal	1990	1092.4	Normal	2002	546.3	Moderate
1979	632.5	Moderate	1991	954.9	Normal	2003	1184.5	Normal
1980	1236.7	Normal	1992	786.4	Normal	2004	491.3	Moderate
1981	712.7	Mild	1993	868.1	Normal	2005	457.2	Moderate
1982	1193.3	Normal	1994	558.0	Moderate	2006	385.3	Severe
1983	1268.8	Normal	1995	802.9	Normal	2007	538.2	Moderate
1984	768.4	Normal	1996	952.7	Normal	2008	1307.5	Normal
1985	1137.9	Normal	1997	985.4	Normal			
1986	739.8	Mild	1998	1005.6	Normal			

Analysis of seasonal rainfall variability

A year is divided into three seasons like monsoon or *kharif* season (June- September), winter or *rabi* season (October to February) and in pre monsoon or summer season (March–May). The mean seasonal rainfall and its variability during *kharif, rabi* and summer season are presented in Table 3. The highest rainfall was received during *kharif* season (791.6 mm) a major part of it generally lost through runoff water which can be stored through *In- situ* management or *ex situ* water harvesting structures and used during *kharif* /*rabi*/summer seasons for growing crops like urdbean, moongbean, sesame, gram, lentil, pea, mustard, linseed, barley, wheat, fodder and vegetables etc. it can also be utilized as life saving irrigation particularly during dry periods because a dry period of one or two week in rainy season also adversely affect the standing *kharif* crops in the region. The mean of rainfall received during *rab*i season is 63.6mm and lowest in summer season (23.1 mm). The percentage contribution of seasonal rainfall to the total was 90.3, 7.1 and 2.6 during kharif, *rabi* and summer season, respectively with the lowest coefficient of variation during *kharif* (32.3%) followed by summer (96.0%) and *rabi* (179.4%) seasons. Summer season (March–May) rainfall amounting to 23.0 mm would be helpful for summer ploughing operation.

Table 3 : Seasonal rainfall pattern of Jhansi (1975-2008).

S.No.	Season	Average rainfall (mm)	Percent	SD	CV
1	Summer or pre monsoon (March -May)	23.1	2.6	21.1	96.0
2	Monsoon or *Kharif* (June - Sept)	791.6	90.3	258.1	32.3
3	Winter or *rabi* (Oct.-February)	63.6	7.1	56.4	179.4
5	**Annual**	**878.3**	**100.0**	**254.7**	**29.0**

Analysis of monthly rainfall variability

The monthly rainfall probability at 10, 25, 50, 75, and 90 per cent confidence level was computed using incomplete gamma distribution model (Fig. 2). It is observed that at 50 per cent level of confidence 96.2 mm rainfall received during the month of June while at 75 per cent level of confidence it was only 21.1 mm. The rainfall received during June in Bundelkhand at least assured level which can be utilized for ploughing and land preparation for *kharif* crops. The assured rainfall or at 75 per cent chances received during the month of July (172.7 mm) and August (185.5 mm). July and August are major (62% of the annual) rainfall contributing months and having lower variability while minor average rainfall received during November (4.2 mm) and April (2.9 mm) months with high variability (Table 4). Gupta *et. al.* (1975) also suggested that the rainfall at 80% probability could safely be taken as assured rainfall while 50% is the medium limit for taking risk.

Table 4 : Monthly variability of rainfall of Jhansi in bundelkhand zone of U.P. (1975-2008).

Month	Mean	Percent	SD	CV	mean+sd	mean-sd
Jan	9.3	1.1	13.9	149.6	24.2	-4.6
Feb	13.1	1.5	24.5	187.6	39.0	-11.4
March	6.2	0.7	9.3	150.0	16.2	-3.1
April	3.1	0.3	7.1	232.0	10.5	-4.0
May	13.8	1.6	18.3	132.5	33.6	-4.5
June	96.4	11.0	116.6	121.0	223.9	-20.2
July	242.3	27.6	99.6	41.1	369.6	142.7
Aug	286.0	32.6	156.6	54.8	475.1	129.4
Sep	166.9	19.0	164.7	98.6	350.6	2.3
Oct	26.3	3.0	57.2	217.2	86.6	-30.9
Nov	4.5	0.5	12.3	271.8	17.4	-7.8
Dec	10.4	1.2	28.9	277.6	40.5	-18.5
Annual	**878.3**	**100.0**	**254.7**	**29.0**	**1233.0**	**623.7**

Analysis of weekly rainfall variability

The weekly rainfall probability and pattern of deficit and surplus rainfall are very useful for suitable crop planning and scientific use of rainwater in rainfed areas for getting maximum production. The weekly initial wet probability P(W) and conditional wet-wet probability of rainfall P(W/W) were worked out by using Markov Chain Model is given in Table 5. Study revealed that the weekly initial and conditional probabilities of getting 25 mm rainfall were more than those at 50% expected during 27th to 37th week. The weekly rainfall of 20 mm or more at 75% probability level can be expected during 29th, 31st, 32nd and 33rd SMW, respectively (Fig. 3). It is observed that the south-west monsoon normally commenced during 25th (18-24 June) week but earliest 23rd week and latest by 32nd week and withdrawal during 38th (17-23 September) week but earliest at 35 week and latest by 42nd week at Jhansi. The mean weekly rainfall during 27th to 37th SMW varied from 48.4 to 70.9 mm and considered as a most stable period and total average length of growing period is 11 weeks at Jhansi. An expected 75 per cent rainfall to the total average of annual is received during 27th to 37th SMW. The collection of surface runoff during the excess rainfall period and its use during the subsequent dry period. The probability of getting 20 mm or more rainfall exceeded 50 per cent level of confidence during 25th SMW and continued till 37th (10-16th September) SMW.

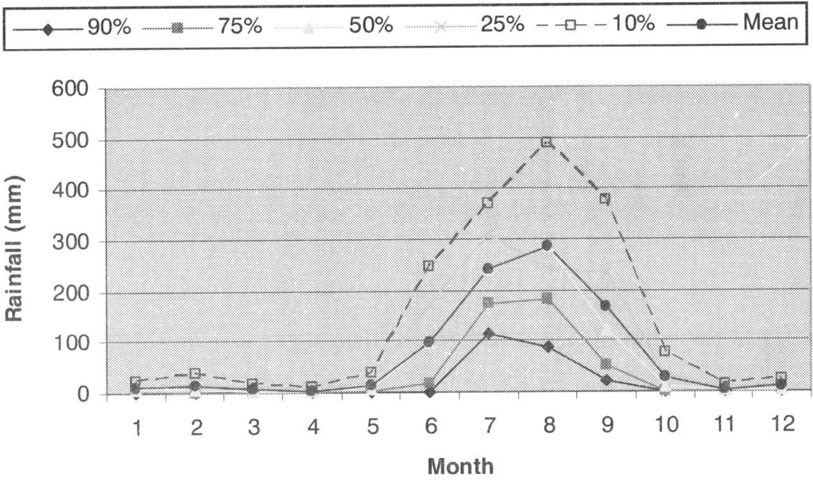

Fig. 2 : Monthly rainfall probability at different levels of Jhansi in bundelkhand zone of U.P. (1975-2008).

Initial and conditional probability of wet and dry weeks

The analysis revealed that the occurrence of 70 per cent of initial probability of a dry week from 1 to 24th week in the zone and the conditional probability of wet week preceded by a wet week is also high from 26th to 37th week. The probability of occurrence more than

50 per cent of getting 20 mm rainfall for two consecutive wet weeks is from 27th to 36th week, but it is varied, in case of occurrence of three consecutive weeks is never above 55 per cent even during the peak season of south west monsoon (July and August), this shows that the live saving or supplemental irrigation facilities could be created to mitigate the effect of dry spells during reproductive stages of crop growth and also select short duration and drought tolerant varieties of growing crops.

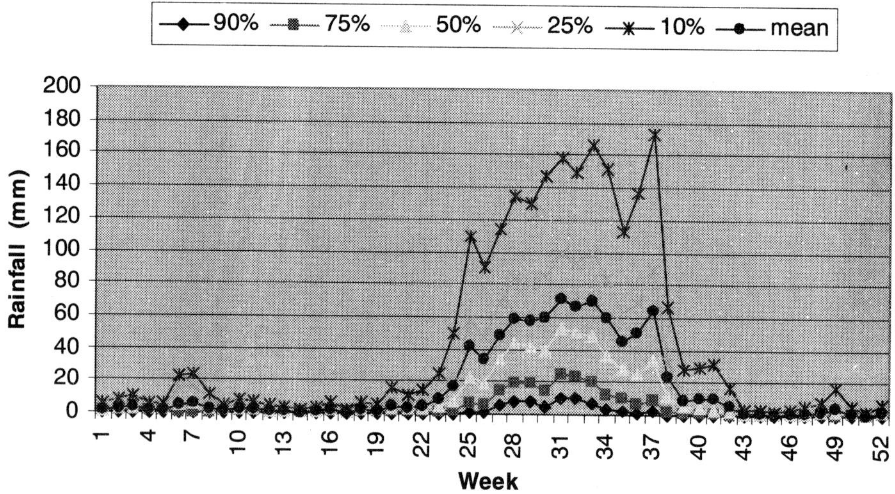

Fig. 3 : Weekly rainfall probability at different levels of Jhansi under bundelkhand zone of U.P. (1975-08).

Crop planning strategies for *kharif* season

In view of high erratic rainfall, runoff and erosion due to rolling topography, frequently occurrence of early, intermittent and late season droughts and decrease in decadal rainfall pattern in bundelkhand zone of Uttar Pradesh is the main reason of low productivity of crops than the national average. Rainfall is the major deciding factor for success and failure of agriculture in the region. Region comprises about 56 per cent of red soils (*Rakar* and *Parwa*) found in upland situation having course texture, shallow in depth, poor in organic carbon content and moisture retention capacity, due to rolling topography and high rainfall intensity causing high runoff and severe erosion problem. The probability of getting onset of rainfall occurred during 25th to 27th week between 50 per cent. All the tillage practices and seed bed preparations should be done within the 25th to 26th week to avail the subsequent weekly rainfall. Sowing of early and long duration crops like groundnut, maize, jowar and soybean should be done immediately so that germination will take place with available moisture at root zone, but on other hand rainfall at 75 per cent probability level during 26th to 27th week is less than the PET value which indicates shortage of moisture. After 28th week, the rainfall is more and runoff is likely to occur so there is huge scope to harvest excess amount of rain water and recycling or live saving irrigation at critical stages of crop growth during drought period. The feasibility of growing recommended and drought tolerant short duration varieties of these crops like groundnut (TG 37 A and Kaushal), sorghum (CSV-13 and 15, Bundela, CSH-23 and SPV 1616), soybean (P.K. 475 and J S 335) and cowpea with intercropping of pigeon pea by 26 to 27th SMW and in case of late situation sesame (Shekhar, Pragati, Tarun and T-78), urdbean (Uttara, PU-31, Azad Urd-3 and N. Urd -1), moongbean (Samrat, Narendra moong-1), maize (Ganga-1 , 2 and 5), pearl millet (BJ-104, WCC 75), cowpea (UPC 4200 and UPC 5287) and clusterbean in *kharif* season could also be sown by 28 to 29th SMW in this zone which can be harvested by 40th SMW.

Adoption of suitable moisture conservation techniques such as mulching, use of anti-traspirants, effective control of weeds, adequate plant stands per unit area helps in better crop production under moisture stress environment or dry spells periods. Conservation of every drop of water to produce more food with less water or more crop per drop of water is very important. Proper land leveling and grading ensures easy and equal distribution of irrigation water resulting increase the crop yields. Flooding method of irrigation having low water application efficiency and require more water, It should be apply high tech methods like sprinkler and drip for achieving higher water application efficiency, gradually these methods are gaining more popularity among the farmers of Bundelkhand region because these are water saving techniques. Contour farming, conservation tillage, trenching, mixed and intercropping and agroforestry techniques and addition of organic matter through crop residue management and green manuring can conserve moisture, minimize evaporation losses and increase water holding capacity of soil will increase the crop yields as fulfill the food, fuel, fodder and fiber needs of the local people to check migration during drought periods.

Table 5 : Weekly initial, conditional, consecutive dry and wet probability of 25 mm rainfall of Jhansi in Bundelkhand zone of U.P. (1975-2008).

Week	Initial		Conditional		Consecutive		Consecutive		Mean
	P(W)	P(D)	P(W/W)	P(D/D)	P(2D)	P(3D)	P(2W)	P(3W)	(mm)
1	0.030	0.970	0.000	0.970	0.937	0.906	0.000	0.000	1.3
2	0.029	0.971	0.000	0.970	0.938	0.907	0.000	0.000	2.4
3	0.029	0.971	0.000	0.970	0.938	0.938	0.000	0.000	3.1
4	0.029	0.971	0.000	0.970	0.969	0.908	0.000	0.000	1.5
5	0.000	1.000	0.000	1.000	0.938	0.875	0.000	0.000	1.4
6	0.059	0.941	0.000	0.941	0.875	0.875	0.000	0.000	4.4
7	0.059	0.941	0.000	0.938	0.938	0.938	0.031	0.000	5.1
8	0.029	0.971	0.500	1.000	0.969	0.939	0.000	0.000	2.7
9	0.000	1.000	0.000	1.000	0.969	0.969	0.000	0.000	1.2
10	0.029	0.971	0.000	0.971	0.969	0.969	0.000	0.000	2.3
11	0.000	1.000	0.000	1.000	1.000	1.000	0.000	0.000	1.8
12	0.000	1.000	0.000	1.000	1.000	1.000	0.000	0.000	1.0
13	0.000	1.000	0.000	1.000	1.000	1.000	0.000	0.000	0.6
14	0.000	1.000	0.000	1.000	1.000	1.000	0.000	0.000	0.3
15	0.000	1.000	0.000	1.000	1.000	1.000	0.000	0.000	0.7
16	0.000	1.000	0.000	1.000	1.000	1.000	0.000	0.000	1.7
17	0.000	1.000	0.000	1.000	1.000	1.000	0.000	0.000	0.3
18	0.000	1.000	0.000	1.000	1.000	0.938	0.000	0.000	2.1
19	0.000	1.000	0.000	1.000	0.938	0.938	0.000		1.5
20	0.059	0.941	0.000	0.941	0.938	0.877	0.031	0.000	5.0
21	0.029	0.971	0.500	1.000	0.906	0.785	0.000	0.000	3.6
22	0.059	0.941	0.000	0.939	0.813	0.784	0.000	0.000	4.8
23	0.118	0.882	0.000	0.875	0.844	0.495	0.063	0.042	8.5
24	0.147	0.853	0.500	0.900	0.531	0.325	0.063	0.027	17.2
25	0.441	0.559	0.600	0.586	0.344	0.109	0.188	0.101	41.3
26	0.441	0.559	0.467	0.579	0.188	0.313	0.219	0.153	34.2
27	0.647	0.353	0.600	0.316	0.063	0.156	0.438	0.310	48.4
28	0.765	0.235	0.727	0.167	0.625	0.278	0.531	0.370	59.3
29	0.677	0.324	0.654	0.250	0.125	0.568	0.500	0.429	57.1
30	0.647	0.353	0.696	0.455	0.156	0.586	0.563	0.422	60.4
31	0.735	0.265	0.864	0.500	0.094	0.021	0.563	0.416	70.9
32	0.706	0.294	0.760	0.444	0.063	0.023	0.531	0.310	67.0
33	0.735	0.265	0.750	0.300	0.094	0.065	0.438	0.345	70.5
34	0.588	0.412	0.560	0.333	0.281	0.173	0.469	0.271	59.7
35	0.588	0.412	0.800	0.714	0.250	0.156	0.344	0.279	45.0
36	0.500	0.500	0.600	0.643	0.313	0.216	0.406	0.150	50.5
37	0.588	0.412	0.765	0.588	0.281	0.255	0.219	0.040	64.8
38	0.324	0.677	0.350	0.714	0.594	0.530	0.063	0.000	23.9
39	0.118	0.882	0.182	0.913	0.781	0.727	0.000	0.000	9.6
40	0.088	0.912	0.000	0.900	0.844	0.786	0.031	0.000	9.9

contd...

41	0.088	0.912	0.333	0.936	0.844	0.844	0.000	0.000	10.0
42	0.059	0.941	0.000	0.936	0.938	0.938	0.000	0.000	5.2
43	0.000	1.000	0.000	1.000	1.000	1.000	0.000	0.000	0.7
44	0.000	1.000	0.000	1.000	1.000	1.000	0.000	0.000	0.6
45	0.000	1.000	0.000	1.000	1.000	1.000	0.000	0.000	0.4
46	0.000	1.000	0.000	1.000	1.000	0.969	0.000	0.000	0.6
47	0.000	1.000	0.000	1.000	0.969	0.938	0.000	0.000	1.3
48	0.029	0.971	0.000	1.000	0.938	0.907	0.000	0.000	2.2
49	0.029	0.971	0.000	1.000	0.938	0.938	0.000	0.000	5.0
50	0.029	0.971	0.000	1.000	0.969	0.969	0.000	0.000	1.5
51	0.000	1.000	0.000	1.000	1.000	0.000	0.000	0.000	0.2
52	0	1.000	0.000	0.000	0.000	0.000	0.000	0.000	2.0

Crop planning strategies for *rabi* season

The probability of getting 30 mm rainfall at 30 per cent level of confidence between 39 to 41th week. Therefore limited chances of receiving adequate amount of moisture at the root zone level in red soils for proper germination of *rabi* season crops. It is only possible by pre sowing irrigation through water harvesting ponds. Crops like barley, duram wheat, mustard, linseed, lentil, and chickpea with intercropping can also be taken up with residual moisture and with one and two live saving irrigation during critical growth stages in black soils. The above results also indicate that the rainfall occurred during 39th SMW or later will allow the soil to be recharged that will help in sowing rainfed *rabi* crops in first fortnight of October. Wheat is the major *rabi* crop growing in the zone and with available assured irrigation thiough water harvesting reservoirs starting from first fortnight of November.

Conclusion

On the basis of probability analysis of rainfall, it is concluded that the deep summer ploughing against the slope should be done immediate after harvesting of *rabi* season crops to take advantage of soil exposed to high temperature, aeration, incorporation of crop residue and conserve moisture available from summer season rainfall under rainfed condition of Bundelkhand zone of U.P. It is also advisable to planting early short duration, recommended and drought resistant varieties of groundnut, maize, pearl millet and sorghum by 25 to 27th SMW while in case of late condition urdbean, moongbean, sesame, intercropping with pigeon pea should be planted by 29th to 30th SMW. It would be better to make ridging and compartmental bunding against the slope after immediate sowing of *kharif* season crops for *In-situ* water harvesting in place of unbunded method where the harvesting of rain water is nil. The collection of surface runoff during the excess rainfall period (27 to 36th week) and its use during the subsequent dry period. Addition of FYM, green manuring of dhaincha, crop residues management would improve the health of soils and increase the yield of crops.

References

Gupta, R.K.; Rambabu and Tejwani K.G. (1975). Weekly rainfall of India for crop planning programme. *Soil Cons. Digest,* **3** :31-39.

Kar, G. (2003). Initial and conditional probabilities of rainfall and wet and dry spells for red and laterite zone of West Bengal using Marcov Chain model. *Indian J. Soil Cons.* **31**(3): 287-290.

Panigrahi, B. and Panda, S.N. (2002). Dry spell probability by Marcov Chain model and its application to crop planning in Kharagpur. . *Indian J. Soil Cons.* **30**(1): 95-100.

Ray, C.R., Senapati, P.C. and Lal, R. (1980). Rainfall analysis for crop planning at Gopalpur, Orissa. *J. Agric. Eng.* I.S.A.E., **17** : 384.

Robertson, G.W. (1982). Frequency and probability of dry and wet spells. WMO Tech. Note. **179**: 149-153.

Tom, H.C.S. (1958). A note on the Gamma Distribution. *Monthly Weather Review* **86** (4): 117-122.

Virmani, S.M. (1994). Climatic resources characterization in stresses tropical environment: Constraints and opportunities for sustainable agriculture. In: Stressed ecosystem and sustainable agriculture. (Eds. Virmani, S.M., Katyal, J.C. Eswaran, E. and Abroll, I.P.), pp 149-160.

□□□

Section - 2

Management and Efficient Utilization of Water Resources

Chapter 8

A Decision Support Tool to Explore and Promote Conjuctive use Options in Canal Command

A. UPADHYAYA[1], A.K. SIKKA[2], A.K. SINGH[1] AND P.R. BHATNAGAR[3]

[1] Senior Scientist, ICAR Research Complex for Eastern Region, ICAR Parisar, Patna-14 [2] Technical Expert (Watershed Development), National Rainfed Area Authority, New Delhi [3] Project Co-ordinator (APA), CIPHET, Ludhiana

Survival of the agrarian economy of world largely depends upon the sustainable use of land and water resources. India is perceived to be endowed with large and reasonably good land and water resources, but with alarming increase in population since last 30 to 40 years per capita resource availability is progressively decreasing. On global basis the irrigated agriculture covers only 16% of the total land under cultivation while it contributes about 36% of the total food production. This supports the fact that irrigation is the most potential and effective means of increasing agricultural productivity. It is however realized that water is a scarce resource, which goes into many competing and conflicting processes and requires a comprehensive planning for appropriate utilization. Development of these resources by individuals has not only created unfavorable conditions for crop production and environmental sustainability but also raised the energy uses in agriculture for irrigation. At the same time intensive irrigation in various canal commands has created the problems of water logging and salinity, making the soil unproductive while overexploitation of groundwater in certain parts of the country has resulted in lowering of water tables and causing decline in yield and productivity of wells, saltwater intrusion along the coast, drying of springs and shallow dug wells, increase in cost of lifting etc. Hence, it is advantageous to look water as a single resource irrespective of its source of development, rain, surface or groundwater. Planning for conjunctive use requires basic understanding of the processes and principles involved in water resource management. A well-formulated conjunctive use programme, must consider fulfilling the irrigation requirement of crops during lean period by other sources of water when supply through canal is not enough to meet the irrigation requirement. In addition it must also take in to account the conservation of rainwater along with augmenting the groundwater storage through recharge.

Conjunctive use models developed earlier may be classified as simulation models, dynamic programming models, linear programming models, hierarchical optimization models, nonlinear programming models and others as reported by Vedula *et. al.,* (2005). The conjunctive use models developed and employed by various researchers like Bredehoeft and Young (1983), Latif and James (1991), Belaineh *et. al.,* (1999), Marino (2001) Barlow *et. al.,* (2003), Rao *et. al.,* (2004) can be classified under a single or combination of categories mentioned above. Vedula *et al.* (2005) attempted to model conjunctive use and developed a stable operating policy for optimal allocation of surface and ground waters for irrigating multiple crops in a canal command area.

Lakshminarayana and Rajagopalan (1977) studied the problem of optimal cropping pattern and water releases from canals and tubewells in the Bari Doab basin, India using a linear programming model. The model is deterministic and the dynamic response of ground water aquifer is not considered. In most of the cases of conjunctive use programmes in India, what is being practiced in the country at the moment does not really envisage the optimal use of rain, surface and groundwater resources (Prasad, 1993). None of the states, so far have formulated clear-cut operational plan for conjunctive use in any irrigation projects though bulk of attempts through research and extension agencies have been undertaken and developed technologies have been disseminated in the country. Reviews indicate that no irrigation project, however, has either been planned, or designed and operated, on conjunctive use principles so far, which can incorporate the institutional framework, socio-economic factors of wide range of stakeholders and policy guidelines for development of water resources along with the technical aspects of canal management due to unresolved technological issues, absence of a methodological framework, and certain inhibiting policy and socioeconomic factors. Upadhyaya *et. al.*, (2004) developed a conceptual framework for a decision support system based on prevailing constraints faced by farmers and opportunities available to them.

In this paper attempts have been made to present the constraints in conjunctive use identified through survey and interaction with farmers to explore and promote conjunctive use options in the canal command with the help of an interactive decision support tool.

Study area

In order to understand the constraints related to conjunctive use and explore possibilities for its promotion RP Channel V distributary of Patna Main Canal under Sone Canal system in Bihar was selected. In this area conjunctive use of rain, surface and ground water is poorly developed, despite the chronic unreliability and shortage of canal water. Tubewell irrigation is practiced mostly at the tail end of RP Channel-V, where canal water supply is poor. Although many wells and tubewells have been created, as shown in Table 1, yet farmers use them rarely.

Table 1 : Wells and Pumpsets in RP channel V.

Reach	Number of wells			Pump sets	
	Open	Tubewells	Open+ Tubewells	Motor	Diesel
Head	57	26	2	-	28
Middle	69	46	4	2	45
Tail	131	101	14	17	98

The cost of tubewell water varies from Rs. 40/- to Rs. 70/- per hour depending on location, time and amount of water supply, whereas cost of canal water supply in the command varies from Rs. 217 to Rs 370.50 per ha depending on the crop. The cost of canal water supply is given in Table 2 below.

Table 2 : Cost of canal water (Rs/ ha per crop season).

S. No.	Crop	1985-95	1995-2001	2001-02 till date
1	Rice	89.41	172.9	217.36
2	Wheat	51.13	148.2	185.25
3	Sugarcane and vegetables	157.59	296.4	370.50

Low cost of canal water misleads cultivators into avoiding use of water sources that are more expensive. Hence they avoid the cost of paying for water from wells or tubewells in the belief that this would be unnecessary in the event that canal water becomes available, or rain occurs. Because canal water is inadequate, irregular, untimely and uncertain, many farmers wait to buy water until it is too late and their crops fail.

Constraints in conjunctive use based on farmers' survey

Before propagating the concept of utilizing rain, surface and ground water conjunctively in the canal command farmers were interviewed through a questionnaire and their responses towards conjunctive use were analyzed. Major constraints in adoption of conjunctive use as reported by farmers are:

No land consolidation

Fifty five percent farmers in the head reach, 35% farmers in the middle reach and 32% farmers in the tail reach of RP Channel V distributary reported that small and fragmented land holding and non-existence of land consolidation were the major problems in owning a tubewell and utilizing groundwater along with canal and rain water conjunctively.

Uneconomical because it needs high initial investment

All the farmers reported that utilization of ground water has become uneconomical because initial investment ranges from Rs. 25,000/- to Rs. 40,000/- depending on the selection of pump, engine/motor and other accessories, whereas canal water though irregular and unreliable, but available at much cheaper rate.

High recurring expenditure due to increase in price of diesel

All the farmers were of the opinion that due to hike in price of diesel recurring expenditure increases and it discourages the farmers to withdrawal ground water and utilize it for crop production.

Frequent power failure

Power supply in the villages is totally unreliable and there is a frequent failure so electric motors are very limited in use and farmers are compelled to use diesel engine in place of electric motors.

Lack of awareness about selection of pump, motor/engine as per farmers' requirement

Sixty percent farmers reported that they don't have knowledge about selection of pump, motor/engine and other accessories and they purchase either on the basis of experience of other farmers or on the advice of supplier. Due to this they don't get the expected output of pump and face problem due to early failure.

Lack of trained manpower to efficiently operate and timely maintain the pump, engine/motor and other attachment

All the farmers reported that trained and efficient mechanic is not easily available and accessible in affordable price. Due to this repair and maintenance of pumping unit is not done at proper time, which results in frequent loss of time and money.

Transportation from one place to another is difficult

Some people have boring in their fields but don't have pumping unit, whereas some people have pumping unit but no boring. These people face problem in transportation of pumping unit due to heaviness of whole assembly.

Advantages of conjunctive use quoted by farmers

When farmers were asked about the advantages of conjunctive use, they reported that ground water is assured source of water supply and we can use it efficiently and economically as per requirement of the crop. Sixty three percent farmers in the head reach, 60% farmers in the middle reach and 95% farmers in the tail reach, who are also using ground water, reported that use of ground water only during nursery period, helped them in attaining higher yield of rice. The farmers requested to demonstrate the concept of conjunctive use and conditions under which it is economically beneficial to them.

This response of farmers inspired for development of a decision support tool in order to analyze the possibility of ground water utilization and promotion of conjunctive use in the canal command.

Decision support tool for conjunctive use

An interactive decision support tool in Visual Basic was developed in english and hindi versions. A pictorial view of english version with example is given below.

Economical Analysis of Tubewell and Canal Irrigation

Fixed Cost of Pumping

	Tubewell Own	Tubewell Rented
Cost of Boring (Rs)	6200	0
Cost of Pump (Rs)	2000	0
Cost of diesel engine/electric motor (Rs)	12000	0
Cost of fittings and accessories (Rs)	1200	0
Total Cost (Rs)	21400	0

	Tubewell Own	Tubewell Rented
Salvage value of total system (Rs)	5000	0
Interest Rate(%)	8	0
Annual interest cost (Rs)	656	0

Canal water chages Rs/Katha

Rice	Wheat	Other crops
2.75	2.34	3.75

Depreciation cost

	Tubewell Own	Tubewell Rented
Pump Life (years)	15	0
Pump salvage value (Rs)	400	0
Depreciation Cost of Pump (Rs)	107	0
Life of Diesel engine (years)	12	0
Salvage value of diesel engine (Rs)	4000	0
Depreciation cost of diesel engine (Rs)	667	0

	Tubewell Own	Tubewell Rented
Life of fittings and accessories (years)	3	6
Salvage value of fitting & accessories (Rs)	300	0
Depreciation cost of fitting & accessories (Rs)	300	0
Total depreciation cost (Rs)	1074	0
Total Fixed Cost (Rs)	1730	0

Cost of Irrigation

	Tubewell Own	Tubewell Rented	Canal
Operating Cost (Rs/hr)		60	
Fuel Consumption (L/hr)	1.5		
BHP of Engine	8		
Specific Fuel Consumption (L/BHP-Hr)	0.1875		
Irrigation during Kharif			
Area irrigated during Kharif Nursery (Katha)	8	8	8
Hours of operation to irrigate one Katha nursery once	0.6	0.6	0.6
No. of irrigations during nursery	2	2	2
Area irrigated during Kharif Rice (Katha)	80	80	80
Hours of operation to irrigate per Katha of Kharif Rice once	0	0	0.5
No. of Irrigations for Kharif Rice	0	0	7
Area irrigated during Kharif Crop1 (Katha)	0	0	0
Hours of operation to irrigate one Katha of Kharif Crop1 once	0	0	0
No. of Irrigations for Kharif Crop1	0	0	0
Area irrigated during Kharif Crop2 (Katha)	0	0	0
Hours of Operation to irrigate one Katha of Kharif Crop2 once	0	0	0
No. of Irrigations for Kharif Crop2	0	0	0
Total Area during Kharif Season (Katha)	80	80	80

Cost of Irrigation

Irrigation during Rabi & other crops	Tubewell Own	Tubewell Rented	Canal
Area irrigated during Rabi Crop1 (Katha)	60	60	60
Hours of operation to irrigate one Katha of Rabi Crop1 once	0.6	0.6	0.6
No. of Irrigations for Rabi Crop1	2	2	3
Area irrigated during Rabi Crop 2 (Katha)	0	0	0
Hours of operation to irrigate one Katha of Rabi Crop 2 once	0	0	0
No. of Irrigations for Rabi Crop 2	0	0	0
Area irrigated during Rabi Crop 3 (Katha)	0	0	0
Hours of operation to irrigate one Katha of Rabi Crop 3 once	0	0	0
No. of Irrigations for Rabi Crop 3	0	0	0
Total Area during Rabi atleast once Season (Katha)	60	60	60
Area irrigated during other Crop 1 (Katha)	0	0	0
Hours of operation to irrigate one Katha of other Crop 1 once	0	0	0

Exit | Save | Call Rec. | |< | << | >> | >| | Move Between Stored Rec.

Name	Location	Outlet	Distributory
Unknown Singh	RamNarayanPur	123	

Next Sheet

Cost of Irrigation

	Tubewell Own	Tubewell Rented	Canal
No. of Irrigations for other Crop1	0	0	0
Area Irrigated during other Crop 2 (Katha)	0	0	0
Hours of operation to irrigate per Katha of other Crop 2 once	0	0	0
No. of irrigations for other Crop 2 (Katha)	0	0	0
Total area irrigated under other crops atleast once (Katha)	0	0	0
Rate of fuel (Rs/litre)	36.2		
Total fuel consumed (litre)	122.4		
Total Annual Cost of Fuel (Rs)	4431		
Pump and engine maintenance and repair charges (Rs)	1000		
Operator's wages per day (Rs)	40		
Days of operation of pump in a year	11		
Annual Operator's Wages (Rs)	440		
Total Operational Cost (Rs)	5871		
Total Fixed and Operational Cost (Rs	7601	4920	360
Do you pay for Canal Water also? ☑	7961		5280
Excees expenditure in irrigation as compared to canal (Rs)	7601	4920	
Yield of Kharif Rice (Kg/Katha)	60	60	45
Yield of Kharif Crop1 (Kg/Katha)	0	0	0
Yield of Kharif Crop2 (Kg/Katha)	0	0	0
Total yield of Kharif Rice (Kg)	4800	4800	3600
Total yield of Kharif Crop 1 (Kg)	0	0	0
Total yield of Kharif Crop 2 (Kg)	0	0	0
Cost of Kharif Rice (Rs./Kg)	4.65	4.65	4.65
Cost of Kharif Crop 1 (Rs./Kg)	0	0	0

Cost of Irrigation

	Tubewell Own	Tubewell Rented	Canal
Cost of Kharif Crop 2 (Rs./Kg)	0	0	0
Yield of Rabi Crop 1 (Kg/Katha)	40	40	35
Yield of Rabi Crop 2 (Kg/Katha)	0	0	0
Yield of Rabi Crop 3 (Kg/Katha)	0	0	0
Total yield of Rabi Crop 1 (Kg)	2400	2400	2100
Total yield of Rabi Crop 2 (Kg)	0	0	0
Total yield of Rabi Crop 3 (Kg)	0	0	0
Cost of Rabi Crop 1 (Rs./Kg)	7	7	7
Cost of Rabi Crop 2 (Rs./Kg)	0	0	0
Cost of Rabi Crop 3 (Rs./Kg)	0	0	0
Yield of other Crop 1 (Kg/Katha)	0	0	0
Yield of other Crop 2 (Kg/Katha)	0	0	0
Total yield of other Crop 1 (Kg)	0	0	0
Total yield of other Crop 2 (Kg)	0	0	0
Cost of other Crop 1 (Rs/Kg)	0	0	0
Cost of other Crop 2 (Rs/Kg)	0	0	0
Cost of total produce (Rs/Katha)	559	559	454.25
Total cost of produce (Rs)	39120	39120	31440
Req. increase in rice equivalent yield (Kg/Katha) to compensate cost of Irrigation	11.68	7.56	

Comparison of costs considering own tubewell

Conjunctive use of Tubewell and Canal is Economical - Go ahead

Comparison of costs considering rented tubewell (Rs/hr)

Conjunctive use of Tubewell and Canal is Economical - Go ahead

Exit Save Back

This tool calculates:

- Annual fixed and operational costs of irrigation from tubewells and canal
- Yield and total cost of produce
- Excess expenditure incurred in tubewell irrigation over and above the canal water charges
- The required yield increase needed to compensate for the additional cost of irrigation from tubewells.

The tool was demonstrated to water users and data was collected through the specified format from 150 farmers (50 farmers each at head, middle and tail reaches). The tool was found capable of making the required analyses and of convincing them to use it to make decisions about ground water use under the prevailing constraints they face. Along with canal irrigation the economic viability of tubewell irrigation under three situations were also analyzed to understand why conjunctive use is not so popular in this region?

These three situations are:

- Own tubewell
- Renting pumping sets to run tubewells for irrigation and
- Purchasing water from tubewell owners.

Analysis showed that farmers applied 2 to 3 irrigation during rice nursery and 2 to 3 irrigations during rabi season to irrigate wheat, the required yield increase to compensate the additional cost of irrigation from tubewell in case of the water users owning tubewell varied in the range of 0.069 to 1.28 t/ha in head, 0.067 to 1.51 t/ha in middle and 0.13 to 1.32 t/ha in tail reach of the canal, whereas for the water users renting pumping sets to run tubewells for irrigation the required yield increase varied in the range of 0.078 to 0.8 t/ha in head reach, 0.052 to 0.81 t/ha in middle reach and 0.16 to 0.78 t/ha in tail reach, and in case of water users purchasing water from tubewell owners it varied in the range of 0.13 to 0.26 t/ha in head reach, 0.16 to 0.35 t/ha in the middle reach and 0.18 to 0.52 t/ha in tail reach of canal. The results reveal that compared to owning tubewell, water purchasing is the most economical option followed by getting only pumping sets on rent basis to run tubewells for irrigation.

Conclusion

Conjunctive use is defined as operation of surface and groundwater in such a way, which enhances their combined output. Since long, researchers and planners are trying to recommend the concept of conjunctive use in canal command as well as salt affected areas and many theoretical studies have been conducted in the past, but so far there is hardly any evidence about wide adoption and acceptance of conjunctive use practices among water users due to many constraints and mainly due to tremendous difference in cost of canal water and ground water. Since canal water is available at much cheaper rates, water users don't prefer ground water utilization unless they feel that their production will reduce drastically in the absence of water.

A user interactive decision support tool was developed both in hindi and english versions using Visual Basic platform and demonstrated among water users in order to create awareness and explore the possibility of conjunctive use in the canal command. Farmers found this tool quite helpful in understanding and analyzing the concept of conjunctive use and taking appropriate decision for its adoption under the prevailing constraints.

References

Barlow, P.M., Ahlfeld, D.P. and Dickerman, D.C. (2003). Conjunctive-management models for sustained yield of stream-aquifer systems. *J. Water Res. Plann. Manage.* **129** (1): 35-48.

Belaineh, G., Peralta, R.C. and Hughes, T.C. (1999). Simulation/optimization modeling for water resources management. *J. Water Res. Plann. Manage.* **125**(3): 154-161.

Bredehoeft, J.D. and Young, R.A. (1983). Conjunctive use of groundwater and surface water for irrigated agriculture:risk aversion. *Water Resour. Res.* **19** (5): 1111-1121.

Lakshminarayana, V. and Rajagopalan, P. (1977). Optimal cropping pattern for basin in India. *J. Irrig. Drainage Eng.* Div., ASCE **103** (1): 53-70.

Latif, M., James, L.D. (1991). Conjunctive water use to control waterlogging and salinization. *J. Water Resour. Plann. Manage.* Div., ASCE **117**(6): 611-628.

Marino, M.A. (2001). Conjunctive Management of Surface Water and Groundwater, Issue **268**: IAHS-AISH Publication, pp. 165-173.

Prasad, R.K. (1993). Conjunctive use of surface water and groundwater. Proc. of National workshop on action for optimum utilization of water resources, WAPCOS, New Delhi, India: 33-49.

Rao, S.V.N., Bhallamudi, S.M., Thandaveswara, B.S. and Mishra, G.C. (2004). Conjunctive use of surface and groundwater for coastal and deltaic systems. *J. Water Resour. Plann. Manage.* **130** (3): 255-267.

Upadhyaya, A., Singh, A.K., Bhatnagar, P.R. and Sikka, A.K. (2004). Is conjunctive use a feasible option? A simple tool for participatory evaluation of the economic benefits of groundwater exploitation. Poster presented in the Workshop on Realizing Potential: Livelihood, Poverty and Governance at NASC Complex, Pusa Campus, New Delhi, August 3-4, 2004.

Vedula, S., Majumdar, P.P. and Chandra Sekhar, G. (2005). Conjunctive use modeling for multicrop irrigation. *Agricultural Water Management* **73** : 193-221.

❏❏❏

Chapter 9

Ground Water Quality for Irrigation in Datia District, Madhya Pradesh — An Assessment

H. BISWAS, P.P. ADHIKARY, S.P. TIWARI AND SANT RAJ
Central Soil and Water Conservation Research and Training Institute, Research Centre, Datia-475 661 (M.P.)

Irrigation water quality, which refers to its suitability for agricultural use is an important factor influencing crop yields and soil quality. The concentration and composition of dissolved constituents in irrigation water determine its quality. The quality of ground water is the resultant of all the processes and reactions that act on the water that is considered in the atmosphere until the time it is discharged by a well or spring (Kesavan and Parameswari, 2005).

Owing to the occurrence of droughts in the recent years, the use of ground water for irrigation has picked up in Datia district of M.P. state. As per Department of Water Resources, Government of Madhya Pradesh, the ground water development in Datia is about 32% of its ultimate potential (93689.79 ha m), rendering it safe for usage as of now. However, studies on irrigation water quality have not been done on an extensive basis in the district. Keeping this in view, an assessment study was taken up at CSWCRTI Research Centre, Datia (M.P.) the documentation of which will benefit district planners and policy makers.

Materials and Methods

Datia, one of the smallest districts of Madhya Pradesh is spread over an area of 2.691 lakh ha, and is situated between 25°28′ and 28°20′ N latitude, and between 78°10′ and 78°45′ E longitude. It consists of three Blocks, *viz.* Datia, Bhander and Seondha. Water samples were collected in 1 litre clean poly-bottles from open and tube wells from 66 randomly selected sites (Table 1) from all three blocks of Datia district after cessation of monsoon season during October, 2007. The samples were analysed for pH, electrical conductivity, total dissolved solids (TDS), cations (calcium, magnesium, sodium and potassium) and anions (carbonate, bicarbonate and chloride) using standard methodology as outlined in APHA (1985). Residual sodium carbonate (RSC) and sodium adsorption ratio (SAR) were computed using the following relationships:

$$RSC = (CO_3^{-} + HCO_3^{-}) - (Ca^{++} + Mg^{++})$$

$$SAR = \frac{Na+}{[(Ca++ + Mg++)/2)]^{1/2}}$$

Where, the ionic concentrations are expressed in mili-equivalents per litre (meq l^{-1})

Results and Discussion

The range of values of different water quality parameters are presented in Table 2. A brief discussion of the analysis is being made below:

pH, EC and TDS: The range of pH values covering all the 66 sites is 7.1 to 8.4, which is within the desirable limits for irrigated agriculture (Ayres and Westcot, 1985). The electrical conductivity (EC) of ground water varied from 0.45 to 2.65 dS m^{-1}, which according to CSSRI, Karnal classification (Yaduvanshi, 2007) comes under good to marginally saline. The TDS values of the ground water samples varied between 189 and 1024 mg l^{-1}, which again under the CSSRI classification can be categorized under good to marginal. This suggests that at present there is no threat to the current cropping pattern in vogue in the district.

Table 1 : Location of sample sites in Datia District.

Sample No	Sample ID*	Location	Sample No	Sample ID	Location
1	S1	Murera-1	34	S34	Basai
2	S2	Murera-2	35	S35	Udguan
3	S3	Nayagaon-1	36	S36	Dang Karera
4	S4	Nayagaon -2	37	S37	Dheerpura
5	S5	Sitapur-1	38	S38	Chikau
6	S6	Sitapur-2	39	S39	Uchaad
7	S7	Maheba-1	40	S40	Bhadol
8	S8	Maheba-2	41	S41	Laxmanpura
9	S9	Bhiti-1	42	S42	Jonia
10	S10	Bhiti-2	43	S43	Laanch
11	S11	Bargain-1	44	S44	Indergarh
12	S12	Bargain-2	45	S45	Senguwa
13	S13	Rajghat	46	S46	Tharet
14	S14	Sita Sagar	47	S47	Charokhera
15	S15	Lala ka tal	48	S48	Bhaguwapura
16	S16	CSWCRTI (Richhari)	49	S49	Rampura
17	S17	CSWCRTI Farm	50	S50	Barokhari
18	S18	Sirol	51	S51	Khanjapura
19	S19	Bajni 1	52	S52	Dariavpur
20	S20	Bajni 2	53	S53	Maharajpura
21	S21	Bajni 3	54	S54	Seondha
22	S22	Prakash Nagar	55	S55	Magrol
23	S23	Chirula	56	S56	Madhopur
24	S24	Khairi	57	S57	Chandrol
25	S25	Unao	58	S58	Bisalpura
26	S26	Beekar	59	S59	Pandokher
27	S27	Dursada	60	S60	Udina
28	S28	Uprain	61	S61	Imliya
29	S29	Sindhwari	62	S62	Bhander
30	S30	Badonkala	63	S63	Bharroli
31	S31	Badoni khurd-1	64	S64	Dalpatpur
32	S32	Badoni khurd-2	65	S65	Sohan
33	S33	Himmatpur	66	S66	Bichhodana

*S1 to S36 – Datia Block; S37 to S55 – Seondha Block; S56 to S66 – Bhander Block

Cations (Ca^{++}, Mg^{++}, Na^{+} and K^{+}): The Ca^{++} and Mg^{++} contents of ground water from all the sampled sites of Datia district were within the permissible limits. Seventy three per cent of the samples from Bhander Block contained Na^{+} content above the permissible limits, while in Seondha Block, 11 and 42 per cent of the samples contained Na^{+} and K^{+} contents, respectively above the permissible limit. The K^{+} content of ground water exceeded the threshold limit in 42% of the samples. Therefore, in Bhander Block, the sodium sensitive crops, especially legumes should be grown with care.

Anions (CO_3^{-}, HCO_3^{-} and Cl^{-}): Carbonate was detected in traces in all the samples, while the bicarbonate (HCO_3^{-}) content was within the safe limits for all the samples. Similar was the case with chloride (Cl^{-}).

Irrigation water quality: The most important characteristic which determine suitability of ground water for irrigation purpose are salinity, residual sodium carbonate (RSC) and sodium adsorption ratio (SAR). Salinity is broadly related to total dissolved solids (TDS) and electrical conductivity (EC). High concentration of TDS and electrical conductivity in irrigation water may increase the soil salinity, which affect the salt intake of the plant. The salts present in the water, besides affecting the growth of the plants directly, also affect the soil structure, permeability and aeration, which indirectly affect the plant growth. The results of the ground water analysis of Datia district indicate that as of now there is not threat of a salinity hazard, although in some pockets the EC value of water is slightly marginal.

Although the bicarbonate (HCO_3^{-}) ion concentration was found to be within safe limits in the whole of the district, in 30% of the cases, the RSC was above the permissible limits (2.5 meq l^{-}) , indicating that the water must be used with caution for irrigation. Ground water containing high concentration of carbonate and bicarbonate ions tends to precipitate calcium and magnesium as carbonate. As a result, the relative proportion of sodium increases and gets fixed in the soil thereby decreasing the soil permeability.

The one index which is considered as effective for most water used in irrigated agriculture is the sodium adsorption ratio (SAR) (Ayers and Westcot, 1985). If the proportion of sodium is high, the alkali hazard is high; and conversely, if calcium and magnesium predominate, the hazard is less. As indicated above, more than 70% of the water samples in Bhander Block contain high amounts of sodium.

However, this is adequately balanced by the calcium ion, which is reflected in the SAR values being well within the safe limits (0-10).

Table 2 : Ranges in values of physico-chemical parameters (post-monsoon) of ground water in Datia during 2007.

Parameter	Range of values	Standard deviation	CV (%)	Maximum Permissible range/value*	% samples over permissible value
		Datia Block			
pH	7.1 - 8.0	0.22	2.9	6.0 - 8.5	-
EC (dS m^{-1})	0.51 - 2.09	0.55	50.4	3.0	-
TDS (mg l^{-1})	356-1024	397	54.1	1500	-
Ca^{++} (meq l^{-1})	2.0 - 6.4	0.99	25.7	20	-
Mg^{++} (meq l^{-1})	0.3 - 2.1	0.44	35.5	5	-
Na^{+} (mg l^{-1})	89 - 191	25.6	20.3	200	-
K^{+} (mg l^{-1})	0.3 - 5.1	1.96	72.3	2.0	25
CO_3^{-} (meq l^{-1})	Traces	-	-	1.0	-
HCO_3^{-} (meq l^{-1})	6.1 - 9.3	0.89	12.0	10.0	-
Cl^{-} (meq l^{-1})	0.6 - 6.5	1.54	59.9	30	-
RSC (meq l^{-1})	0.7 - 4.3	1.03	44.3	2.5	42
SAR	0.3-4.0	0.81	77.3	15	-
		Seondha Block			
pH	7.2 - 8.4	0.37	4.9	6.0 - 8.5	-
EC (dS m^{-1})	0.45 - 1.91	0.37	40.1	3.0	-
TDS (mg l^{-1})	189 - 964	200	43.0	1500	-
Ca^{++} (meq l^{-1})	2.9 - 7.2	1.10	27.1	20	-
Mg^{++} (meq l^{-1})	0.3 - 2.1	0.47	40.4	5	-
Na^{+} (mg l^{-1})	123 - 221	26.4	17.1	200	11
K^{+} (mg l^{-1})	0.3 - 2.6	0.63	49.9	2.0	21
CO_3^{-} (meq l^{-1})	Traces	-	-	1.0	-
HCO_3^{-} (meq l^{-1})	5.9 - 9.3	1.01	13.6	10.0	-
Cl^{-} (meq l^{-1})	0.9 - 3.5	0.79	37.9	30	-

RSC (meq l^{-1})	0.7 – 4.7	1.17	52.6	2.5	42
SAR	0.3 – 4.7	0.50	44.3	15	-
		Bhander Block			
pH	7.2 – 8.0	0.25	3.3	6.0 – 8.5	-
EC (dS m^{-1})	0.54 – 2.65	0.59	60.5	3.0	-
TDS (mg l^{-1})	215 – 512	106	28.5	1500	-
Ca^{++} (meq l^{-1})	2.9 – 5.0	0.56	13.9	20	-
Mg^{++} (meq l^{-1})	0.5 – 1.2	0.24	29.2	5	-
Na^{+} (mg l^{-1})	190 – 257	20.0	9.2	200	73
K^{+} (mg l^{-1})	0.3 – 1.6	0.46	54.3	2.0	-
CO_3^{-} (meq l^{-1})	Traces	-	-	1.0	-
HCO_3^{-} (meq l^{-1})	5.9 – 7.5	0.55	8.1	10.0	-
Cl^{-} (meq l^{-1})	0.9 – 3.9	1.02	55.7	30	-
RSC (meq l^{-1})	1.0 – 2.5	0.49	26.9	2.5	-
SAR	1.0 – 2.4	0.35	21.7	15	-

*Compiled from Richards (1954); Ayers and Westcot (1985); Anonymous (2003); Yaduvanshi (2007)

Conclusion

From the study it can be concluded that there is no salinity or sodicity related problems in the ground water in Datia district in more than 70% of the area. Further, the SAR values being within the safe limits in all the locations render the ground water safe for irrigation for the prevailing cropping pattern in this area.

References

Anonymous (2003). Saline water management for irrigation. *International Commission on Irrigation and Drainage*, New Delhi

APHA (1985). Standard methods for the examination of water and wastewater, 16th Edn. Washington DC: APHA.

Ayres, R.S. and Westcot, D.W. (1985). Water Quality for Agriculture. FAO Irrigation and Drainage Paper No. 29. Food and Agriculture Organization of the United Nations, Rome. pp. 1-117.

Kesavan, K.G. and Parameswari, R. (2005). Evaluation of ground water quality in Kancheeppuram. *IJEP*, **25** (3), 235–239.

Richards, L.A. (1954). Diagnosis and improvement of saline and alkali soils U.S. Salinity laboratory staff. USDA Handbook **60**: 160.

www.mp.gov.in/WRD/HIS/GW. Department of Water Resources, Government of Madhya Pradesh, Bhopal, India.

Yaduvanshi, N.P.S. (2007). Diagnosis of Soil and Water Quality. *In*: Kaledhonkar, M.J. *et. al.* Eds. On-Farm Land and Water Management. CSSRI, Karnal, India. 178 p.

□□□

Chapter 10

Economic Water Productivity of Drum Kit Drip Irrigation Systems for Vegetable Production

MALI S.S.[1], AND S. KUMAR[2]
[1] Scientist, ICAR Research Complex for Eastern Region, Research Center Ranchi. [2] Principal Scientist, ICAR Research Complex for Eastern Region, Research Center, Ranchi

Availability of water is a major concern for sustainable agricultural production in Jharkhand. The irrigation sector of this state of eastern plateau region is grossly underdeveloped with only 10% of the net sown area enjoying irrigation facilities as against the national average of 40 per cent (FAO, 2009). On an average state receives 1350 mm of rainfall in a year but the undulating topography of the region does not favor perennial streams as the streams dries up immediately after withdrawal of monsoon. The undulating terrain combined with higher rainfall intensities gives rise to high runoff production which, owing to hilly lands, flows out of the area rapidly. Consequently, only one crop is taken during the year and most of the land remains uncultivated for eight months from October to May. Limited water is available during this period and farmers are cultivating vegetables on smaller areas by using furrow irrigation system of water application in most parts of the region. These smaller areas can be increased by the adoption of drip irrigation practices. Economic status of the farmers of this region is such that they cannot afford big-high pressure drip irrigation systems. Also land holdings are not suitable for this. Small scale drip irrigation systems like Gravity Drum Kit System (GDKS) can be promoted for the adoption by the farmers. These systems are low cost and are within the reach of small farmers.

Recently, a number of innovations have been made for small-scale farmers to manage water effectively during the dry season, and increase on-farm water efficiency using simple, affordable micro-irrigation systems (Postel *et al.*, 2001; Polak and Yoder, 2006). One of such low cost micro irrigation systems is Gravity fed Drum Kit Drip (GDKD) irrigation system. This technology integrates simple, low-cost micro-irrigation techniques into vegetable farming systems for resource-poor farmers. To promote such small scale systems, number of demonstrations needs to be carried out and the farmers should be made aware about the additional economic gains they are going to get by the adoption such systems.

In order to make the efficient use of available limited water supplies, the available water should be used effectively, i.e., to increase water productivity without compromising food security. Water productivity is the ratio of mass or value of product to volume or value of water depleted or diverted. The water productivity can be increased by increasing crop yields or by reducing the quantity of water. Several previous studies (INCID, 1994; Singh 2005) have showed water saving of 36 to 63% and yield increase of 10 to 40% under drip irrigation system. So drip irrigation systems are very effective in increasing crop water productivity.

Seckler *et. al.*, (1998) expressed the crop water productivity in three ways; in physical terms as quantity of the product divided by the amount of water depleted or diverted; in combined physical and economic terms as gross or net present value of the crop divided by the amount of water diverted or depleted; and in economic terms as gross or net present value of the product divided by the value of the water diverted or depleted. In the present study the second approach of combined physical and economic terms has been used to estimate the crop water productivity because the net income per unit of water consumed will be more illustrative, as it considers economic value of yields, inputs and all other expenditures.

In the present study the economic water productivity of Gravity Drum Kit drip irrigation system was compared to that of traditional furrow irrigation system. The major objective of the study is to assess the performance of low head (1.2 m) Gravity Drum Kit (GDK) drip irrigation system with respect to economic water productivity and water and labor savings.

Methods

Treatments

Two irrigation methods, GDKD and furrow, were considered for the study on two crops namely cabbage and cauliflower. These two crops in two irrigation methods were imposed with four treatments viz. T_1-Control, T_2-application of Boron, T_3-Application of Lime and T_3- application of Boron + lime. There were 16 plots of size 10 x 10 m, out of which 8 were under furrow and remaining under GDKD irrigation system.

Irrigation Systems

The study was carried out at the research farms of RC Ranchi, located in the Chota nagpur region of the eastern plateau. The soils of the study area are fine sandy loams which have moderate water holding capacity. The GDKD system consisted of two major component parts - a water source tank and water distribution and application system along with control valves and filters. Other than a network of pipes and water tank, it had a shut off (control) valve and a small screen filter. The black plastic drums of 200 lit capacities were installed at a height of 1.2 m. The integral drip lateral with rated emitter discharge of 2 lph spaced at 30 cm was used to irrigate 10 paired rows of plants in a 10 x 10 m land area. These laterals were connected to drum using LLDPE main line of 16 mm diameter. The system also had a screen filter of 3 m^3/hr discharge capacity. The layout of the system is shown in Fig 1(a) and the cross sectional view of the field is shown Fig 1 (b).

Traditional furrow irrigation had, in 10 x 10 m area, 16 numbers of furrows having 10 m length. The furrows constructed were having depth of 15 cm, top width of 25 cm and bottom width of 5 cm. The ridge width at the top, for the planting of crops, was 35 cm. There was one crop row between tow furrows. Two crops, cabbage and cauliflower, were taken during winter season of year 2007 and 2008 on sixteen plots of 100 m^2 area. Irrigation of vegetable, in case of furrow system was done according to the practice fallowed by the farmers of this region. This included irrigating the vegetables at an interval of 6 to 7 days. Volume of water applied each irrigation was taken as the volume of water stored in all the furrows. The schematic of the furrows used is shown in Fig 2.

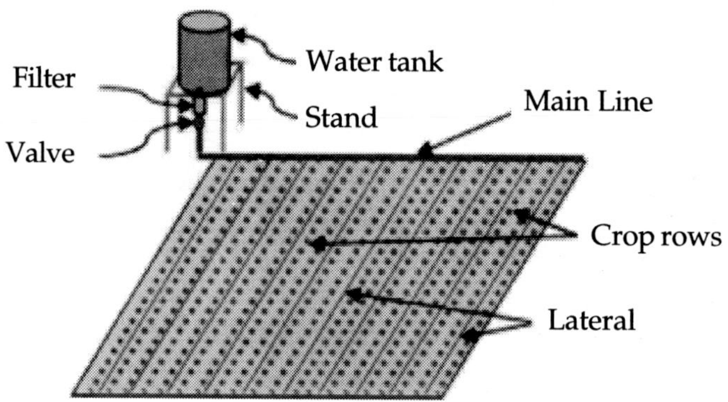

Fig 1(a) : Layout of the GDKD irrigation system.

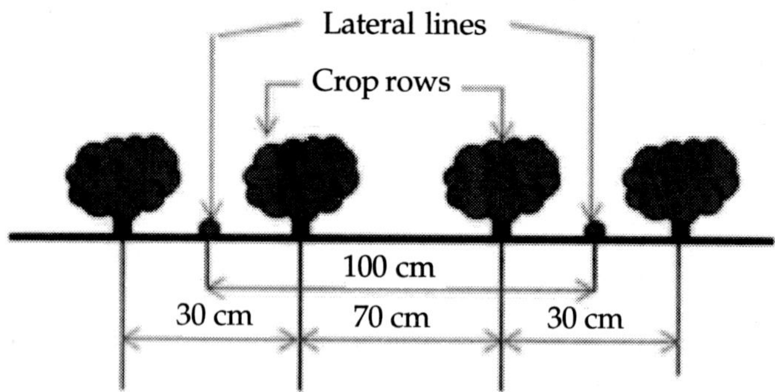

Fig 1 (b) : Cross sectional view of GDKD irrigation layout.

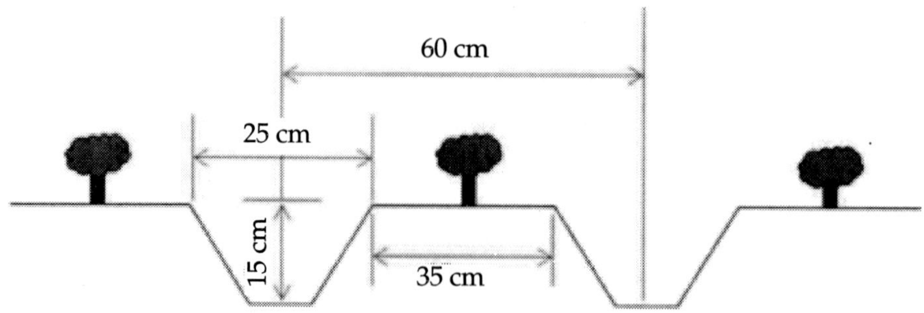

Fig. 2 : Schematic of furrow irrigation method.

In case of furrow system of irrigation the crop spacing was 0.60 x 0.45 m and at this spacing, 100 m^2 area accommodated 370 plants (0.27 m^2 area per plant) while in case of GDKD irrigation the paired row spacing was 0.3 x 0.45 x 0.7, this geometry gave plant population of 445 plants in 100 m^2 area (0.22 m^2 area per plant). Irrigation scheduling for GDKD irrigation system was done using past pan evaporation data and taking Kc values for both the crops as 0.45, 0.75, 1.05 and 0.9 for initial, development, mid and late growth stages respectively. Daily irrigation was given to both the crops. Data on water usage, labor requirement and yield was recorded throughout the growing season.

Results and Discussion

Labor requirement

Labor requirement for various field operations for cultivation of cabbage and cauliflower under both the irrigation systems is given in Table 1. Adoption of GDKD irrigation system reduced the labor usage by an average value of 40.25 % in cabbage and cauliflower cultivation.

Table 1 : Usage of labor (hr/100^2) in different operations for cabbage and cauliflower throughout the growing season.

Operation	GDKD		Furrow	
	Cabbage	Cauliflower	Cabbage	Cauliflower
Land Preparation	6.00	6.00	15.00	15.00
System Installation	3.00	3.00	0.00	0.00
Planting	1.50	1.50	1.50	1.50
Inter culturing	8.00	6.00	21.00	18.00
Spraying	1.40	1.40	1.40	1.40
Irrigation	6.93	5.73	6.25	5.00
Harvesting	4.50	4.50	7.50	6.00
Total labor use	**31.33**	**28.13**	**52.65**	**46.90**
Percent labor saving under GDKD	**40.50**	**40.00**	**0.00**	**0.00**

The major savings in labor requirement were on land preparation, inter culturing, irrigation and harvesting operations. Uniform maturity of curds under GDK system required only 3 harvestings for both the crops while that under traditional furrow system required 5 and 4 harvestings for cabbage and cauliflower respectively. Also, irrigation labor use in GDK was only for opening and closing of valves while that for furrow required continuous presence and monitoring of irrigation for 25 minutes of irrigation of single plot.

Water usage

Irrigation scheduling was done using past pan evaporation data and the cumulative volume of water applied in GDK system for cabbage and cauliflower are given in Fig. 3. In case of furrow method the water was applied at weekly interval and amount of water application was same every irrigation. The difference in irrigation between cabbage and cauliflower was only in number of irrigations and time of start of irrigation as cauliflower was planted one week after cabbage plantation.

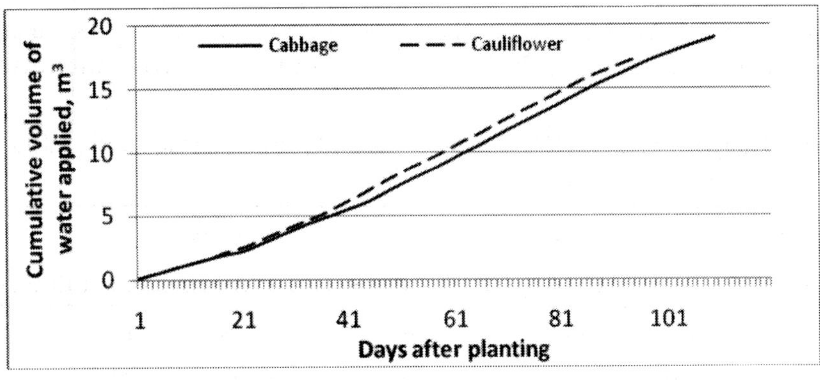

Fig. 3 : Cumulative volume of water applied in GDKD system.

Since the geometry of plantation in case of cabbage and cauliflower was same quantity of water applied in single irrigation was same for both the crops in case of furrow system. In total cabbage required 15 irrigations while cauliflower required 12 irrigations. Volume of water stored in furrow in single irrigation was 2.17 cubic meter. The amount of water required for cabbage and cauliflower under furrow and GDKD irrigation system is given in Table 2.

In GDK system, application of calculated amounts of water on daily basis reduced the water usage considerably over furrow irrigation system. This resulted in saving of irrigation water to the extent of 36.4 % and 33.5 % in cabbage and cauliflower, respectively.

Table 2 : Seasonal water usage ($m^3/100\ m^2$) for cabbage and cauliflower under GDKD and furrow irrigation system.

Particulars	Cabbage	Cauliflower
No of irrigations	15	12
Volume of water stored per irrigation in furrows, cu m	2.17	2.17
Seasonal water use in furrow system, cu m	32.55	26.04
Seasonal water Used in GDKD, cu m	20.70	17.31
Per cent saving in GDKD System	**36.4**	**33.5**

Crop Yield

The yields obtained under four treatments viz T_1-Control, T_2-application of Boron, T_3-Application of Lime and T_4- application of Boron + lime are presented in Table 3. In all the treatments and in both the crops GDKD system outperformed the traditional furrow irrigation system. Considering the highest yields under all the treatments, the yields under GDKD system were 42.86% and 50.45% higher than furrow system of irrigation for cabbage and cauliflower, respectively.

Table 3 : Yields ($kg/100\ m^2$) obtained under GDKD and furrow systems.

Particular	GDKD		Furrow	
Treatment	Cabbage	Cauliflower	Cabbage	Cauliflower
Control	394.2	317.15*	251.75	210.8*
Boron	716.82*	237.3	249.75	119.7
Lime	679.65	159.05	501.75*	107
Boron + Lime	639.15	169.5	242.5	148.8

* Yield values considered in productivity analysis

In case of GDKD system, cabbage responded positively to the application of boron while cauliflower yields were higher for control treatment. In furrow system cabbage yields were higher for lime application while increase in yield was not observed for cauliflower under application of lime, boron or both. The yield values obtained under the treatments resulting in highest yield were used for the economic analysis and water productivity computations.

Income and expenditure statement

The gross income and expenditures were calculated at the prevailing market prices. For the first season GDKD showed lower net income than furrow system in case of cabbage. Also, in case of cauliflower GDKD system showed negative net income for the first season. This is because the material cost of GDKD system is more than traditional furrow system.

Table 4 : Income and expenditure (Rs/100 m^2) statement for GDKD and furrow irrigation systems.

Particular	GDKD		Furrow	
	Cabbage	Cauliflower	Cabbage	Cauliflower
A. Income				
Gross Income	4300.9	3171.5	3010.5	2108.0
B. Expenditures				
System Cost	3150.0	3150.0	0.0	0.0
Seed	145.0	170.0	145.0	170.0
Fertilizer	38.9	38.9	38.9	38.9
Pesticide	176.0	160.0	189.0	215.0
Labor	454.3	407.9	763.4	680.1
Fuel charges	19.3	16.1	45.5	39.8
C. Net Income (A-B)	**317.4**	**-771.4**	**1828.7**	**964.2**
Net Income excluding GDKD system cost	3467.4	2378.6	1828.7	964.2
Percent increase in Net income of GDK system over Traditional furrow	89.6	146.7	0.00	0.00

Excluding the cost of system, an increase of 89.6% and 146.7% was observed in net income for cabbage and cauliflower under GDKD system. Though the cauliflower yields are lower than that of cabbage, the higher *per cent increase* in net returns is due to high market price of cauliflower produce and lower cauliflower yields under traditional furrow irrigation system.

Water use efficiency, labor returns and economic water productivity

Combined effect of reduced water application and increased yields resulted in considerably higher water use efficiency of the GDKD system for both the crops. In case of GDKD system, the water use efficiency expressed in terms of marketable yield per unit of water applied was found to 34.6 t/ML and 29.0 t/ML while that under traditional furrow system was 9.7 t/ML and 8.1 t/ML for cabbage and cauliflower respectively. Reduction in labor usage under various operations for GDKD system resulted in high net returns per labor day of 8 working hours. On an average the net labor returns per working day were 781.25 Rs/day and 223.25 Rs/Day for GDKD and furrow systems respectively showing an impressive increase in EWP of the GDKD system.

Table 5 : Crop water productivity, net returns on labor and economic water productivity.

Particulars	GDKD		Furrow	
	Cabbage	Cauliflower	Cabbage	Cauliflower
Yield (t/ha)	71.7	50.2	31.7	21.1
Water use (ML/ha)	2.07	1.73	3.26	2.60
Net Income, (Rs/ha)*	346,738	237,856	182,867	96,424
Labor usage (man days/ha)	391.7	351.7	658.1	586.3
Water use efficiency (t/ML)	34.6	29.0	9.7	8.1
Net returns on labor (Rs/man day/ha)	885.3	676.4	277.9	164.5
Economic water Productivity, (Rs/m^3)	167.53	137.43	56.18	37.03

* Excluding GDKD system cost.

The net returns on per unit of water consumed were highest for cabbage under GDKD system. Drum kit systems' average economic water productivity was 152.45 Rs/m^3 while that for furrow system was found to be 46.60 Rs/m^3. If cost of the GDKD system is considered in calculation of net income then actual benefits under cauliflower cultivation will start to accrue from the second season (Table 6). In case of cabbage cultivation under both the irrigation systems and cauliflower cultivation using furrow method of irrigation, the system becomes profitable from the first season itself. Assuming current levels of input and output if we project the data for consecutive three seasons

(Table 6), it may be observed that for the second season GDKD systems' profit under cauliflower cultivation is less than furrow systems' profit, but it becomes more profitable than furrow in third season. Likewise, the GDKD system for cabbage cultivation is more profitable than furrow from the second season.

Table 6 : Projected *cumulative income* for three cropping seasons assuming current levels of gross income and expenditures (including system cost).

System	Crop	1st season	2nd season	3rd season
GDKD	Cabbage	317.4	3784.8	7252.1
Cauliflower	-771.4	1607.1	3985.7	
Furrow	Cabbage	1828.7	3657.3	5486.0
Cauliflower	964.2	1928.5	2892.7	

Conclusion

The gravity fed drip irrigation system was found to be more effective than furrow system of irrigation in all the aspects of water and labor saving, economic water productivity, water use efficiency and net returns on labor. The technology behind a drum system is quite easy to explain as well, and gives the field a measured volume of water. Drum kits provide affordable entry points for small farmers to irrigate horticultural crops with drip irrigation, opening the possibility of rapid increase in the adoption of drip irrigation. This in turn would generate significant improvements in irrigation productivity. The small drum kit systems will enable small-scale and poor farmers to utilize efficiently marginal quantities of water and to cultivate vegetables and other horticultural crops. With drum kit drip irrigation, water sources previously considered too small for irrigation can now be utilized, and the same limited amount of water can be used to irrigate twice as much land as is possible under furrow system in the dry season. Such low cost irrigation systems are within the reach of small farmers and should be promoted for adoption.

References

INCID, (1994). Drip Irrigation in India, Indian National Committee On Irrigation And Drainage, New Delhi.

F.A.O. (2009). "Global Map of Irrigation Areas" FAO's Information System on Water and Agriculture, (© FAO, NRL 2009), http://www.fao.org/ date of citation 15/07/2009).

Polak P. and Yoder R., (2006). Creating wealth from groundwater for dollar-a-day farmers: Where the silent revolution and the four revolutions to end rural poverty meet. *Hydrogeology J.* **14**: 424-432.

Postel S., Polak P., Gonzales F. and Keller J., (2001). Drip irrigation for small farmers. A new initiative to alleviate hunger and poverty. *Water Institute* **26**: 3-13.

Seckler D., Molden D. and Sakthivadivel R., (1998). The concept of efficiency in water resource management and policy. Sri Lanka, Colombo: International Water Management Institute.

Singh H.P. (2005). '*Micro irrigation – an option for sustainability*' In: *Souvenir issue National seminar on Micro irrigation* June 3-5 at G.B. Pant university of Agril. & Tech. Pantnagar, page: 1-15.

□□□

Chapter 11

Groundwater Use Efficiency and Price Equilibrium of Groundwater Use in Central Plain Zone of Uttar Pradesh

S.K. SRIVASTAVA[1], RANJIT KUMAR[2] AND R.P. SINGH[1]
[1] Research Scholar and Former Professor & Head, respectively, Division of Agricultural Economics, Indian Agricultural Research Institute (IARI), New Delhi-110012. [2] Senior Scientist (Agricultural Economics), Indian Institute of Soil Science (IISS), Nabibagh, Bhopal- 462038.

Irrigation development has always been the priority sector for the policy makers for food security and sustainability of ever growing population. With the passage of time, groundwater irrigation has emerged as a vital factor in the modern agriculture due to its well established comparative advantages over canal irrigation and the development of groundwater took a major stride through private modern water extraction mechanisms (WEMs) which are highly skewed towards large farmers due to huge capital investment needed and relatively better consolidation of land holdings among them (Dhawan, 1982 and Shah, 1993). Thus small and marginal farmers and even large farmers with fragmented holdings have to depend on owners of modern WEMs to irrigate their uneconomic land holdings. This has led to emergence of an informal groundwater market which has been well documented (Patel and Patel, 1970, Shah, 1985, Kolavalli *et. al.*, 1993, Pant, 2004, Singh and Singh, 2003; 2006) and has various productivity considerations too. The possibility has been shown of increasing the productivity in major crops like sugarcane and wheat by reducing the excessive water- use on self-users farms, which in turn would increase the availability of water on the buyers' farms (Singh and Singh, 2006). Though the groundwater markets have facilitated to mitigate inequality in the physical access to the irrigation water among the resource-rich large and resource- poor small farmers, the sellers of the water are charging exorbitant prices from buyers (Sharma and Sharma, 2006). The present paper is an attempt to measure the productivity of groundwater at micro level and to examine the price equilibrium of groundwater use in Central Plain Zone of Uttar Pradesh which has well developed groundwater market and exclusively engaged in growing water intensive crops.

Methodology

Data source and sampling framework

The study is primarily based on primary data which was collected from randomly selected two representative districts of Central Plain Zone of Uttar Pradesh through multistage random sampling in which Central Plain Zone was selected purposively due to well developed water market and water intensive cropping pattern. Further, within the zone, two districts namely Lucknow and Sitapur, were selected randomly. Finally, from each of selected blocks, a cluster of two to three villages were selected randomly out of which twenty five sample farmers were drawn randomly from each block. Thus, total sample size from two selected districts was one hundred.

Conceptual framework and Statistical analysis

Water markets are informal institutions, which provide access to groundwater irrigation particularly to the small and marginal farmers who can not afford to invest themselves to construct water extraction structure for irrigating their small land holding. The water markets are very crucial, where agriculture increasingly depends on groundwater for crop production and has various efficiency and productivity aspects among different strata of the agricultural community. In the present study, groundwater productivity has been estimated for the different water market regimes such as such as (a) Self-users, (b) Self-user + Buyer, (c) Only Buyer and, (d) Self-user + Seller.

Extent of Groundwater Extraction and Accessibility

The amount of groundwater extracted and accessibility to irrigation water were studied through personal interview with farmers. Implication of groundwater accessibility on cropping pattern was examined through comparing the choice of crops under different water regimes. The volume of groundwater exploitation/ extraction (in litres) was estimated by using a pre-tested estimation model (Eyhom *et. al.*, 2005) given below:

$$Q = t*129574.1 * BHP / [d + ((255.5998 * BHP^2) / d^2 * D^4)] \quad ...(1)$$

where,

Q = Quantity of groundwater extracted (in litres)

t = Total duration of irrigation (in hours)

BHP = Engine power of pump (in HP)

d = Average depth of the well (in metres)

D = Diameter of the suction pipe (in inches)

Water use efficiency: Production function approach

For estimating water use efficiency, production function approach was used and out of several functions, log linear production function (Cobb-Douglas version) was used as it gave best fit to the observed data. Functional form of the log linear production function can be given as:

$$Y = A\ X_1^{b1}\ X_2^{b2}\ X_3^{b3}\ e^{Ui} \quad ...(2)$$

where,

Y = Annual gross return per farm (Rs/ha)

X_1 = Quantity of water used on the farm for all the crops (liters/ ha)

X_2 = Value of seed used in irrigated crops (Rs/ha)

X_3 = Amount of fertilizer (NPK) applied in irrigated crops (kg/ ha)

Ui = Random error term

Since the production function was fitted to whole farm data, in order to avoid aggregation problem of output for different crops, gross farm income was taken into consideration and variables showing multi-co linearity problems were dropped out from the model. Aggregation of all the outputs was done due to the insufficient number of observations for all the crops under four different water regimes for doing functional analysis. For example, if a farmer is growing paddy, wheat and sugarcane in a particular water regime, then the number of pooled observation of output-input data for the same farmer became three. In this case, revenue from each crop has been taken as dependent variable, while other inputs were taken in physical term to estimate the VMP, except for seed as improved seed costs very high but required less in quantity as compared to traditional varieties and thus underestimated while taking in physical term.

The value of marginal product (VMP) of water was computed at its geometric mean level of both dependent (Y) and independent variable (X_i) and the elasticity coefficient b_1 by using the formula:

$$VMPx_{i} = b_i * Y / X_i \quad .... (3)$$

where,

$VMPx_i$ = Value marginal product of X_i coefficient.

B_i = The regression coefficient (elasticity coefficient) of X_i variable, water

Y = Geometric mean of gross return.

X_i = Geometric mean of X_i variable.

In order to examine the Water Use Efficiency, the $VMPx_i$ was compared with its factor cost. For the resource to be optimally allocated necessary condition is,

$$VMPx_i = Px_i \text{ or } VMPx_i / Px_i = 1$$

If the ratio is equal to unity, the resource will be said to be optimally allocated. If the ratio is more than one, it implies that the particular resource is under used, and *vice versa*. Water use efficiency estimated for different water market regimes indicate whether the particular group of farmers under respective water regime is efficient water user or not. Certainly, it would not explicitly explain in which crop, water is being used efficiently by the farmers.

Price equilibrium for groundwater use

Price equilibrium for groundwater use was examined in terms of bargaining power of the Buyer and sellers for which Nash equilibrium framework was used to model the bargaining power of water sellers and Buyers. The model assumes that water Buyers and sellers are highly rational with equal bargaining skills, each has full knowledge of the taste and preferences of the other, and each individual wishes to maximize his utility. Bargaining power of seller is assumed in terms of gross irrigated area (GIA) of sellers and total water extracted by them, while bargaining power of Buyer is assumed in terms of GIA of Buyers. Water price per unit tends to remain uniform and this was chosen as the dependent variable after it was standardized by amortized cost incurred by sellers for extracting every unit of water.

Amortized cost of irrigation

The amortized cost represents the annual fixed cost component of irrigation water. The amortized cost varies with the type of well, status of the well, year of construction, average age of well and interest rate chosen. The amortized cost of pump-set, bore well and over ground structures was calculated separately as fallows:

$$A = \frac{CB * (1+i)^{n} * i}{(1+i)^{n} - 1} \quad \ldots (4)$$

where,

A = Amortized cost

CB = Cost of respective asset at current price.

i = Interest rate

n = Average life of the asset.

Average life of the pump set and bore well was assumed to be 15 and 10 years respectively which shows the number of years, pump set and bore well have already functioned and no longer functioning. Compound interest rate of 2 per cent had been used to amortize the irrigation cost as it offered a close approximation to the growth of investment in well irrigation (Diwakara H. and M.G. Chandrakanth, 2007)

Total amortized cost was the sum of the amortized cost of pump set, bore- well, over ground structures and annual maintenance cost. After standardization with amortized cost, price-cost ratio (margin ratio) was assumed as a surrogate for bargaining power.

Nash's bilateral bargaining model

Nash equilibrium assumes that bargaining power among the Buyers and sellers of groundwater is the key determinant of the groundwater markets. Accordingly, bargaining power of the seller is measured in terms of two variables namely gross irrigated area (GIA) of sellers and total water extracted by the sellers. The bargaining power of Buyer is measured by gross irrigated area (GIA) of Buyers. For performing regression analysis, the variation in groundwater price is reduced by considering the margin ratio as equal to price per acre inch groundwater charged by sellers divided by the amortized cost per acre inch of groundwater of the seller (Deepak *et. al.*, 2005). This is taken as a surrogate for market price. Nash bargaining model is thus constructed as,

$$Y= f (X_1, X_2, X_3) \quad \ldots (5)$$

where,

Y = Margin ratio
X_1 = Gross irrigated area of seller
X_2 = Total water extracted by seller
X_3 = Gross irrigated area of Buyer.

Standard conversion units like, one acre-inch (hectare cm) equivalent to 3630 cubic feet and one cubic feet equivalent to 28.32 liters, were used for transforming water quantities in to comfortable figures.

Empirically, the Nash Bargaining model followed the quadratic function;

$$Y = a + b_1X_1 + b_2X_2 + b_3X_3 + C_1X_1^2 + C_2X_2^2 + C_3X_3^2 + U_i \quad \ldots (6)$$

The elasticity of price-cost ratio with respect to each of the explanatory variables was considered as,

$$(bi + 2C_iX_i) X_i/Y_i \quad \ldots (7)$$

This elasticity shows that for one per cent increase in GIA of Seller or Buyer or total water extracted by Seller, how much increase will be there in price - cost ratio. The ratio thus explored bargaining power of Seller.

Results and Discussion

Type of groundwater market and farm size in study area

Average size of the farm holding in the study domain was 1.46 hectares, which approximately matches to small farm size category (Table 1). Average land holding was highest (2.75 hectare) for self-user + buyer category, who were relatively large farmers.

Table 1 : Type of groundwater market and average size of holding among Sample Households (hectare).

Category of water market regimes	Farm category			Overall
	Marginal (0-1 ha)	Small (1-2 ha)	Others (> 2 ha)	
Buyer	0.46	1.29	4.11	0.82
Self-user	0.67	1.42	4.02	1.64
Self-user + buyer	0.80	1.47	4.19	2.75
Self-user + seller	0.78	1.73	4.56	2.00
Total	0.55	1.47	4.22	1.46

But, due to fragmented land holding, these farmers had to buy water for the fields where they couldn't have own water extraction installation. Buyer category farmers were having uneconomic land holding with the average size of 0.82 hectare. They did not have their own water extraction devices due to uneconomic nature of land holding, and thus became an important agent of water market. Farmers included in Self-user + Seller category were having 2 hectares of average land holding and were the net sellers of groundwater.

Cropping pattern of selected farm under different water regimes

Cropping pattern within an agro-climatic area is mainly influenced by a number of factors such as soil type, size of holding, availability of irrigation facilities and other resources, level of investment, public policy, marketing facilities, etc. Cropping pattern followed by the farmers under different types of water market in study area is presented in Table 2. In the study area, although, there was not any significant trend for choice of crops by the farmers under different water regimes, but number of farmers growing different crops varied very much across the groups (Appendix).

Table 2 : Cropping pattern of selected farm under different water regimes.

Crops	Buyer	Self-user	Self-user + buyer	Self-user + seller	Overall
Paddy	7.51	8.12	4.20	3.20	6.43
Wheat	36.26	34.37	35.47	38.93	38.28
Maize	0.71	3.01	3.66	n.a.	2.00

Sorghum	4.43	n.a.	n.a.	2.00	1.93
Pearl millet	3.43	0.60	2.74	4.53	2.95
Mustard	2.14	3.01	2.65	1.87	2.58
Pulses*	8.49	8.42	4.93	4.4	6.51
Sugarcane	20.34	32.36	19.47	19.20	18.80
Potato	4.28	3.51	15.54	8.27	8.18
Vegetable**	6.27	4.21	10.15	13.20	8.50
Others	6.14	2.40	1.19	4.40	3.87
Total	100 (1.10)	100 (2.10)	100 (2.27)	100 (2.14)	100 (1.60)
Cropping intensity	**130.05**	**115.98**	**100.43**	**114.05**	**120.29**

Figures within parentheses are the gross cropped area for the respective group of farmers.

n.a. = Respective crops are not being grown by the farmers in corresponding group.

* Pulses include Pigeon pea, Lentil, and Black gram. **Potato is not included in vegetable group.

For all the farmers, Wheat has dominated the cropping pattern, while Paddy seems to be grown only for household consumption. It was also surprising to note that Self-user or tube-well owners, who were also relatively large farmers, were also growing vegetables on significantly large areas. This may be due to assured irrigation facilities available to them. On the other hand, buyer and exclusive self-user were allocating about 8 per cent of area to pulse crops.

Inputs used and crop yield harvested by sample farmers

In the study, crop-wise detailed analysis has been done for only four crops- Paddy, Wheat, Sugarcane and Potato, as these were the major irrigated crops. Input use pattern and corresponding yield achieved in selected water intensive crops namely wheat, paddy, sugarcane and potato was examined and details are presented in Table 3.

Out of four crops, potato followed by sugarcane was found to be most water intensive crop requiring maximum amount of irrigation water. Potato being short duration cash crop, farmers applied maximum irrigation to the crop. In all the four crops, highest amount of water was applied by either self-user or self-user+buyer category, while farmers under buyer category applied minimum amount due to constraint in irrigation water. In addition to water, farmers under Buyer category also used minimum amount of fertilizer than other categories in case of wheat crop resulting into lower crop yield realized by these small farmers as compared to their counterparts. Interestingly, Banerji *et. al.* (2007) in his case study of sugarcane growers in Western UP observed that tube well owners wanting to maximize profits from water sales could choose to sell little water, using most of it on their own plots, given the fairly low water price and the relative shortage of water (due to paucity of electricity). Instead, they sell substantial volumes of water even though it would make better economic sense to use the water to boost the productivity of their own land.

Table 3 : Application of critical inputs and yield realized by selected farms.

Particulars	Irrigation (No./ha)	Quantity of irrigation water (m^3/ha)	Fertilizers applied (kg/ha)			Yield (Kg/ha)
			N	P_2O_5	K_2O	
		WHEAT				
Buyer	2.99	2784	140.11	71.90	25.5	3665
Self user	3.53	4084	195.30	98.91	30.0	3767
Self-user + buyer	3.47	4091	157.23	84.35	37.5	3777
Self-user + seller	3.32	3057	163.23	93.64	-	3525
Overall	**3.33**	**3504**	**157.03**	**82.40**	**28.8**	**3681**
		PADDY				
Buyer	2.75	2658	130.50	56.00	-	2657
Self user	2.71	2577	155.39	73.50	7.5	3010
Self-user + buyer	3.00	3648	165.84	62.71	-	3417
Self-user + seller	3.00	3422	145.13	80.50	12.0	4556
Overall	**2.87**	**2826**	**153.65**	**66.72**	**8.5**	**3159**
		SUGARCANE				
Buyer	2.26	2384	169.44	59.37	9.0	59200
Self user	3.86	5639	251.66	89.10	-	62875
Self-user + buyer	3.11	3766	198.00	82.80	-	55018
Self-user + seller	3.33	3647	237.60	86.25	-	61111
Overall	**3.14**	**3859**	**211.71**	**77.49**	**9.0**	**59868**

		POTATO				
Buyer	5.00	2886	196.88	138.89	135.38	18783
Self user	5.50	4955	228.75	153.33	150.00	21086
Self-user + buyer	5.14	4585	185.22	149.91	131.82	27434
Self-user + seller	5.00	3928	223.17	148.06	135.00	19771
Overall	**5.02**	**4089**	**150.11**	**147.33**	**136.02**	**22643**

Source : From field survey, 2007.

Regression estimates for irrigated crop production

Cobb-douglas production function was fitted for the whole farm under different water regimes to see the major determinants and factors explaining the variation in value of crop output. Revenue from output of various crops was chosen as dependent variable and seed cost, fertilizer quantity and water uses were dependent variables. Among all the four major factors explaining the variation in the dependent variable (revenue), seed cost was found to be highly significant for all the water regimes except self-user + seller, where it was found insignificant with negative value (Table 4). Fertilizer quantity was also observed to be significant for self- user + buyer and self-user + seller category. For small and marginal farmers, who mainly came in buyer category, water- use was most significant factor as compared to other. For self-user and self-user + seller, the coefficient was negative showing water as not an important variable for farmers in this category. For self-user + buyer water regime also, water was not a significant input. However, it is worth to note that less number of observations in this category could not capture the variations in the pooled data of different crops, giving rise to low value of R^2.

Water use efficiency for different category of water market regimes

Regression coefficients estimated from the regression analysis for irrigation water for different group of farmers were further used to estimate the water use efficiency under different water market regimes with the details presented in Table 5. For buyers, ratio of VMPx and Px (price per unit of irrigation water) obtained was 1.361 (more than one) which indicated that buyers were under utilizing water resource. For them water was a significant and most limiting factor for the farming due to their small and uneconomic land holding. For self-user, having their own water extraction facilities, water was found to be least significant and the negative VMPx to Px ratio of -1.140 showed over utilization of the water resource due to injudicious use of it.

Table 4 : Regression estimates for irrigated crop production under different water market regimes.

Variables	Buyer	Self user	Self user + Buyer	Selfuser + Seller
Intercept	2.607 (2.1370)	12.57 (2.4797)	0.7724 (4.5402)	9.7657 (6.3300)
Seed cost (Rs/ha)	0.312*** (0.0512)	0.531*** (0.0734)	0.369*** (0.0859)	-0.1675 (0.1225)
NPK (Kg/ha)	0.0208 (0.1737)	-0.340 (0.2920)	0.534* (0.3194)	1.0917** (0.4449)
Water-use (litre/ha)	0.342** (0.1436)	-0.301* (0.1658)	0.245 (0.3418)	-0.2944 (0.4369)
Adjusted R^2	0.4175	0.5941	0.5298	0.1178
No. of observations	68	36	32	26

Dependent variable: Revenue from output of different crops (Rs/ha)

***, ** and * indicate significance at 1, 5 and 10 per cent probability level, respectively.
Source: From field survey, 2007

Table 5 : Water use efficiency for different category of water market regimes.

Category	VMP_X	P_X (Rs/litre)	VMP_X/P_X	Remark
Buyer	0.00365	0.00268	1.361	Under utilisation
Self user	-0.00247	0.00216	-1.140	Over utilisation
Selfuser + buyer	0.00234	0.00225	1.039	Almost optimal utilisation
Selfuser + seller	-0.00211	0.00239	-0.881	Over utilisation

Source: From field survey, 2007

Price equilibrium for the groundwater use

To examine the price equilibrium of the irrigation water in the study domain. Nash Bilateral Bargaining model was used as bargaining power is the most important factor for the price equilibrium. Details of the findings regarding pricing mechanism of irrigation water are presented in the following sub sections.

Trends in irrigation charges

Both diesel and electric operated tube-wells were observed in the study domain but due to irregularity of electricity and administrative bottlenecks for government operated electric tube-wells, majority of the sample households were found to be using diesel operated tube-wells. In the region, irrigation charges from diesel operated tube-wells have increased from Rs. 66.44/hr to Rs. 84.55/hr during the last four years (on farmers' recall basis). When enquired from the respondent households regarding the fixation of the irrigation charges, about 97 per cent of them confirmed that irrigation charges were decided by Seller and varied mainly according to cost of diesel proportionately and to some extent, plot size of buyer as well as personal relations between buyers and sellers. In case of electric operated tube-wells, irrigation charges were found to be increased from Rs. 50/hr to Rs. 58.57/hr during last four years. In this case, increase in prices was slow as compared to diesel operated tube-wells.

Table 6 : Trends in irrigation charges (Rs/hr).

Type of tube-well	2003	2004	2005	2006
Diesel operated	66.44	71.82	78.03	84.55
Electric operated	50.00	51.79	55.71	58.57

Amortized cost of irrigation water extraction structure for seller

To know the annual fixed cost component of irrigation water, amortized cost of the water extraction structure was estimated using discounting formula. Amortized cost of pump set, bore well plus over ground structures were estimated as Rs. 1937 and 424 per annum, respectively (Table 7). Amortized cost of conveyance was included in pump set and average number of years for pump set and bore well were assumed as 15 and 10 years, respectively. Finally total annual amortized cost of irrigation (Rs. 3404 per annum) was calculated by the summation of amortized cost pump set, bore well and over ground structures and annual maintenance cost.

Table 7 : Amortized cost of irrigation water extraction structure for seller.

Particulars	Cost (Rs)	Average no. of years
Amortized cost of pump set*	1936.66	15
Amortized cost of bore well and over ground structures	424.40	10
Annual maintenance cost	1042.86	-
Total annual amortized cost	3403.92	

* Amortized cost of conveyance is included in pump set.

Nash bilateral bargaining power estimates

Empirically Nash Bilateral Bargaining Power model follows quadratic function. This function was fitted with price-amortized cost ratio as dependent variable and water extracted (acre-inch), gross irrigated area of seller (ha), and gross irrigated area of buyer (ha) as explanatory variables. Results obtained are presented in Table 8. Water extracted, gross irrigated area of seller and gross irrigated area of buyer were found to be significant at 1 per cent level of significance level with the coefficients 0.042, 0.136 and 1.173 respectively. R^2 value was 0.86 showing that 86 per cent variation in dependent variable, price-cost ratio (margin ratio) explained by these three explanatory variables.

Table 8 : Nash bilateral bargaining power estimates.

Variables	Coefficient	Standard error
Intercept	-1.889	0.508
Water extracted (acre inch)	0.042*	0.012
Gross irrigated area of seller (ha)	0.136*	0.086
Gross irrigated area of buyer(ha)	1.173*	0.188
Sq. of Water extracted (acre inch)	-0.0002*	0.00006
Sq. of Gross irrigated area of seller (ha)	-0.052*	0.035
Sq. of Gross irrigated area of buyer (ha)	-0.154*	0.025
R square	0.86	
No. of observations	15	

Dependent variable: Price-cost ratio for per litre of irrigation.
* Indicates significant at 1 per cent significance level.

Elasticity coefficients for pricing of irrigation water

Using the coefficients of the Nash bilateral Bargaining model, elasticity coefficients for all the three explanatory variables were estimated which explains the bargaining power of sellers and Buyers. The result of the model given in Table 9 showed that for one per cent increase in water extracted by seller, price-cost ratio increased by 1.243 per cent. Similarly, one per cent increase in gross irrigated area of the Seller, price-cost ratio increased by only 0.019 per cent while, one per cent increases in gross irrigated area of buyer, the price-cost ratio increased by 0.971 per cent. Thus, it can be interpreted that water extracted and gross irrigated area of Buyer were the key explanatory variables in price determination in groundwater market. Total water extracted by Seller gave greater bargaining power to the seller in price determination than gross irrigated area of buyer and gross irrigated area of seller. On the other hand, gross irrigated area of the seller was not found to be major factor for price determination as explained by small value (0.019) of elasticity coefficient.

Table 9 : Elasticity coefficients for pricing of irrigation water.

Variables	Elasticity coefficient	Remark
Water extracted (acre inch)	1.243	Most significant factor
Gross irrigated area of seller (ha)	0.019	Not a significant factor
Gross irrigated area of Buyer (ha)	0.971	Less significant factor

Conclusion and Policy options

Central Plain Zone of Uttar Pradesh is exclusively engaged in growing water intensive crops (like Wheat, Paddy, Sugarcane and Potato) and thus depends on groundwater as a reliable source of irrigation to a great extent. However, due to income and resource disparities in the zone, the ownership of water extraction mechanisms (WEMs) was highly skewed towards large farmers. Thus, groundwater market emerged as an informal institution and provided the resource poor small and marginal farmers physical access to irrigation water, but with differential water use efficiency. Farmers included in Buyer category were found to be under-utilizing irrigation water because of their small and uneconomical land holding. Self-users were over-utilizing groundwater because of injudicial use of it. For self-user + buyer category, almost optimal utilization of water resource was observed while self-user+seller were over-utilizing groundwater because of their business behaviour and thus injudicious use of water. In the price determination (equilibrium) of the groundwater use, water extracted and gross irrigated area of buyer were found to be key explanatory variables and total water extracted by seller gave greater bargaining power to the seller in price determination than gross irrigated area of Buyer and gross irrigated area of Seller.

From the study, following policy options emerges:

- With development of groundwater market, small and marginal farmers also got benefited as they could access to irrigation. However, due to lack of assured electric supply, all the farmers have to largely depend on diesel-operated *tube-wells*, which cost them dearly. Therefore, attempts should be made to provide assured electricity for irrigation purposes with economic electricity charges.
- The region has witnessed cropping pattern upheavals especially in the past two decades favouring water intensive crops at the cost of pulses, which needs to be corrected, not only for reducing reliance on import of pulses for domestic supply but also to arrest the depletion of groundwater resources. This requires adequate policy support for bringing required changes.
- Finally, quantity of water extracted and GIA of buyer being the major contributing factors for price equilibrium of groundwater, assured irrigation to these categories of farmers (Buyer) will help in ameliorating the price mis-match in the groundwater market regime.

References

Dhawan, B.D. (1982). "Development of Tubewell Irrigation in India", New Delhi. Agricole Publishing Academy

Deepak, S.C., M.G. Chandrakanth and Nagraj N. (2003). "Groundwater markets and water use efficiency: A case of Karnatka" *International Water Management Institute*. Water Policy Program, Report No. 12, www.iwmi.org.

Diwakara H. and Chandrakantha M.G. (2007). "Beating Negative Externality through Groundwater Recharge in India: A Resource Economic Analysis", *Environment and Development Economics* **12**: 271-296, Cambridge University Press.

Eyhom E., Mader, P. and Ramakrishnan, M. (2005). The impact of organic cotton farming on the livelihoods of smallholders: Evidence from the Maikaal bioRe project in central India, *Research Report,* Research Institute of Organic Agriculture FiBL, Frick, Switzerland.

Kijne, J.W., Barker R. and Molden D. (2003). "Improving Water Productivity in Agriculture: Editor's Overview", International Water Management Institute, Columbo, Srilanka. www.iwmi.org

Kolavalli, S., G. Naik and Kalro AH. (1993). Groundwater Utilization in Eastern Uttar Pradesh. Centre for Management in Agriculture, IIM, Ahmedabad.

Patel, S.M. and Patel K.V. (1970). Economics of Tubewell Irrigation, Indian Institute of Management, Ahmedabad (mimeo).

Pant, Niranjan (2004). "Trends in Groundwater Irrigation in Eastern and Western Uttar Pradesh", *Economic and Political Weekly,* July **31**: 3463-3468.

Shah, Tushar (1985). Transforming Groundwater Markets in to Powerful Instruments of Small Farmers Development: Lesions from Punjab, Uttar Pradesh and Gujrat, Overseas Development Institute, London.

Shah, Tushar (1993). "Groundwater Markets and Irrigation Development: Political Economy and Practical Policy", Bombay: Oxford University Press.

Singh, D.R. and Singh R.P. (2003). "Groundwater Market and the Issues of Equity and Reliability to Water Access" *Indian Journal of Agricultural Economics,* **58**(1): 115-127.

Singh, D.R. and Singh R.P. (2006). "Structure, Determinants and Efficiency of Groundwater Markets in Western Uttar Pradesh", agricultural economics research review, **19**: 129-144.

Sharma, P. and Sharma R.C. (2006). "Factors Determining Farmers' Decision for Buying Irrigation Water: Study of Groundwater Markets in Rajasthan", *Agricultural Economics Research Review,* **19**: 39-56.

□□□

Chapter 12

Participatory Action Research for Irrigation Water Management in Sugarcane Cultivation

D.V. YADAV, R.P. VERMA, KAMTA PRASAD, A.K. SAH, RAJENDRA GUPTA AND K.P. SINGH
Indian Institute of Sugarcane Research, Lucknow-226 002

Irrigation efficiency is as low as 30 per cent at farmers fields. Availability of water for irrigation is also showing decreasing trend as it progressively being diverted for other usages e.g. industry, power generation, sanitation, domestic use etc. (Table 1). Even per capita availability of water is also decreasing and already reached to a water scarcity line (Fig. 1).

Sugarcane being one year crop, it has to survive through all the seasons of the year. Formative phase of the crop, in which the crop is young and tender, coincides with hot and desiccating summer. For this reason, optimum soil moisture in the formative phase of the sugarcane crop is utmost important to have the good yield of the crop. Water requirement of sugarcane is 1500–2500 mm/year depending upon soil, climate and crop conditions. As more than 90 per cent of

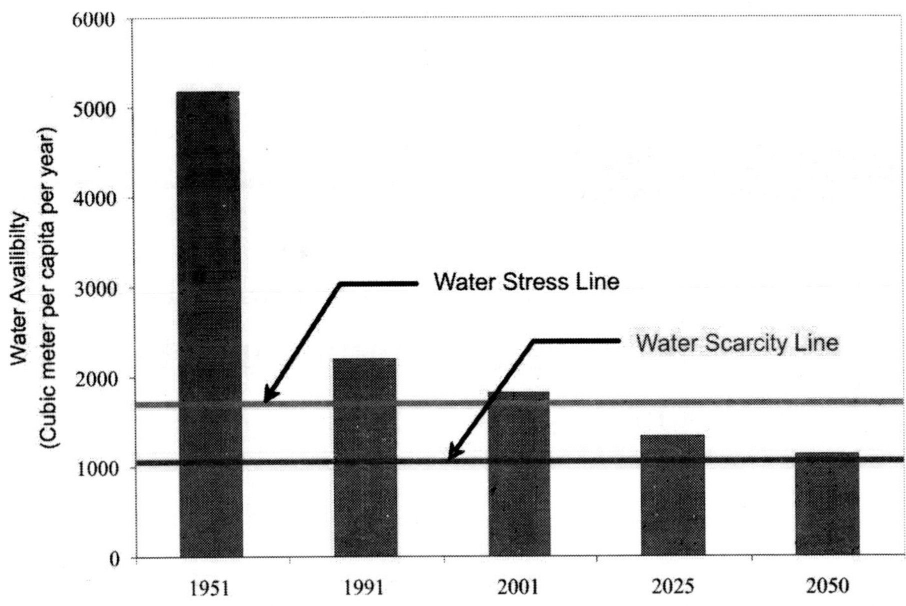

Fig. 1 : Per capita availability of water over a period of time.

Table 1 : Sectoral Requirement of Water Resources.

Sector	Requirement in M ha m			
	1990	**2000**	**2025**	**2050**
Irrigation	460	630	770	800
Domestic & Livestock	25	34	52	60
Industries	15	36	120	130
Thermal Power	30	50	160	160
Total	**530**	**750**	**1100**	**1120**
Population (M)	800	1000	1400	1700

the sugarcane crop is irrigated by the surface irrigation methods, which have 35-45 per cent irrigation efficiency. To increase irrigation efficiency of surface irrigation methods, voluminous work has been carried out by the research institutions of National Agricultural Research System, WALMI etc. but the water saving technologies developed by these institutions could not find way to the farmers fields due to lack of resources and technology transfer mechanism. Scientists have worked to enhance irrigation efficiency by adopting advance irrigation methods or by modifying existing surface irrigation methods. Prasad *et al.* (1987) found higher water use efficiency with skip furrow irrigation as compared to check basin irrigation methods. The water use efficiency for plant crop obtained by them was 2.97 and 2.31 t / ha cm with skip furrow and check basin irrigation methods, respectively. Singh *et. al.*, (1994) reported higher water use efficiency with alternate furrow irrigation methods. The observed irrigation water use efficiency with alternate furrow irrigation was 0.75 t/ha cm as compared to check basin and every furrow irrigation method, where the irrigation water use efficiency was 0.47 and 0.68 t/ha cm, respectively. Hapse, *et. al.*, (1991) observed highest water use efficiency (0.79 t/ha cm) with sub surface drip irrigation as compared to 0.33 t/ha cm with conventional furrow irrigation method. Parikh *et. al.*, (1992) observed highest irrigation water use efficiency of 2.42 t/ha- cm when irrigation was applied through drip method at 0.3 IW/CPE ratio, while with furrow irrigation methods, irrigation water use efficiency achieved was only 0.90 t/ha-cm. To increase effectiveness of irrigation by reducing evaporation loss of moisture, trash mulching had been advocated by many researchers. Yadav and Prasad (1988) observed highest sugarcane yield and irrigation water use efficiency when irrigations are applied at 25 per cent deplition of available soil moisture. Motiwale & Singh (1977) observed higher irrigation water use efficiency with trash mulching. They reported 3.7 t/ha cm and 2.3 t/ha-cm irrigation water use efficiency in the fields which received trash mulch and no-mulch, respectively. The crop if irrigated scientifically, irrigation water use efficiency of sugarcane crop can be enhanced by 1.5 to 2.5 times. Therefore, there is a need to improve irrigation efficiency at the farmer's field to save the most precious natural resource i.e., water through use of water efficient technologies. In this regards some of the technologies developed by the Indian Institute of Sugarcane research, Lucknow (U.P.) were transferred to the farmer's fields trough participatory action research programme. The results are presented in this paper.

Methodology

Study Locale and Sampling

Two districts; Barabanki and Sitapur in U.P. were selected purposively as these were in the list of the backward districts of the country as declared by the Planning Commission. From each selected district two blocks and from each block 2 villages thus, 2 districts,

4 blocks and 8 villages were selected by following suitable sampling procedure. A total of 35 demonstrations of one hectare area each were conducted during crop season 2008-09 on farmers' fields.

Criteria for farmers' selection

- The farmer should have at least one ha area under sugarcane cultivation.
- He must be growing sugarcane from last 4-5 years.
- He should be ready to spare his land for conducting demonstration as well as agree to actively participate.

Observations recorded

- Cane yield
- Irrigation water applied
- Irrigation water use efficiency
- Economics of crop production under demonstrated technologies

Data Collection and analyses

On above-mentioned parameters, data were collected periodically by following standard procedures and techniques and were analyzed with the help of suitable statistical tools.

Table 2 : Year wise break up of demonstrations.

Irrigation Technology	No. of Demonstrations			Total
	I Year	II Year	III Year	
Ring-Pit	8	8	0	16
Skip Furrow Method	14	14	6	34
Drip Irrigation	0	8	8	16
Trash Mulching	8	16	10	34
Total	30	46	24	100

Description of technologies

Trash Mulching : Sugarcane trash is a waste material available after harvesting of the crop. Trash is spread @ 10 t ha^{-1} in the inter row spaces in ratoon crop at the time of its initiation. Because of trash mulching, effectiveness of irrigation is increased as the evaporation losses of moisture from soil surface reduced considerably. Sugarcane crop yield and water use efficiency increases by 26 and 40 per cent, respectively due to trash mulch as the trash mulch keep the soil moisture at the higher level for a longer time as compares to uncovered soil surface. Increase in sugarcane yield due to trash mulch is attributed to favourable moisture condition, increased microbial activities and addition of water-soluble nutrient of the trash. In the long run, soil organic carbon content is also improved.

Ring Pit Planting Technique: The field is marked at a regular distance of 120 cm, leaving 1 m space in the beginning, both length and width wise. Nearly 6900/9000 pits per ha of 90/75 cm diameter and 45 cm depth are made by pit digger. The soil duged up from the pit is kept in the periphery of the ring in 30 cm space left in between the two pits. In every pit 3 kg farmyard manure or compost or press mud cake is mixed uniformly before placing the setts for planting. In addition to this manure 8 g urea, 20 g DAP, 16 g MOP and zinc sulphate 2 g are also added in the pit. Twenty, two buded setts are placed in a circular manner in each pit. The chloropyriphos solution is applied on the setts and 2-5 cm soil cover is made over the setts. One irrigation just after planting and blind hoeing prompt germination. Thirty days after germination, 16 g urea is applied in the pit and half of the soil remaining at periphery is filled back in the pit. In the month of April-May the remaining soil is filled back in the pit and 16 g urea is also applied. The filling of soil is completed when all the mother shoots have emerged. The crop under ring planting consists mainly of mother shoots, which are thicker and heavier. Ratoon yields are also higher because of deeper planting of plant crop. As the irrigation water is applied in the pits, more than 40 per cent surface area remains dry. For this reason, with the ring planting system 25-30 per cent irrigation water is saved. Higher sugarcane yield and reduced quantity of irrigation water results in 30-40 per cent higher irrigation water use efficiency.

Drip Irrigation: The sugarcane crop is planted in ring pits or in paired row planting systems. Drip laterals are laid down and 3-4 emitters of 4 litres discharge per pit are provided. In paired row planting system, drip tape or in line emitter laterals are provided in between the paired rows of sugarcane. Water-soluble fertilizers and other chemicals are also applied through drip system. Since soil moisture and nutrient supply is maintained at the optimum level, the crop yield is improved significantly. Due to no deep percolation, losses of water and nutrient through leaching can be avoided This results in higher water and nutrient use efficiency. This technology saves 40-60 per cent irrigation water.

Table 3 : Locational details of demonstrations conducted during 2008-09.

S.No	Technology	Sugar Mill Zone									
		Biswan Village		Kamlapur Village		Rauzagoan Village		Hydergarh Village		Total	
		Maizuddiripur	Piperakala	Muzafferpur	Devipur	Said Khanpur	Udvatnagar	Garhi Satrahi	Bijapur	Target	Achievement
1.	Trash Mulching	1	2	1	1	1	1	1	2	8	10
2.	Skip Furron Method	4	2	2	1	1	2	1	1	10	14
3.	Irrigation at C G S	2	2	1	-	1	1	1	1	6	9
4.	Ring- Pit planting Method	1	1	-	-	-	-	-	-	8	2
	Total	8	7	4	2	3	4	3	4	32	35

Irrigation at critical growth stages: In the areas of limited water supply, ensuring irrigation at critical period of water need of the crop and deferring the same at somewhat less critical period, improves yield and irrigation water use efficiency. These critical stages for sugarcane are emergence, first order of tillering, second order of tillering and third order of tillering. Depending upon the availability of water, the crops irrigated at these stages. If only one irrigation is available then it is provided at third order of tillering. If two irrigations are available then the irrigations are provided at emergence and at third order of tillering. If three irrigations are available then the irrigations are provided at emergence, first order of tillering and third order of tillering. If four irrigations are available then the irrigations are provided at all the four critical stages.

Results and Discussion

Increase in Yield and Water Saving

During the crop 2008-09, four technologies namely skip furrow method of irrigation, ring-pit method of planting, trash mulching and irrigation at critical growth stages were demonstrated in 8 villages of 4 blocks from Sitapur and Barabanki districts of Uttar Pradesh. Technology wise locational details are provided in Table 3. A total 35 demonstrations (Table 5) were conducted during 2008-09 crop season. The results of demonstation reveals that there is significant increase in crop yield, irrigation water saving and irrigation water use efficiency (Table 6).

With skip furrow method of irrigation there was 30per cent increase in cane yield, 33 per cent saving in irrigation water with corresponding increase of 70 per cent in irrigation water use efficiency (IWUE). In case of ring-pit method of planting there was 40 and 128 per cent increase in cane yield and irrigation water use efficiency. The water saving in ring-pit method was observed to be 38 per cent. Trash mulching increased sugarcane yield by 38 per cent and IWUE by 94 per cent. However, irrigation water saving was recorded 42 per cent. Irrigation at critical growth stages resulted in 17per cent water saving and 31 per cent increase in yield and 57 per cent increase in irrigation water use efficiency. It was observed that ring-pit method was best in terms of increase in cane yield, saving in irrigation water and increase in IWUE. However, the lowest increase in cane yield (30 per cent) was observed in skip furrow method. The lowest saving in irrigation water (17 per cent) and lowest increase in IWUE (57 per cent) was recorded in irrigation at critical growth stages.

Table 4 : Details of Technologies.

S. No.	Particulars	Technologies				
		Ring Pit Planting Technique	Irrigation at critical growth stages	Skip furrow method of Irrigation	Drip irrigation	Trash mulching
1.	Places where the technologies have already been in use	Punjab, Haryana and Western U. P.	Some parts of U. P., Punjab, Haryana and M. P.	Punjab, Haryana and Western U. P.	Western U. P, Gujrat, Maharastra, Karnataka, Tamilnadu & A. P.	U. P., Uttaranchal, Punjab and Haryana.
2.	Cost/ha (Rs.)	Rs. 44, 500	Rs. 30, 000	Rs. 33, 000	Rs. 55, 000	Rs. 15, 430
3.	Crops/farming system for which suited	Sugarcane Based cropping system	Sugarcane Based cropping system	Sugarcane Based cropping system	Sugarcane Based cropping system	Sugarcane Based cropping system
4.	Water use efficiency/water conserved	30-40% increase in WUE (WUE=4200 Kg $ha^{-1}cm^{-1}$) & 25-30 % saving of irrigation water	Water saving is 33.3% and (water use efficiency is 893 kg $ha^{-1}cm^{-1}$)	36.5 % irrigation water is saved 64 % increase in WUE. (WUE=2970 kg $ha^{-1}cm^{-1}$)	40-60 % water saving (WUE=775 kg $ha^{-1}cm^{-1}$)	Saves 33.3 and 40 % irrigation water at 25 and 50 % ASM, respectively. (WUE=3671 kg $ha^{-1}cm^{-1}$)
5.	Increase in yield & other benefits	142 t ha^{-1} (67 more than that of 75 t ha^{-1} under flat method)	The water saved under this system can be utilized for cultivation of other crops	3-5 t/ha more yield than conventional method (66.9 t ha^{-1})	30-35 % yield increase (94.3 t ha^{-1}) along with 0.5-1.0 unit increase in sugar recovery.	26 % increase in cane yield (82.6 t ha^{-1})
6.	Cost: Benefit Ratio	1:1.85	1:1.91	2.04	1:3.7	2.28

Table 5 : Technology wise demonstration conducted (2008-09) and Target (2009-10).

S. No.	Technologies	Demonstrations Conducted during 2008-09	Demonstrations to be conducted in 2009-10
1.	Ring Pit Planting Technique	2	14
2.	Skip Furrow Method	14	6
3	Drip Irrigation	0	8
4	Trash Mulching	10	8
5	Irrigation at critical growth stages	9	5
	Total	35	41

Economics of Sugarcane Production

In comparison to Farmers' practices, the cane yield recorded for all the four demonstrated technologies was higher (table 7). The highest yield was recorded in case of ring-pit method (116 t/ha), followed by irrigation at critical growth stages (ICGS) and skip furrow method of irrigation (80 t/ha) and trash mulching (70 t/ha). Although the cost of cultivation was highest in case of Ring-pit method, the highest return accrued for this technology well compensated the increased cost. The highest net return (Rs. 74363/ha) was recorded in case of Ring-pit method, followed by skip furrow method of irrigation (Rs.57169/ha), trash mulching (Rs.54966/ha) and ICGS (Rs.53210/ha). However, among the demonstrated technologies, the highest B/C ratio was observed for Trash mulching technology (1.228) followed by skip furrow method of irrigation (1.31), ICGS (1.91) and Ring-pit method of planting (1.71). The highest increase in B/C ratio over conventional method was recorded for trash mulching.

Conclusion

The results of this study reveal that crop performance in demonstration field clearly established the benefit of these water saving sugarcane production technologies in increasing yield and saving irrigation water. The sugarcane cultivation with demonstrated technologies were found useful not only in raising profit margin of the farmers also proved beneficial in saving most precious natural resource i.e. irrigation water.

Table 6 : Effect of technologies on sugarcane yield, irrigation water applied and irrigation water use efficiency.

S.No	Name of Technology	No. of Demonstration	Average Yield (t/ha)			Irrigation water Applied			IWUE (kg/ha cm)		
			Demonstration	Conventional	% increase	Demonstration	Conventional	% decrease	Demonstration	Conventional	% increase
1	Skip furrow Method of Irrigation	14	82.80	63.70	30.00	48.70	63.90	33.20	1700.20	996.90	70.60
2	Ring-pit Method of Planting	2	98.50	70.00	40.70	37.50	61.00	38.50	2626.70	1147.50	128.90
3	Trash Mulching	10	75.92	54.92	38.20	42.50	59.60	28.70	1786.40	921.50	93.90
4	ICGS*	9	82.80	63.40	30.60	46.00	55.30	16.80	1800.00	1146.50	57.00

Table 7 : Economics of sugarcane production in plant cane.

S.No.	Particulars	Farmers' practice	Demonstrated technologies Trash mulching	Ring pit	ICGS	Skip furrow
1.	Cost of production (Rs/ha)	56721	43034	88037	58790	54831
2.	Yield (t/ha)	62	70	116	80	80
3.	Gross return (Rs.)	86800	98000	162400	112000	112000
4.	Net return (Rs.)	30079	54966	74363	53210	57169
5.	B:C ratio	1.53	2.28	1.85	1.91	2.04
6.	% increase in B:C ratio over farmers' practice	-	49.02	20.26	24.18	33.33

References

Hapse, D.G., Deshmukh, A.S. and Shinde, P.P. (1991). Techno-economic valuation of drip irrigation system in sugarcane crop. Proceeding of National Seminar on Modern Irrigation Techniques in sugarcane Farming Systems held at VSI, Manjari, Pune on Sep. 4-5, 1991.

Motivale, M.P. and Singh. A.B. (1971). Effect of trsh mulch in economy of water and fertilizer use in sugarcane, Annual report IISR, Lucknow, pp.23-24.

Parikh, M.M., Shrivastava, P.K., Savani, N.G. and Raman, S. (1992). Response opf sugarcane crop to drip method of irrigation. *Cooperative Sugar* **23** (10):673-677.

Prasad, S.R., Alam, M. and Singh, R.N. (1987). Role of skip furrow irrigation in increasing waster use efficiency in sugarcane under Staggered Row Planting, *Indian Sugar* **36** (11):569-573

Singh S.N., Singh, S.B. and Singh, N.P. (1994). Effect of moisture regimes and methods of irrigation on growth, yield, quality and water economy in sugarcane crop. Proceeding of All India Seminar on Water Managent held at Jalgaon, Maharastra on March 13, 1994 pp. 7-12.

Yadav, R.L. and Prasad, S.R. (1988). Moisture use characteristics of sugarcane genotypes under different available soil moisture regimes in alluvial entisols, *J. Agric. Sci. Camb* **110** (1): 5-11.

□□□

Section - 3

Conservation Technologies for Food and Environmental Security

Chapter 13

Agro-Forestry Practices in Central India : Prospects and Potentials

R.K. DUBEY AND S.K. DUBEY
Central Soil and Water Conservation Research and Training Institute, Research Centre, Agra, Uttar Pradesh, India

Agro-forestry, role of tree from local watershed to regional level, finds a vast scope and high relevance in Central India (area: 8,56,467 Km^2 and human population: 35,70,22,157 equivalent to 19.90 and 34.76per cent of national area and population, respectively); in the states of Madhya Pradesh, Chhattisgarh, Uttar Pradesh, Bihar and Jharkhand; from natural resource management (NRM) point of view. Central India has a sparse forest cover as well as low agricultural productivity. As a result farmers face acute shortage of fodder, fuel, timber and non timber forest produce, hence there is considerable potential of tree based farming system. Agro-forestry, traditional and commercial practices, tree species etc. differ widely in Central India in accordance with edaphic factors, rainfall, physiographic features (Vindhyan and Satpura ranges and Narmada valley, Indo-Gangetic plains and Malwa and Chhotanagpur plateaus), socioeconomics and regional production constraints viz. highly erratic onset, precipitation and distribution of annual rainfall (500 to 1500 mm); land degradation and erosion on sloping and ravine lands; heavy runoff and soil loss in hilly tracts; declining ground water table in plain areas; drainage problem in low lying areas of Malwa plateau; presence of hard calcium pan or gravels in sub soils; soil salinity, alkanity and water logging; injudicious irrigation water use in Indo-Gangetic plains; increasing soil nutrient deficiencies mainly N, P and Zn; droughts of variable duration and intensity; impeded sub soil drainage and cracking of black soils due to expanding and shrinking clay i.e. montmorillonite; floods in 41per cent of geographical area and droughts in area equivalent to 40per cent of geographical area in South Bihar of Bihar; shifting cultivation in Santalpargana of Bihar by Mal Paharia and Samia Paharia tribes on about 162 Km^2 area and in Bastar of Chhattisgarh by hill Maria tribes on area of 81 Km^2. Agro-forestry, combining agriculture with forestry, is a technology from large scale riparian buffers to family scale forestry. Agro-forestry is a viable climate change mitigation measure and way of carbon storage in agro-ecosystems (IPCC, 2000). This is an alternative land use for more diverse, productive, profitable, healthy and sustainable agriculture. If properly designed and integrated, agro-forestry can protect crops and improve crop yields, shelter livestock, reduce animal stress while improving weight gain and enhance resource stewardship and land conservation. Scientific and systematic research on tree crop interactions in India started in 1970 which got a major impetus with launch of All India Coordinated Research Project (AICRP) on Agro-forestry in 1983. Re-vegetating rangelands lying currently in dwindling state need scientific inputs involving tree, grass and legume components to alleviate wide gap between demand and supply of food, fuel wood, fodder, fiber and medicine etc. (Kundu *et. al.*, 2008). India has 14224 to 24602 million trees are outside the forests (Ravindranath and Hall, 1995 and Prasad *et. al.*, 1999) spread over an area equivalent to 17 M ha that is supplying 49per cent of 201 M tons of fuel wood and 64 M M^3 timber consumed annually in the country. Major constraints in promoting agro-forestry in Central India are lack of awareness about benefits and wisdom of selection of tree species, higher priority to food crops, non availability of planting material, high cost of establishment, late economic returns and poor marketing facility. Trees in agro-forestry may reduce rate of oxidation of carbon due to shading and canopy i.e. micro-climate. Trees are known to ameliorate soil acidity, alkalinity and toxicity.

In Uttar Pradesh, agro forestry practices vary considerably according to the agro climatic zones, socio-economic conditions and site-specific tree species. Due to assured market of agro-forestry produce created by flourishing wood based industries (WIMCO etc.), *Eucalyptus* and *Poplar* (introduced in 1970s) are preferred in western U.P. whereas *Dalbergia sisso* and *Tectona grandis* are preferred species in eastern parts. Zomer *et. al.* (2007) reported that growth of large scale agro-forestry with poplar on irrigated lands mostly having rice-wheat cropping system in western U.P.; though it is also practiced with crops like sugarcane, berseem, clover, pulses, teak and mango; is mainly driven by economic incentives i.e. wood production is more valuable or less labor intensive; as well as due to National Forest Policy that aims to limit further deforestation. Fruit trees also have considerable share in agro forestry particularly in western part. Research circle of U.P. forest department has worked out suitable agro forestry models with preferred timber, fodder, fuel and fruit tree species for different agro climatic zones of U.P. in a workshop held on dated 5-6th May 2001 with priority to reduce soil erosion and environmental pollution, developing waste and degraded lands, reducing biotic pressure on existing forests, increasing soil fertility by planting nitrogen fixing tree species and creating availability of raw material for wood based industries.

Acacia nilotica on rice bunds of small farmers is traditional agro-forestry system in Madhya Pradesh beset with sub humid climate and hot summer. In M.P., about 19.31per cent of total geographical area (5.923 ha) is degraded and this state alone occupies 17per cent of wasteland in the country, therefore, the state Government is promoting and investing on agro-forestry and agro-based industries in areas of bio-fuel trees, pulp trees and medicinal plants. *Jatropha curcas*, a hardy and quick maturing bio-fuel plant, not browsed by cattle, grows from arid to humid climate is specifically being promoted by M.P. government on farmers field and boundaries, under stocked, public and denuded lands etc. Chhattisgarh is one of the richest bio-diversity habitats in India with 44per cent area under forests i.e. equivalent to 12per cent of Indian forests. Chhattisgarh has the densest forests, rich wild life, more than 200 non timber forest products and nearly ten thousand industrial units depending on forests (saw mills, furniture units, bidi, silk and kattha) besides, mineral mines of gold (Sonakhan area of Raipur), iron ores (Bailadila hills and Dalhi-Rajhara), coal, corundum, buxite, dolomite, lime (Raipur, Bilaspur, Durg and Bastar), tin, granite etc. Agro-forestry project (1994) promoted by Kitply Industries Ltd involving *Gmelina arbora* (white teak) and Khamar on degraded waste lands in Raipur and Durg districts of Chhatisgarh having an annual rainfall of 800 to 1500 mm on a red medium soils poor in nutrient status utilizing industrial waste like fly ash and sewerage sludge has been successful by virtue of planting more than 8.5 lakh trees planted over 1600 acres provided employment to 1200 local people together with free fuel and fodder, improvement in ground water regime aswell as The state of Chhattisgarh is clearly divided in to five river basins i.e. Mahanadi, Godavari, Ganga, Narmada and Brahmani.

Poplar based agro-forestry is also popularizing in Bihar. Champaran district in Bihar represents typical moist deciduous forests that also consist of scrub, grass and reeds. Climatologically luxuriant sal forests are favored in Bihar. Important tree species in Bihar comprise of Sal, Shisham, Cedrela, Toona, Khair and Semal. Bihar is rich in both above and underground water and Ganga and its tributaries like Sarju, Gandak, Budhi Gandak, Bagmati, Kamla Balan, Mahananda, Sone, Uttari Koyal, Punpun, Panchani and Karmanasha constitute natural drainage system. Jharkhand with considerable tribal population (22.5per cent) and tribe and an annual average rainfall of 1370 mm has severe problem of water erosion due to light textured soils with low availability of phosphates, poor soil water retention and undulating topography. Development based on watershed approach with diversified production systems including agro-forestry is highly recalled in the state. Low land areas mostly having rice mono crop constitute 40per cent of total cultivated area in Jharkhand. Intercropping with multi-purpose trees (boundary plantation, alley cropping with *Leucaena latisiliqua, Sesbaena grandiflora* and Albezia *lebbeck* and random mix) as well as agri-horti-silvi systems together with horti-pastoral systems with grasses like *Pennisetum purpureum, Pennisetum pedicellatum* and *Cenchrus ciliaris* and home gardens prevalent with tribes may be adopted in Jharkhand state.

Eucalyptus introduced in India during 1782-90 at Nandi hills in Mysore grows widely in all five states of Central India. India occupies 8per cent of global coverage under Eucalyptus (0.8 million ha). It is safe to grow on marginal agricultural and degraded lands in agro-forestry. *Eucalyptus* with short cycle of 8 to 10 years produces 10-15 M^3 biomass ha^{-1} yr^{-1} as compare to 0.5 M^3 biomass productivity of Indian forests ha^{-1} yr^{-1} and improves the fertility of site (Agriculture

Tribune, 10.08.1999). Medicinal and aromatic plants (MAPs) have promising potential in Central India. Lemon grass, menthe, java, citronella and sarpagandha have good potential in Central India specifically in U.P. and Bihar. Many MAPs are shade tolerant thus, agro-forestry offers a convenient strategy for promoting their cultivation and conservation through several feasible approaches (integrating shade tolerant MAPs as lower strata species in multistrata systems; cultivating short cycle MAPs as intercrops in existing stands of plantation tree-crops and new forest plantations; growing medicinal trees as shade providers, boundary markers, and on soil conservation structures; inter planting MAPs with food crops; involving them in social forestry programs etc.) (Rao *et al.*, 2004).

Agro-Forestry in Central India

I. Carbon sequestration

Average carbon sequestration potential in agro-forestry in India is estimated to be 25 ton $ha^{-1}yr^{-1}$ over 96 M ha but it considerably varies in different regions depending upon biomass production (Sathaye and Ravindranath, 1998 and Maikhauri, 2000). Land use options that enhance uptake of CO_2 or reduce its emission may sequestrate atmospheric C. Agro-forestry sequesters C in to vegetation and possibly in soil to a tune of 12 to 128 Mg ha^{-1} in tropical climate, thus, 1.1 to 2.2 Pg C can be stored in terrestrial ecosystems through agro-forestry on global basis in next 50 years (Albrecht and Kandji, 2003).

II. Enhancement in soil fertility

Enhancing soil fertility is vital for food security and environmental sustainability. Ecologically sound agro-forestry systems (intercropping and mixed arable livestock) can increase agriculture sustainability simultaneously reducing on site and off site consequences (Rasmussen *et. al.*, 1998). Nitrogen fixing trees grown in the agro forestry systems are capable of fixing about 50 -100 Kg N ha^{-1} $year^{-1}$ (Tewari, 1995). Agro-forestry systems are more effective in soil organic matter restoration (Manna *et. al.*, 2003). They may improve the site in non green revolution belt like many parts of Central India having low fertility. In Indo-Gangetic plains, biomass production potential to a tune of 49 t ha^{-1} $decade^{-1}$ exists which may be used for improving nutrient dynamics in deficient soils and amelioration of toxic soils (Singh, 1998). In Central India at Raipur, *Azadirachta indica* trees produced substantial biomass in N&P deficient soils (Puri and Swamy, 2001). Agro-forestry is an option to address slash and burn practice i.e. shifting cultivation (Semwal *et al.*, 2003). Das and Chaturvedi (2005) at PUSA, Bihar found that total biomass, floor litter mass, litter fall and net primary productivity of tree plantations increased with increase in age whereas biomass of any crop inter planted with poplar decreased with age of plantation. Total plantation biomass increased from 12.08 to 90.59 Mg ha^{-1} and NPP varied from 5.69 to 27.9 Mg ha^{-1} $year^{-1}$. Biomass accumulation ratio ranged between 2.1 and 3.2 while total annual litter fall between 1.95 and 10.00 Mg ha^{-1} $year^{-1}$, of which 92-94per cent was contributed by leaf litter. Appropriate agro forestry systems improve soils physical properties, maintain soil organic matter and promote nutrient cycling. Leaf litter after decomposition forms humus, releases nutrients and improves various soil properties; it also reduces the fertilizer needs (Sanchez, 1987).

III. Improving water use efficience and balance

There is robust evidence that agro-forestry improves water use efficiency by correcting water balance through its effect on runoff, soil evaporation and drainage (Turner and Ward, 2002). Simultaneous agro-forestry systems may double the rainfall utilization than annual cropping systems (Lovenstein *et. al.*, 1991; Droppelmann and Berliner, 2003). Poplar plantations existing on about 10per cent of irrigated area in Western U.P., improved water profitability and productivity and resulted in enhancement of regional water requirement just by 1.6per cent where as bund and boundary plantations were found to have little or no impact on regional water use (Zomer *et. al.*, 2007). There is general impression that *Eucalyptus* and *Prosopis* increase water use and have a negative effect on water balance (Calder, 1992, 2000 and 2002). Jackson *et. al.* (2005) found that plantations decreased stream flow by an average of 227 millimeters (mm) per year (52per cent), with 13per cent of streams dried completely for at least a year. The magnitude of this decrease in water flow is proportional to the percentage of vegetation cover and is due to an increase in actual evapo-transpiration *(AET)*, an increase in the net addition to evaporation from interception losses and the larger volume of the root zone under trees from which water is extracted (Dingman, 1993).

IV. Soil and water conservation

Trees exploit deep soil water (Maidment, 1992) also available during prolonged dry seasons (IPCC, 2000). On site effects trees are reduced runoff and soil erosion, improved micro-climate, increased control over nutrient fluxes, decreased sediment loads and increased water quality while off site effects may be lower base flow (FAO, 2006).

Table 1 : Average runoff and soil loss under different land uses (Young, 1989).

Treatment	Runoff (per cent)	Soil Loss (t/ha.)
Chrysopogon fulvus	12.7	8.65
Maize alone	27.5	28.27
Maize + *Leucaena*	21.4	17.83
Maize + Eucalyptus	20.8	13.51
Leucaena + grass	17.6	10.15
Leucaena alone	2.4	1.74
Eucalyptus + grass	6.3	3.52
Eucalyptus alone	2.1	1.20
Cultivated fallow	38.2	56.58

V. Reclamation of problem soils

Agro-forestry may provide satisfactory drainage control to ameliorate salinity (Dunin, 2002). Eight year plantation of five tree species (*Acacia nilotica, Albizia procera, Leucaena leucocephala, Azadirachta indica* and *Eucalyptus* hybrids) planted in agro-forestry reduced pH and electrical conductivity (EC) from 10.5 and 0.73 to 8.2 and 0.2 mmohs cm^{-1}, respectively in mid Indo-Gangetic alluvium at Kanpur. *Leucaena leucocephala* produced maximum girth (cm) i.e. 50.98 while *Morus alba* produced lowest girth (33.96). *Cassia siamia* and *Acacia nilotica* showed maximum improvement in soil organic carbon (from 0.12 to 0.85per cent) while maximum improvement in available P (kg ha^{-1}) was shown by *Cassia siamia* (6 to 23.4). Maximum improvement in available K (kg ha^{-1}) was recorded under *Azadirachta indica* (85 to 240). Trees showed 17.14 to 31.42per cent reduction in pH, 194.4 to 268per cent reduction in EC, 275 to 608.33per cent improvement in organic carbon, 50 to 290per cent improvement in available P and 47.6 to 182.35per cent improvement in available K (Khan, 2003).

In a study on highly alkaline wastelands in Aligarh, Uttar Pradesh, *Prosopis juliflora* produced maximum biomass (12.05 t ha^{-1}) at 3.5 year age. The survival was *Prosopis juliflora*> *Acacia nilotica*> *Terminalia arjuna*. In an another study, *Leucaena leucocephala* (7500 trees ha^{-1}) was found most promising in rehabilitation of sub standard soils. *Sesbania sesban* and *Tamarix dioca* also showed good adaptability in difficult habitats. All tree species significantly improved the soil properties, enhanced soil organic carbon content and showed marked reduction in pH, EC and ESP (Singh, 2006).

VI. Sustained crop production

Agro-forestry enhances resource capture for instance nutrient and moisture from deeper soil layers as well as radiant energy above ground since trees capture solar radiation three dimensionally while crops two dimensionally (Sachumature, 1974). Although, some traditional agro-forestry systems increase crops yields near trees but fast-growing trees may also reduced crop yield at short run. Long-term studies are required to resolve the issue of yield reduction near trees. Reduction of crop yield in agro-forestry may be observed due to several reasons but it may also be compensated in the long run by microclimate modification (Kohli and Saini, 2003) and residual nitrogen benefits from trees. Agro-forestry systems may reduce crop yields but there may be trade off for instance after removal of 15 year *Acacia nilotica* trees, crop yields increased (12.5 t ha^{-1}) that was equal to crop yield reduction during 15 years in agro-forestry (Pandey and Sharma, 2003).

Poplar either on bunds and boundaries planted sequentially with crops or intercrops in agro-forestry configuration can lead to increased land productivity while simultaneously diversifying the farming enterprises (Atta Krah *et. al.*, 2004, Huxley, 1999, Young, 1989). Greater efficiency of tree species for photosynthesis, improved soil structure and fertility with favorable effects on crop yield, reduction in losses due to soil erosion and cycling of organic matter and nutrients etc creates better micro climate for growth of agricultural crops which resulted in about 20per cent higher of grain and wood than pure agriculture in poplar based agro-forestry areas in western U.P. (Dwivedi and Sharma, 1989). Integration of fodder grasses and fodder trees was more productive than pure fodder grass cultivation. *Leucaena leucocephala* intercropped with agricultural crops and fodder grasses increase the total yield of food grains, fodder and fuel (Pathak, 1989). The total production and value of fuel, fodder and small timber in degraded lands were many times more than the coarse grains usually produced (Gupta and Mohan, 1982).

Knowledge of the quantitative assessment and structural development of root systems is essential to improve and optimize productivity of agro-forestry systems. In a study on root biomass of different size roots (2.0, 1.0, 0.5 and 0.25 mm) and root distribution of five agro-forestry trees (*Acacia auriculiformis*, *Azadirachta indica*, *Bauhinia variegata*, *Bombax ceiba*. and *Wendlandia exserta* Roxb) at PUSA, Bihar, Das and Chaturvedi (2008) found that all tree species exhibited a large variation in root depth and horizontal root spread at four years age. *W. exserta* (2.10 m) produced maximum root depth while *B. variegata* (1.00 m) produced minimum root depth. Horizontal root spread was 2.05 m in *B. ceiba* and 8.05 m in *A. auriculiformis*. Root spread exceeded crown cover in all species. Primary roots were more horizontal than the secondary roots. The length and diameter of the main root were highest in *A. indica* (108.3 cm) and *B. ceiba* (23.2 cm), respectively. Highest length and diameter of lateral roots were produced by *B. variegata* (201.6 cm) and *A. indica* (1.8 cm), respectively. Total root biomass of tree species accounted for 18.2–37.9per cent of total tree biomass. This study suggests that although all species have potential to conserve moisture and improve fertility status of the soil but *A. auriculiformis* may be most effective for promoting soil fertility while deep rooted *W. exserta* and *A. auriculiformis* may be preferred in agro-forestry as they may reduce competition for nutrients and moisture with crops as they may derive resources from deeper soil layers. In a silvo-pastoral study at Jhansi with 18 year old fodder trees (*Acacia tortilis* + *Cenchrus ciliaris*, *Leucaena leucocephala* + *Panicum maximum*, *Hardwickia binata* + *Sehima nervosum*, *Albizia amara* + *Cenchrus ciliaris* and *Albizia lebbeck* + *Sehima nervosum*), Suresh and Khan (2007) reported that root density significantly influenced due to different tree species, five radial distances (45^0, 90^0, 135^0, 157^0 and 170^0) and three depths (0-10, 10-20, 20-30 cm). In general, the root density decreased significantly ($P<0.01$) with increase in depth from 10-30 cm. The root length density of fodder trees decreased with increase in radial distance (angle) from the tree trunk (45-170 ^{0}). The pattern of biomass accumulation in roots varied significantly among species and depths.

VII. An alternative to shifting cultivation

Agro-forestry can be effectively practiced on shifting cultivation lands in mountain ecosystem usually denuded of vegetation due to biotic causes as well as in arid and semi-arid tracts with aims to conserve and improve the site and simultaneous optimization of the combined production of trees and crops. Agro-forestry had a lead role in success of integrated area development approach for control of shifting cultivation since tribal prefer to raise agricultural crops and trees simultaneously to cater their domestic needs for food, timber, bamboos etc for house construction, agricultural implements, fencing, etc. Agro-forestry also caters to the wish of tribes to obtain an economically beneficial way of life from their land. However, methods of agro-forestry has to be tuned in accordance with tribal customs, rituals and festivals, socioeconomics and cultural life, plantation crops raised such as coffee, large cardamoms and medicinal plants and acceptance of tribes. For instance in Bihar. 61,000 tribal people practice shifting cultivation and grow rice, millets, pulses and maize, and often rear tassar silkworms on sam (*Terminalia tomentosa*) while in Madbya Pradesh tribal people grow rice, millets, beans, pulses, maize, sweet potatoes, chillies, oil seeds and vegetables. They often protect trees of mahua (*Bassia latifolia*), palas (*Butea monosperma*) semal (*Bombax ceiba*), kirni (*Mimusops indica*), ber (*Zizyphus jujube*), munga (*Moringa oleifera*), palms (*Caryota urens, Phoenix sylvestris, P. farinifera*), and siari (*Baubinia vahlii*). Therefore, it is necessary to design agro-forestry practices involving such trees in order to achieve higher adoption in these areas.

VIII. Bio-diversity conservation

Deforestation has reduced the extent of forests globally, in general and across tropics, in particular (FAO, 2006) due to increasing population and standard of living, demand for food and wood led to conversion of large forest areas in to agriculture, industries, urban and other land uses. Role of trees and forests in terrestrial biogeochemical cycle urgently recalls for stopping further deforestation and conserving remaining forests. Synergy in livelihood security and biological diversity is of immediate concern due to continuous deforestation. Though, agro-forestry may not entirely reduce deforestation but may act as an effective buffer to deforestation by mitigating demographic pressure on forests for fuel wood, fodder, fiber, silk, gum, resins, fruits etc. Trees support cavity nesting birds which acts as refuse to bio-diversity after catastrophic events as fire (Griffith, 2000). Agro-forestry leads to a more diversified and sustainable production system and increases social, economic and environmental benefits of land users. In some cases agro-forestry supports as high as 50-80per cent bio-diversity of comparable natural systems (Noble and Dario, 1997), acts as buffer to parks and protected areas, acts as a corridor for dispersal and gene flow in wild life population (Pande, 2002). Trees populate area with a number of birds and mammals.

IX. Breaking poverity and food insecurity

Traditional *Acacia* and upland rice agro-forestry in Central India is characterized by on an average 20 *Acacia nilotica* trees of 1 to 12 year age ha^{-1} and provide illuminating economics equivalent to 10per cent of annual farm income, benefit-cost ratio of 1.47 and internal rate of return (IRR) of 33per cent at 12 per cent discount rate in a ten year cycle (Vishwanath *et al.*, 2000). A study on 8 year old community agro-forestry in Palamu district of Jharkhand showed that farmers dependent on rain-fed farming and animal husbandry definitely gained from agro-forestry (Minj and Quli, 2000). Suitable dual purpose (silk, lac and honey) community plantation of marketable non timber products may improve livelihood status of local people apart from conservation of useful species (Singh *et al.*, 1994; Quli, 2000; Dwivedi, 2001 and Jaiswal *et al.*, 2002). Agro-forestry supports wood carving industry, a source of income to local artisans worldwide, therefore, strengthens livelihood (Cunningham *et al.*, 2005). Future food security of small and marginal farmers, landless labors etc is becoming a major issue in view of declining productivity, soil erosion, land degradation, loss of bio-diversity, climatic change etc. Along with mitigating the climate change, agro-forestry systems can at least partially meet the energy needs of one billion people in India through bio-energy options, by a prudent use of agricultural residues and biomass generated in agro-forestry systems. Biomass energy based supply options can create rural wealth and employment necessary for livelihood improvement and sequester large amount of carbon in a decentralized manner.

X. Economic feasibility

Poplar based agro-systems were found to be not highly risky, economically viable, more profitable than many prevalent crop rotations and provided on farm employment opportunities in Uttar Pradesh (Jain and Singh, 2000). *Azadirachta indica, Acacia nilotica, Dalbergia sisso* and *Eucalyptus spp* are the most dominant species of traditional system while Popular and *Eucalyptus* are main species in commercial agro-forestry in western Uttar Pradesh. Major driving force for agro-forestry adoption were fuel wood (50.6 per cent), additional income (24.4 per cent) and shade (17.5 per cent) in traditional system while additional income (71.3 per cent) was major factor in commercial agro-forestry in the region. Yearly net return from tree produce ha^{-1} in traditional system was Rs. 989, 541 and 440 for marginal, small and medium farmers, respectively. In commercial agro-forestry, B: C ratio was higher (3.00) for Poplar than *Eucalyptus* (2.68). Although traditional system was less promising than commercial agro-forestry but it was relevant to the farmers' livelihood (Dwivedi *et al.*, 2007). *Populus deltoides* increases the farm return by 50per cent in Tarai region of Uttar Pradesh (Chaturvedi, 1981). Poplar as tree component in agro-forestry increased economic security for small farmers in U.P. (Russel and Franzel, 2004). Growing arable crops in association with multi-purpose trees (MPTs) was beneficial in northern alluvial plains of Madhya Pradesh comprising the districts of Gwalior, Bhind and Morena (Kushwaha *et al.*, 2000).

Importance Insect Pests in Agro-Forestry

Clostera cupreata, C. fulgurita and *Apriona cinerea* have been identified as 'major pests' in Poplars as their incidence exceeded 50per cent and led to extensive economic loss during outbreaks, coupled with tree mortality which persisted for several years in succession over large area of western Uttar Pradesh. Five insects were identified as 'potential pests', as they also caused heavy infestation and economic loss only in some years on a localized scale with no large scale plant mortality, but need to be checked to improve productivity (in plantations: *Ascotis selenaria*) and in nurseries: *Eucosma glaciata, Phalantha phalantha, Nodostoma waterhousie* and white grubs) with >50per cent incidence). Integrated pest management practices i.e. cultural methods, use of bio-pesticides, tolerant poplar varieties/clones, natural enemies, recommended safe insecticides and avoiding alternate host plants of the pest were main control measures (Singh *et. al.*, 2004).

Government Promotion to Agro-Forestry

Government of India as well as respective state governments have made variety of efforts to enhance tree cover in various states of Central India. Planning Commission, GOI (2001) made following recommendations for promoting agro forestry which briefly comprise of 1. Commercial agro forestry with suitable species (*Acacia nilotica, Bamboo species, Casuarina equisetifolia, Eucalyptus species, Populus deltoides* and *Prosopis cineraria*) in irrigated districts (10 M ha) of the country considering climatic, edaphic and agricultural conditions. 2. Strategy of environmental security, sustainable agriculture (production and economy) and food accessibility in rain fed areas. 3. Modification of export - import policies, putting in place a market regulation system, amending existing laws and executive orders relating to tree felling transport, processing etc and introduction of a suitable market information system to encourage agro forestry product marketing as more and more farmers are taking up agro forestry. 4. Expanding scheme of NABARD for farm/agro forestry and investing Rs. 100 crores per year should be ensured. 5. To cover 18 M ha of rain fed areas on watershed basis under agro forestry for conservation of soil and water and plantation of hardy species (Eucalyptus, Bamboo and Babul) under various schemes of Rural Development, NAEB and 'food for work. 6. Major states may establish Agro forestry Cooperative Federation for increasing bargaining powers of farmers in marketing of agro forestry products. 7. Suitable buy back arrangement after supply of quality planting material to farmers by wood based industries. *Samudai Adharit Samanvit Van Prabandhan Evam Sanrakshan Yojana* of Bihar [Funded by Planning commission Agro-forestry in Vaishali district, North Bihar and Joint Forest Managing (JFM) in Banka and Jamui Division of South Bihar.

Conclusion

Commercial and traditional agro-forestry systems find a vast potential and high relevance both in irrigated and rain-fed production systems of hill, valley, plain and plateau regions in Uttar Pradesh, Bihar, Jharkhand, Madhya Pradesh and Chhattisgarh states of Central India. Owing to increasing demographic pressure on forests due to increasing population and raising standards of living, role of agro-forestry i.e. trees and forests in terrestrial biogeochemical cycle urgently recalls for stopping further deforestation or conserving remaining forests. Agro-forestry in Central India mainly comprise of trees on farms, community forestry and a variety of local forest management and ethno-forestry practices that broadly cover trees on farm boundaries, trees grown in close association with village rain water ponds, crop fallow rotations, variety of agro-forests, silvi-pastoral systems and trees within settlements. Agro-forestry increases resilience and reduce vulnerability of contemporary societies that are fundamental to livelihood improvement through simultaneous production of food fodder and fire wood; conservation of bio-diversity and soil fertility enhancement; improvement in water use efficiency and water balance through affecting runoff, soil evaporation and drainage and enhancement of resource use. Thus, agro-forestry contributes vigorously to ecological, social and economical functions. Central India with area of 85.65 M ha (26.05 per cent of national area) mainly comprising of alluvial, red and black soils; support 34.8per cent human (357.3 million) and 31per cent livestock (150 million) population, however, natural resources are either shrinking (forest cover: 13.95per cent) or degrading lands like ravines: 2.51 million ha apart from other soil related problems like saline and alkali and waterlogged soils, soil erosion in Yamuna Chambal ravines and Narmada valley etc. Agro-forestry is a right answer to farming risk created by rainfall aberrations due to micro-climate benefits as well as for mitigation of land degradation and soil erosion by virtue of providing effective ground cover and soil binding. Agro-forestry may ameliorate soil salinity, alkanity, nutrient deficiencies, fertility of site, marginal, waste and pasture lands due to above and below ground litter, bird droppings, nutrient cycling, nitrogen fixation, improved infiltration of soil. Agro-forestry had a lead role in integrated area approach in stopping shifting cultivation apart from its prospects in mitigating climate change and Chloflurocarbon (CFCs) emissions and sequestrating atmospheric carbon. Agro-forestry systems have more growth potential which became evident during last decade, when agricultural production is stagnant at 2 to 4per cent annual growth rate. Comprehensive agro-forestry management practices suiting different agro-climate regions of Central India are called upon to expand and sustain employment and strengthen livelihood security particularly in drought prone areas. Therefore, this paper mainly discusses various on site and off site effects of agro-forestry and focuses on key issues like ecology and environment agriculture, poverty and livelihood and economical security in different states of Central India.

References

Albrecht, A. and Kandji, S.T. (2003). Carbon sequestration in tropical agro-forestry systems. *Agric. Eco syst. Environ.* **99**: 15–27.

Atta-Krah, K.; Kindt, R.; Skilton, J. N. and Amaral, W. (2004). Managing biological and genetic diversity in tropical agro-forestry. In: *New Vistas in Agro-forestry: A Compendium for the 1st World Congress of Agro-forestry, 1004,* eds. P. K. R. Nair; M. R. Rao; L. E. Buck. Netherlands: Kluwer Academic Publishers. pp. 183-194.

Chaturvedi, A.N. (1981). Poplar for planting. Uttar Pradesh Forest Department Bull. No. 50, Lucknow, **27** pp.

Cunningham, A., Campbell, B. and Belcher, B. (2005). *carving out a Future: Forests, Livelihoods and the International Woodcarving Trade*, Earth scan, London, UK.

Calder, I.R. (1992). Water use of eucalyptus - a review. In: *Growth and Water Use of Forest Plantations,* eds. I. R.Calder; R. L. Hall; P. G. Adlard. New York: J. Wiley. pp. 167-179.

Calder, I.R. (2000). Land use impacts on water resources. Land-Water Linkages in Rural Watersheds Electronic Workshop, Background Paper No. **1**: September 18 October 27, FAO.

Calder, I. R. (2002). Eucalyptus, water and the environment. In: *Medicinal & Aromatic Plants - Industrial Profiles: Eucalyptus.* pp. 36-51. Amsterdam: Harwood Academic Publishers, 2002.

Das, D.K. and Chaturvedi, O.P. (2005). Structure and function of *Populus deltoids* agro-forestry systems in eastern India. Dry matter dynamics. *Agro-forestry systems.* **65**(3): 215-221.

Das, D.K. and Chaturvedi (2008). Root biomass and distribution of five agro-forestry tree species. Agro-forestry systems. **74**(3): 223-230.

Dingman, S.L. (1993). Physical hydrology. New Jersey, USA.

Droppelmann, K. and Berliner, P. (2003). Runoff agro-forestry– A technique to secure the livelihood of pastoralists in the Middle East. *J.Arid Environ.* **54**: 571–577.

Dwivedi, R.P., Kareemulla, K., Singh, Ramesh, Rizvi, RH and Chauhan, J. (2007). Socio-economic analysis of agro-forestry systems in western Uttar Pradesh. *Ind. Res. J. Ext. Edu.* 7(2&3): 18-22.

Dwivedi, M.K. (2001). Apiculture in Bihar and Jharkhand: A study of costs and margins. *Agric. Market.* **44**: 12–14.

Dwivedi, A.P. and Sharma, K.K. (1989). Agro forestry: it's potential. Paper in Seminar on Eucalyptus, FRI, Dehradun.

Dunin, F.X. (2002). Integrating agro-forestry and perennial pastures to mitigate water-logging and secondary salinity. *Agric. Water Manage.* **53**: 259–270.

FAO. (2006). Global Forest Resources Assessment 2005, Main Report. Progress Towards Sustainable Forest Management. FAO Forestry Paper 147. Rome, Italy: FAO.

Griffith, D.M. (2000). Agro-forestry: A refuge for tropical biodiversity after fire. *Conserv. Biol.*, 2000, **14**: 325–326.

Huxley, P. (1999). Tropical Agro-forestry. Oxford, UK: Blackwell Science Ltd. p. 371.

IPCC (2000). IPCC Special Report on Land Use, Land-Use Change and Forestry. A special report of the Intergovernmental Panel on Climate Change. Approved at IPCC Plenary XVI (Montreal, May 1-8, 2000). IPCC Secretariat, c/o World Meteorological Organization, Geneva, Switzerland. Available at http://www.ipcc.ch/

Jackson, R.B.; Jobbágy, E.G.; Avissar, R.; Roy, S.B.; Barrett, D.J.; Cook, C.W.; Farley, K.A.; le Maitre, D.C.; McCarl, B.A. and Murray, B.C. (2005). Trading Water for Carbon with Biological Carbon Sequestration. *Science* 310 (5756): 1944-1947.

Jain, S.K. and Singh, P. (2000). Economic analysis of industrial agro-forestry: Poplar (*Populus deltoids*) in Uttar Pradesh (India). *Agroforestry Systems*. **49** (3): 255-273.

Jaiswal, A.K., Sharma, K.K., Kumar, K.K. and Bhattacharya, A. (2002). Households survey for assessing utilization of conventional lac host trees for lac cultivation. *New Agric.* **13**: 13–17.

Khan, S.A. (2003). Studies on sodic soil (problematic soils) rehabilitation through agro-forestry systems for central plain zone of Uttar Pradesh, India. XII World Forestry Congress, Quebec, Canada.

Kohli, A. and Saini, B.C. (2003). Microclimate modification and response of wheat planted under trees in a fan design in northern India. *Agro for. Syst.* **58**: 109–117.

Kundu, S.S., Chaturvedi, O.P., Dagar, J.C. and Sirohi, S.K. (2008). Environment, agro-forestry and livestock management. International Book Distributing Co., Lucknow.

Kushwaha, H.S., Gupta, K.P. and Raghu, (2000). Economics and uses of agro-forestry in northern alluvium plains of Madhya Pradesh. *Ind. J. Soil Cons.* **28** (3).

Lövenstein, H.M., Berliner, P.R. and Keulen, H. (1991). Runoff agro-forestry in arid lands. *For. Ecol. Manage.*, 1991, **45**: 59–70.

Maidment, D.R. (1992). Handbook of hydrology. New York, USA: McGraw-Hill.

Maikhuri, R.K., Semwal, R.L., Rao, K.S., Singh, K. and Saxena, K.G. (2000). Growth and ecological impacts of traditional agro-forestry tree species in Central Himalaya, India. *Agro for. Syst.*, **48**: 257–271.

Manna, M.C., Ghosh, P.K. and Acharya, C.L. (2003). Sustainable crop production through management of soil organic carbon in semiarid and tropical India. *J. Sustain. Agric.* **21**: 87–116.

Minj, A.V. and Quli, S.M.S. (2000). Impact of agro-forestry on socioeconomic status of respondents. *Indian For.* **126**: 788–791.

Noble, I.R. and Dirzo, R. (1997). Forests as human-dominated ecosystems. *Science*. **277**: 522–525.

Pandey, D.N. (1998). Ethno-forestry: Local Knowledge for Sustainable Forestry and Livelihood Security, Himanshu/AFN, New Delhi.

Pathak, P.S. (1989). Management of subabul for optimizing production. In production of fodder and fuel wood trees (Eds. N.G. Hegde and Others), BAIF Publications, Pune.

Prasad, R., Pandey, D.N. and Kotwal, P.C. (1999). Trees Outside Forests in India: A National Assessment, Indian Institute of Forest Management, Bhopal, 2000.

Pandey, D.N. (2002). Global climate change and carbon management in multifunctional forests. *Curr. Sci.* **83**: 593–602.

Pandey, C.B. and Sharma, D.K. (2003). Residual effect of nitrogen on rice productivity following tree removal of *Acacia nilotica* in a traditional agro-forestry system in central India. *Agric. Eco syst. Environ.*, **96**: 133–139.

Pandey, D.P. (2007). Multi-functional agro-forestry systems in India. *Current Sci.* **92** (1): 155-463.

Quli, S.M.S. (2001). Agro-forestry for NTFPs conservation and economic up lift ment of farmers. *Indian For.* **127**: 1251–1262.

Rao, M.R., Palada, M.C. and Becker, B.N. (2004). Medicinal and aromatic plants in agro-forestry systems: New Vistas in Agro-forestry. *Agro-forestry systems.* **61**(1): 107-122.

Rai, S.N. and Chakrabarti, S.K. (2001). Demand and supply of fuel wood and timber in India. *Indian Farming.* **127**: 263–279.

Ravindranath, N.H. and Hall, D.O. (1995). *Biomass, Energy and Environment: A Developing Country Perspective from India,* Oxford University Press, New York p. 376.

Russell, D. and Franzel, S. (2004). Trees of prosperity: agro-forestry, markets, and the African smallholder. In: *New Vistas in Agro-forestry: A Compendium for the 1st World Congress of Agro-forestry,* 1004, eds. P. K. R. Nair; M.

Rao, R. and Buck L.E.. The Netherlands: Kluwer Academic Publishers. pp. 345-356.

Sathaye, J.A. and Ravindranath, N.H. (1998). Climate change mitigation in the energy and forestry sectors of developing countries. *Annu. Rev. Energy Environ.* **23**: 387–437.

Schumacher, E.F. (1974). Small is Beautiful. Sphere Books Ltd., London.

Semwal, R.L., Maikhuri, R.K., Rao, K.S., S en, K.K. and Saxena, K.G. (2003). Leaf litter decomposition and nutrient release patterns of six multipurpose tree species of central Himalaya, India. *Biomass Bio energy* **24**: 3–11.

Singh, Arun; Bhandari, R.S. and Verma, T.D. (2004). Important insect pests of poplars in agro-forestry and strategies for their management in northwestern India. *Agro-forestry Systems*: **63** (1): 15-26.

Singh, B. (1998). Biomass production and nutrient dynamics in three clones of *Populus deltoides* planted on Indo-Gangetic plains. **2** (3): 15–26.

Singh, B. (2006). Rehabilitation of alkaline wastelands on the Indo-gangetic alluvial plains of Uttar Pradesh, India through afforestation. *Land Degradation and Development.* **1**(4): 305-310.

Singh, M.P., Dayal, N. and Singh, B.S. (1994). Importance of genetic conservation of tasar host plants in agro-forestry program in Chhotanagpur region of Bihar. *J. Palynol.* 30, 157–163.54. Suresh, G. and Khan T.A. (2007). Studies on root distribution pattern of fodder trees in silvopastoral systems. *Range Mgmt. & Agro-forestry* **28** (2): 79-81.

Tewari, D.N. (1995). Agro forestry for increased, productivity, sustainability and poverty alleviation. International Book Distributors, Dehradun.

Turner, N.C. and Ward, P.R. (2002). The role of agro-forestry and perennial pasture in mitigating water-logging and secondary salinity: Summary. *Agric. Water Manage.* **53**: 271–275.

Viswanath, S., Nair, P.K.R., Kaushik, P.K. and Prakasam, U. (2000). *Acacia nilotica* trees in rice fields: A traditional agro-forestry system in central India. *Agro-for. Syst.* **50**: 157–177.

Young, A. (1989). Agro-forestry for Soil Conservation. ICRAF: Science and Practice of Agro-forestry #4. UK: CAB Int.p. 276.

Zomer, R.J.; Bossio, D.A.; Trabucco, A.; Yuanjie, L.; Gupta, D.C. and Singh, V.P. (2007). Trees and water: Smallholder agro-forestry on irrigated lands in Northern India. Colombo, Sri Lanka: International Water Management Institute. 47p. (IWMI Research Report 122).

❑❑❑

Chapter 14

Measures for Enhancing Crop Productivity and Resource Use Efficiency in Rainfed Vertisols of Madhya Pradesh

S.S. TOMAR[1] AND S.K. SHARMA[2]
[1] Director Research Services, JNKVV, Jabalpur
[2] Senior Scientist, College of Agriculture, Indore

The major objectives for improving productivity of rainfed/ dryland agriculture are to enhance moisture retention capacity of the soil and retain rainfall by using agronomic, biological and engineering measures in an integrated way to improve soil to change farming practices and make full use of resources like light, heat, soil, fertilizer, water and improved seeds to increase yield and agriculture productivity in the rainfed areas. The global climate change and its impact in the form of abiotic and biotic stresses are likely to further aggravate the situation. There would be growing pressure on technological needs to make agriculture sustainable, competitive in terms of cost and quality for varying situations and systems. An ideal farming approach should, therefore, aim at increasing the agriculture production, conservation of soils, improving the productivity of water, ensuring livelihood and generating employment. In this endeavour, developing technologies that supports resource use efficiency, improves soil and environmental health and over all returns to the producers. Hence, research efforts are to be seen in system's perspective on a fast changing agriculture environment (Mangala Rai, 2009). A good deal of information has been generated specifically, for rainfed soybean based and rice based production systems of Madhya Pradesh on various aspects of crop production. These include the practices related to cost effective management of crops and cropping systems, rain water management, soil fertility and nutrient management, conservation of natural resources and their efficient management and conclusions have been drawn for direct use by the beneficiaries.

Material and Methods

Series of field experiments were conducted in Vertisols, to generate the technological options for sustainable crop production under rice-wheat and soybean-based cropping systems, at JNKVV, Jabalpur, Madhya Pradesh under AICRP on "tillage requirement of major Indian soils" and AICRP for Dryland Agriculture, College of Agriculture, Indore.

Results and Discussion

A brief account of the technologies developed and recommendations generated has been discussed in this paper in the following sub heads.

Rainfed agro-ecosystem

Rainfed Agro-Ecosystem occupies an important place in Indian agriculture. It covers about 66per cent of the net cultivated area supporting 40per cent of the India's population and contributed about 44per cent to the food basket. The rainfed agro-ecosystem also supports two third of India's livestock population. The ecosystem as a whole is characterized by instability in biological productivity caused by aberrant weather. Farmers are resource poor with inadequate infrastructure and credit support. The major risk and hazards associated with rainfed crops in central part of India are related to weather and associated soil problems. This paper discusses the potential of various technologies developed and tested in the region for sustainable crop production and for maintaining soil fertility in long run under rainfed/dryland condition of Vertisols region. Key issues of rainfed agriculture are : Continued resources degradation and increased weather related risks, Stagnant productivity in major growing districts, Loss of soil fertility, Stagnant/falling cropping intensity, Poor replicability of watershed approach and poor technology uptake by stakeholder.

Soybean based production system

Soybean with coverage of 6 mha and production of 6 mt has occupied an important place in agricultural and oil economy of India. It has revolutionized rural economy and has uplifted socio-economic status of farmers of M.P. Occupying third place in area, production and share in national kitty for edible oils, soybean is the only cost effective alternative to ameliorate pulse protein deficiency in the country. While there had been unprecedented increase in area and production (75per cent and 155per cent respectively during 1986-87 to 1993- 94), the increase in productivity had been slow (91per cent). It is gaining more coverage in adjacent areas to Madhya Pradesh and Maharashtra and even in tribal districts in Andhra Pradesh like Adilabad. The present yield levels in these new areas are more than the Madhya Pradesh. Therefore, it is necessary to identify the areas that are more suitable for growing soybean. The major concern is that the yield at national level has stagnated at 1 t/ha as against 2-3 t/ha in Front Line Demonstrations and 3-4 t/ha achieved in research experiments. For the sustainability of the crop in India, creation of domestic market is inevitable, which in turn calls for development of products based on soya meal and consumer awareness. The Major soybean based production system which are followed in dryland production system are Soybean-gram/wheat/safflower etc.

Rice based production system

At the current rate of population growth in India (1.45per cent), rice requirement by 2025 would be around 120 mt. According to an estimate (Subbaiah *et. al.*, 2006) with increasing population and improving standard of living, additional 2.0 mt of rice is to be produced every year, to maintain the self-sufficiency in rice achieved during the early 80s. In view of little or almost no scope for horizontal expansion, the increased demand has to be met through vertical growth. The area, productivity and production of wheat have increased 119per cent, 236per cent and 634per cent respectively, by the year 1999-2000, as compared with 1965-66 (base year). During 1999-2000, the wheat production in the country touched the highest peak of 76.37 Mt. In wheat production since last seven years, India is maintaining its second position in the world next to China and at present produces more wheat than the United States of America.

The target of 80 Mt appears to be achievable since the yield gap between research stations and farmer's field is around 1.0-1.5 t/ha, this gap being more prominent in the states of Madhya Pradesh, Bihar, Rajasthan and Uttar Pradesh (Mishra, 2006). The 80mt wheat production target by the end of the next five-year plan could be realistic one only if we concentrate on the short term strategies which give more emphasis to the effective transfer of available technologies to the farmers. With the available technologies it is possible to bridge the gap between farmer's practice and FLD yield and also to increase resource use efficiency on sustainable basis. Due to intensive rice-wheat cropping system there is a deterioration in soil physical properties resulting in declining in total factor productivity of rice-wheat system.

Experimental results

A good deal of information has been generated on various aspects of rice - based and soybean-based production systems mainly related to cost effective management of crops and cropping systems, rain water management, soil fertility and nutrient management, conservation of natural resources and their efficient management and conclusions have been drawn for direct use by the beneficiaries. Major options for enhancing crop productivity and resource use efficiency includes:

1. Reduced tillage option for rice-wheat system

In a long-term study conducted (1991 to 2000) under rice wheat cropping system in Vertisols (Typic Haplusters) at Jabalpur, MP, India it was reported that puddling during rice deteriorates the soil physical environment for subsequent wheat crop as compared to under direct seeded plots (Tomar, 2008). It was concluded that in Vertisols paddy should be grown as direct seeded crop, which provides better physical environment for subsequent wheat crop. Significantly higher yield of wheat was recorded in direct seeded paddy plots as compared to transplanted plots, and the total productivity of the system was higher under reduced tillage conditions i.e. Direct seeded paddy followed by zero till wheat (Figs. 1 and 2).

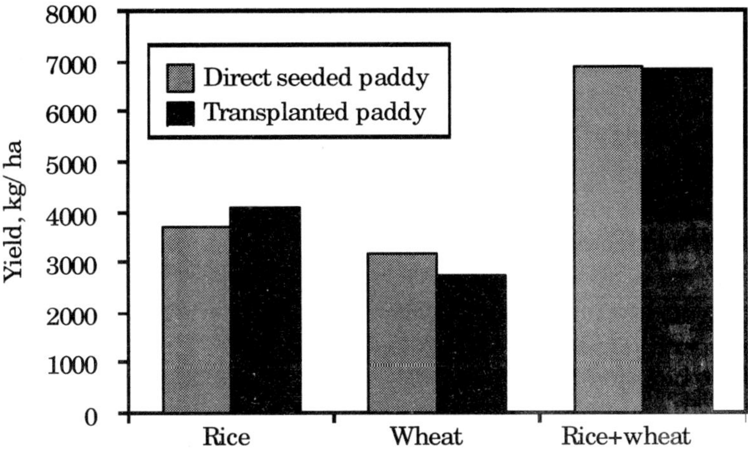

Fig. 1 Grain yield of crops under rice -wheat sequence as influenced by methods of rice cultivation.

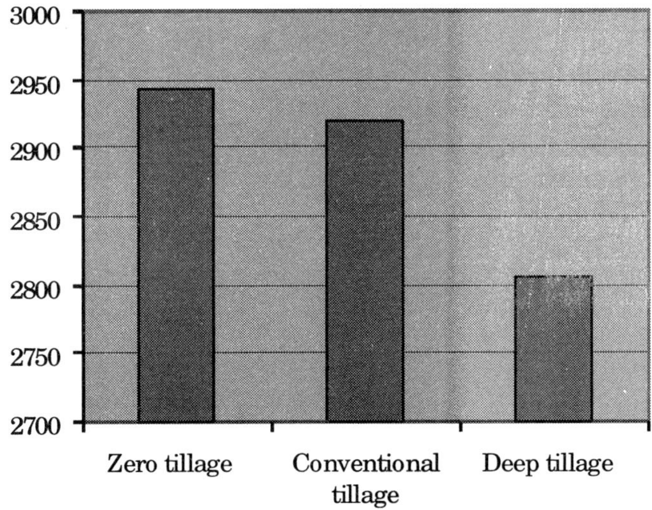

Fig. 2 Effect of tillage on the produvtivity of wheat grown after rice.

Table 1 and Fig.1 shows that the mean yield of transplanted rice (TP) was higher than the yield of direct seeded rice (DS). Outof ten years the grain yield of TP rice was significantly higher in six years than the DS rice, while in other years the yield levels were statistically at par except during 1992 where the DS paddy gave significantly higher yield than that of transplanted paddy. A comparison of method of rice cultivation on root growth showed that total root length upto 45 cm depth was significantly higher in DS rice as compared to that in TP rice Fig 3.The soil organic carbon content was significantly higher under DS rice than the TP rice (Fig. 4). Lal (1986) also reported 2.2 and 1.7 per cent organic carbon in unpuddled and puddled plots, respectively.

Table 1 : Paddy yields ($kgha^{-1}$) as influenced by methods of rice cultivation.

Year	Method of rice cultivation		CD $_{(5\%)}$
	Transplanted paddy		Direct seeded paddy
1991-92	4913	4209	417
1992-93	2715	3502	353
1993-94	5560	4788	562
1994-95	3036	2871	NS
1995-96	3940	2898	116
1996-97	3482	3043	279
1997-98	4540	4208	NS
1998-99	4302	3726	NS
1999-00	3573	2922	521
2000-01	4938	4097	384
Average	4093	3711	-

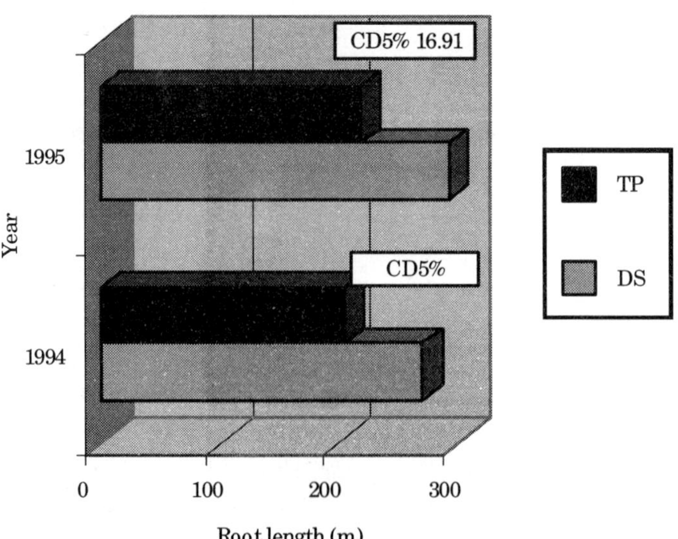

Fig. 3 Root length (m) up to 45 cm depth under Ds and TP rice.

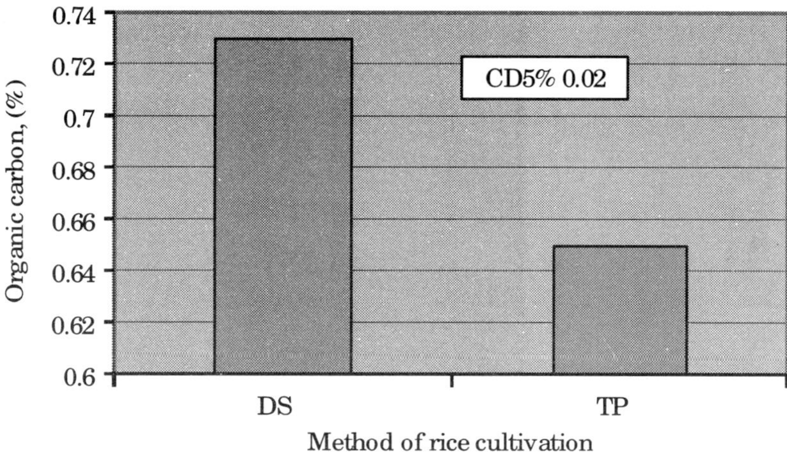

Fig. 4 Organic carbon (per cent) under Ds and TP rice conditions.

On black clay Vertisols, moisture content profile (Fig. 5) at harvest were similar on puddled and unpuddled soils. But drying and the rate of drying were greater an nonpuddled soils. Cultivation of puddled paddy produces cloddy seedbed, resulting in poor seed soil contact, rapid drying of surface soil and reduces germination. Having to wait for a puddled soil to arrive at the optimum moisture content for tillage has another disadvantage. A longer 'turnaround' time resulted in low crop yields. Results of regional experiments showed that delay in wheat planting resulted in decrease in yield potential in the range of 1 to 1.5

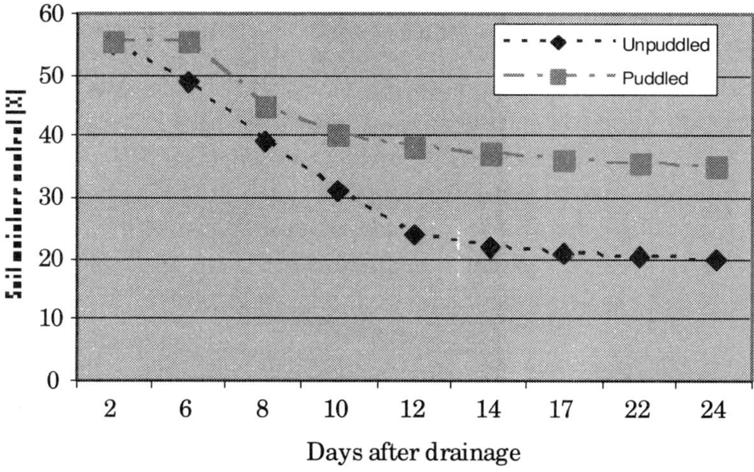

Fig. 5 Drying pattern of puddled and unpuddled vertisols.

percent loss/ha/day when planting occurs beyond last week of November (Tomar and Verma, 1985). Late planting not only results in lower yield but also reduces the efficiency of applied inputs. Fig 6 shows the wheat yield reduction pattern when it is planted beyond November (Tomar and Verma, 1985).

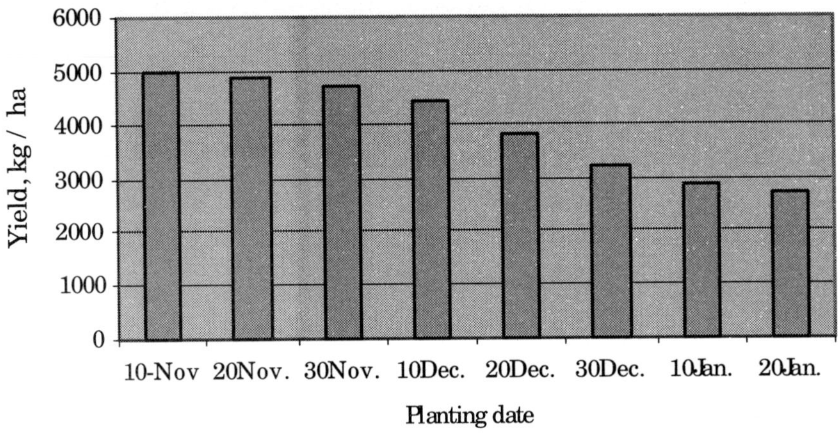

Fig. 6 Wheat yield Reduction trend under late planted condition.

The long term results of rice-wheat cropping system in Vertisols (Typic Haplusterts) at JNKVV, Jabalpur, MP, revealed that zero tillage as a technological option for timely establishment of wheat (Tomar, 2008). In this technique, direct drilling of wheat using Pantnagar Zero till fertilizer seed drill can be done without any tillage operations immediately after harvest of paddy on residual moisture. In rice -wheat system, paddy followed by no till wheat resulted in significantly increased productivity of rice wheat system as compared to the productivity of paddy followed by shallow till or deep till wheat (Fig. 1 & 2). Zero tillage has yield advantage particularly in early sown crop i.e. November planted wheat (Table 2). No till system of wheat and reduced tillage in paddy (avoiding puddling and direct seeding of paddy) helped to reduce the turnaround time and get the wheat planted closure to the optimum date. This system offers the advantages of timely wheat sowing, favouring crop productivity, saving energy (Fig. 7) and turnaround time. This system not only increases productivity of the system but also reduces the cost of cultivation, fuel cost, wear and tear of tractors and increases input use efficiency. The farmers of rice wheat sequence are resource poor and the savings on tillage operations can be used to buy other inputs like weedicides, fertilizer, irrigation which will further enhance productivity. The twin advantage of enhancement of productivity and its profitability under zero till system, will go a long way in enhancing the sustainability of rice wheat system.

Table 2 : Influence of tillage on the grain yield of paddy and wheat under rice – wheat system (Tomar, 2008).

Year	Paddy yield (kg ha^{-1})				Wheat yield (kg ha^{-1})				
	No-tillage	Conventional tillage	Deep tillage	CD (5%)	No-tillage	Conventional tillage	Deep tillage	CD (5%)	Wheat Planting date
1991	4311	4182	4092	NS	3873	3807	3275	251	Nov.13
1992	3042	2922	2919	NS	3545	3137	3646	418	Dec.13
1993	5219	5413	5165	NS	4431	4302	4095	194	Nov.25
1994	2930	2931	2885	NS	4221	3582	3195	101	Nov.11
1995	3378	3049	2670	139	4112	3270	3005	266	Nov.20
1996	3321	3343	3122	NS	2423	2731	2605	NS	Dec.4
1997	4583	4254	4286	268	781	739	730	NS	Jan1
1998	4109	4110	3823	NS	1516	2083	2064	459	Dec.15
1999	3508	3011	3210	NS	2338	2859	2769	270	Nov.25
2000	4722	4558	4372	NS	2195	2674	2669	455	Dec.2
Av.	3912	3773	3654	-	2943	2918	2805	-	-

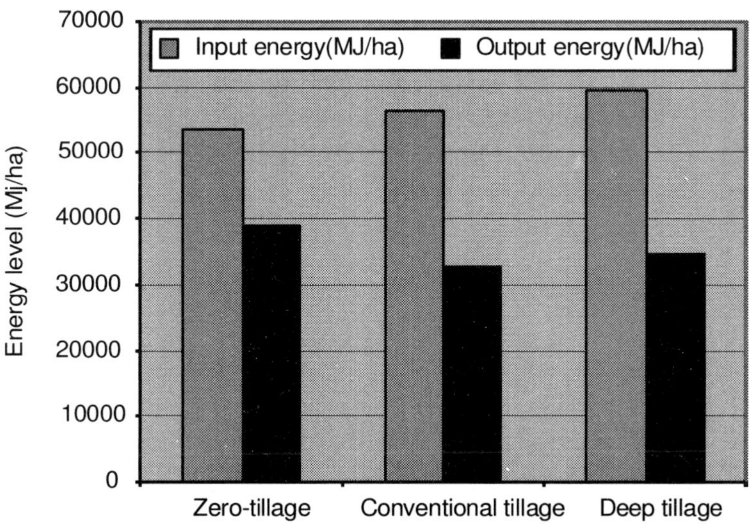

Fig. 7 Input-Output energy requirement in wheat production under different tillage treatments.

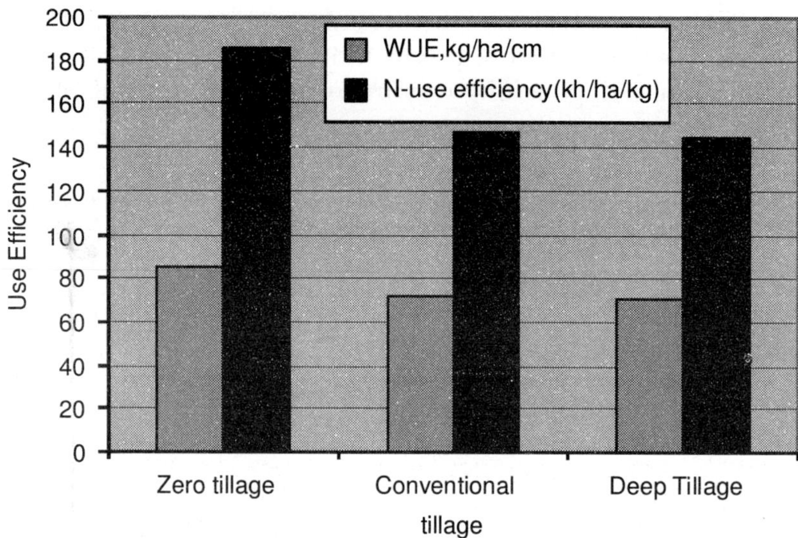

Fig. 8 Effect of tillage on resource use efficiency.

Data presented in table 6 further indicated that after six years of experimentation there is decrease in yield of wheat under zero till plots. Poor yield of wheat recorded during 1997-98 was due to continuous heavy rains received in the month of November and December months.

Recycling of crop residue

Field experiment was conducted in Vertisols at Jabalpur by incorporating of rice and wheat crop residue both in situ or as compost, increased the soil organic carbon content and improved soil health (Fig. 9), while improving the productivity of the rice-wheat system. The return of carbon to soil and yield of crops was maximum, when crop residue was added along with fertilizer N @ 120 kg N ha^{-1}.

Nitrogen management in rice - wheat system

Low N use efficiency in rice – wheat system, especially in rice is a major concerned, recommendation has been evolved for application of FYM along with urea super granules/prilled urea deep placement for rice crop. Data presented in Table 3 revealed that significantly higher yield of paddy was recorded when N was applied through Urea Super granule (USG) along with FYM @ 5t ha^{-1}. All the amendments in combination with USG gave higher grain yield of paddy in comparison to prilled urea combinations. The grain yield of rabi crops was also significantly superior in amendment treated plots irrespective of forms of urea used. Application of organic amendments improved the soil physical environment, which has favourable effect on the grain yield of both paddy and subsequent rabi crops (Fig. 10).

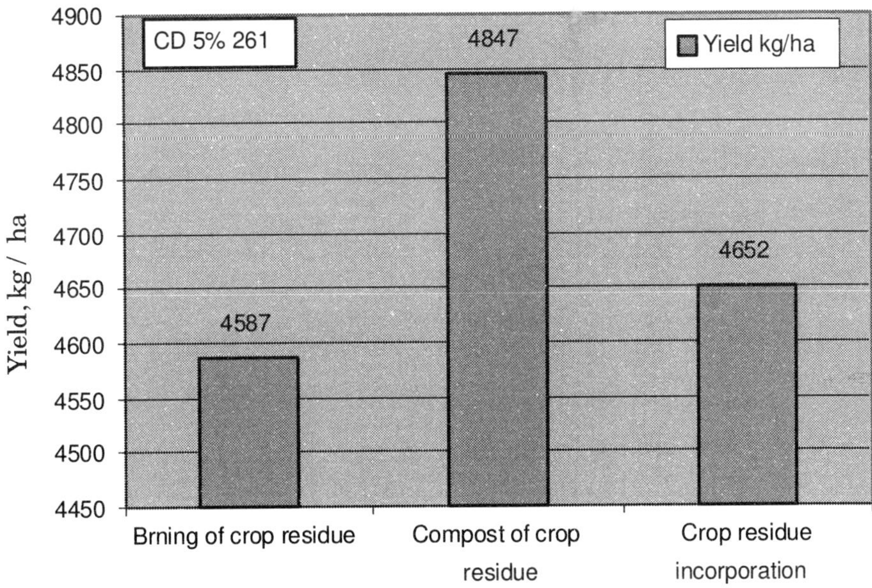

Fig. 9 Wheat yield as affected by crop residue management.

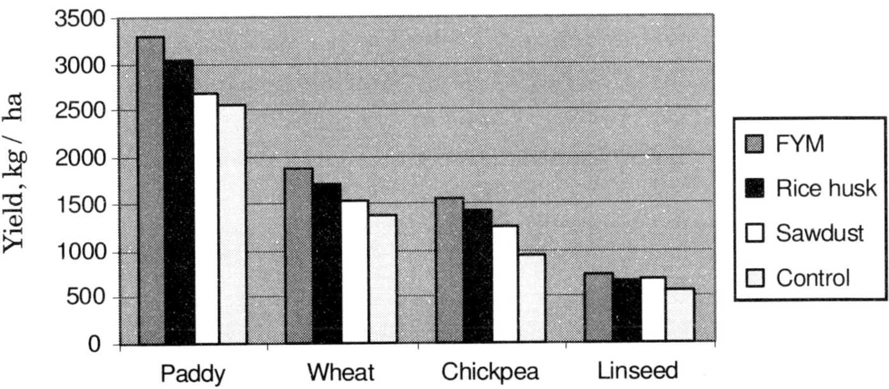

Fig. 10 Effect of organic ammendments on the productivity of crops under rice wheat system.

Table 3 : Effect of organic amendments and forms of urea on the grain yield of crops in rice based cropping system (average of two years) (Tomar *et al.*, 1996).

Treatment	Yield, kgha^{-1}			
	Paddy	Wheat	Linseed	Chickpea
Prilled Urea (PU)	3848	3561	798	1923
FYM+PU	4909	4450	1312	2568
Paddy husk + PU	4079	4112	1123	2314
Paddy Straw+PU	3818	3762	1014	2260
Sawdust + PU	3950	3935	810	2046
UreaSupergranules (USG)	3928	3621	835	2141
FYM+USG	5085	4597	756	1850
Paddy husk + USG	4241	4290	1250	2507
Paddy Straw + USG	4171	3816	1068	2314
Sawdust + USG	4025	4025	949	2200
Control	2214	1966	375	1012
$CD_{(5per\ cent)}$	489	109	361	644

Improving seed soil contact in rice based cropping system

After paddy harvest, the seedbed is quite cloddy, which affects germination of post rainy season crops adversely. Vertisols are having a high water storage capacity which can be very well exploited if we can have a good germination of rabi crops, specifically under rainfed situation (Tomar *et. al.*, 1995). The data depicted in Fig. 10 clearly revealed that highest germination of all the crops was recorded in watering treatment followed by compaction. This was due to the better moisture condition and good seed soil contact in these treatment as compared to control. For better stand of rainfed rabi crops in rice based cropping system either watering in seeded rows or rollering with 20-30 kg roller has been found effective in improving seed soil contact, which resulted in water conservation, higher water absorption and thereby increased germination and over all productivity of rabi crops grown after rice. Data presented in Table 4 revealed that about 40per cent and 26 per cent higher wheat yield, 25 and 18per cent higher chickpea yield, 28 and 13 per cent higher yield of linseed were obtained due to watering and compaction, respectively as compared to control.

Table 4 : Effect of watering and compaction on the productivity of rainfed rabi crops (Average of five years) grown after puddled paddy (Tomar *et al.*, 1996).

Treatment	Yield, kgha^{-1}		
	Wheat	Chickpea	Linseed
Watering in seeded rows	1598 (40per cent)	1438 (25per cent)	724 (28per cent)
Compaction in seeded rows	1436 (26per cent)	1355 (13per cent)	639(13per cent)
Control	1144	1148	566

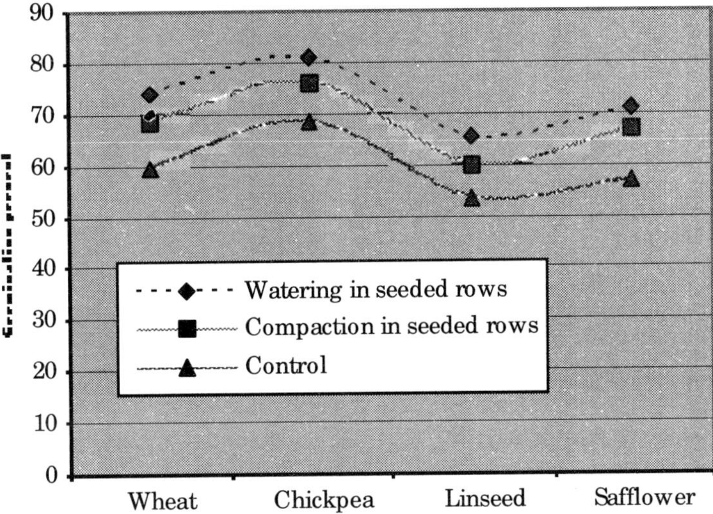

Fig. 11 Germination percent of rabi crops as influenced by water and compaction inseded rows.

Low till farming strategy for improving soybean crop productivity and soil quality in long run

With the objective of exploring potential of conservation tillage in rainfed soybean based production system for enhancing productivity and soil quality, a field experiment was initiated in the year 1999 at JNKVV, College of Agriculture Indore for evolving low till farming strategies for rainfed production system. On the basis of sustainability index it is recommended that the conjoint use of organics with reduced level of tillage and chemicals which are not only cost effective but also enhance the productivity of soybean (Table 5) and build up soil health in the long run. Under rainfed conditions, soybean should be grown with the addition of organics such as compost, straw as a mulch and addition of toppling of gliricidia. With the help of these additives low till farming strategy is highly recommended with hand weeding as the application of herbicide was not found effective and low sustainability index was obtained in the treatments involving herbicide application. Energy requirement of various treatments revealed that conservation tillage requires lesser input energy and gave higher energy use efficiency as compared to conventional tillage and intensive tillage options.

Table 5 : Low till farming strategy for sustainable of soybean productivity.

Treatment	Mean	B:C	SI
T_1:CT + RF (-OT) + HW.	1376	1.45	0.53
T_2:CT + RF (+OT) + HW.	1653	1.52	0.63
T_3:LT + 4 t /ha straw + HW.	1493	2.27	0.55
T_4:LT + 4 t /ha straw + Hb.	1178	1.67	0.43
T_5:LT + 4 t/ha Compost + HW	1462	1.97	0.57
T_6: LT + 4 t/ha Compost + Hb.	1177	1.48	0.46
T_7:LT+2 t/ha Gliricidia green leaves + Hb	1099	1.57	0.42

The major recommendation emerge from this experiment is that low till farming strategy (minimum tillage + use of organics i.e. Compost @ 5 t/ha, Gliricidia leaves @ 2 t/ha) is recommended for enhancing productivity and soil quality in long run under rainfed conditions. Low till farming strategy can be adopted in place of intensive cultivation which includes low tillage + use of organics like compost, straw and gliricidia toppling. The SI > 0.50 revealed that this is one of the most promising technologies for rainfed soybean in Vertisols.

Advantages of low till farming

The sustainability index (SI) calculated for the low farming strategy treatment revealed that this is one of the most promising technologies for rainfed soybean in Vertisols as the Sustainability Index is > 0.50.

Improves soil quality in terms of increased levels of available OC (0.77 per cent), N (278 kg ha^{-1}), P_2O_5 (58.80 kg ha^{-1}), K_2O (710 kg ha^{-1}) and S (12.11 kg ha^{-1}) as compared to control in which case the values of available OC (0.52 per cent), N (207 kg ha^{-1}), P_2O_5 (36.0 kg ha^{-1}), K_2O (692 kg ha^{-1}) and S (7.19 kg ha^{-1}).

Over all results suggested that by the use of organics the tillage intensity can be reduced which will help in increasing soil quality and reducing energy consumption without sacrificing the crop yields.

Integrated nutrient management practices for soybean based cropping system

Maintaining soil health is an important consideration so that productivity potential of a soil is not diminished. Integrated nutrient management envisages the combined application of chemical fertilizers along with organic manures, green manures, biofertilizers and other recycling materials for crop production. The basic concept underlying IPNS is the maintenance or adjustments of soil fertility and plant nutrient supply to an optimum level for sustaining the desired crop productivity through optimization of the benefits from all possible sources, plant nutrients in an integrated manner. IPNS is ecologically, socially and economically viable and environmentally healthy technology, which can be adopted by the farming community to get higher productivity while maintaining soil fertility. Studies on permanent manorial trails at Indore (1993-08) showed that the conjoint use of FYM along with chemical fertilizers helps in increasing soybean yields on sustainable basis (Table 6). Application of FYM has been found to increase water use efficiency, moisture availability to rainfed soybean-safflower sequences and recovery of N, P, K and S from black clay soils. The data presented in the Table 10 indicated that the productivity of soybean can be increased with application of FYM @ 6 t/ha + 50per cent of the recommended dose of N and P. The highest sustainability index of 0.50 was obtained in this treatment as against 0.22 of recommended dose of chemical fertilizers.

Table 6 : Soybean yield and sustainability index under Long term manurial trail in Vertisols of Indore, M.P.

Treatments	Mean yield (16 Years) kg ha^{-1}	Sustainability Index
Control (N0P0)	1261	0.22
N 20 P 13 to each crop	1620	0.33
N 30 P 20 to each crop	1781	0.37
N 40 P 26 to each crop	1886	0.41
N 60 P3 5 to each crop	1994	0.45
FYM @ 6 t/ha+ N 20 P 13	2095	0.50
Residue @ 6 t/ha + N 20 P 13	1790	0.35
FYM @ 6 t/ha	1863	0.39
Residue@ 6 t/ha	1629	0.31

The highest safflower seed yield of 1408 kg ha^{-1} was recorded due to the treatment FYM 6 t + $N_{20}P_{13}$ which was statistically superior to the rest of the treatments. Thus, a combination of 6 t FYM, 20 kg N and 13 kg P is optimum for realizing higher yield of safflower. Decline in levels of fertilizer N and P down to 60 kg N and 35 kg P ha^{-1} resulted in gradual reduction in seed yield of safflower (Table 11). Application of FYM at the rate, of 6 t ha^{-1} along with reduced level of fertilizer (FYM + $N_{20}P_{13}$) resulted 35per cent additional seed yield of safflower as compared to the treatment having RDF. Crop residues application as surface mulch at the rate of 5 t ha^{-1} alongwith reduced level of fertilizer N and P (Residues 5 t + $N_{20}P_{13}$) enhanced the seed yield in the range of about 14.6per cent. The highest sustainable yield index (SYI) of 0.41 was estimated due to conjunctive use of FYM at the rate of 6 t ha^{-1} along with reduced levels of fertilizer N and P at the rate of 20 and 13 kg ha^{-1}.Chemical fertilizer application 60 kg N and 35 kg P ha^{-1} gave sustainable yield index of 0.30 which, emphasized that integrated application of nutrient enhances crop productivity of safflower on sustainable basis when applied in Vertisols under rainfed conditions. Similar results were also reported by Sharma and Dixit (1987), Sharma (1997) and Sharma and Sharma (2005).

Table 7 : Safflower seed yield and sustainability index as influenced by different fertility treatments.

Treatment	Mean Seed Yield, kg ha^{-1} (11 years Av.)	Sustainability index
T1:N0P0 (Control)	628	0.13
T2:N20P13	735	0.12
T3:N30 P20	1042	0.26
T4:N40P26	1166	0.30
T5:N60P35	1182	0.29
T6:FYM 6 t/ha + T2	1408	0.41
T7:Residue 5 t/ha + T2	1194	0.32
T8:FYM 6 t/ha	1211	0.33
T9:Residue 5 t/ha	1022	0.25

Changes in soil quality

Soil analysis at sowing and after harvest of safflower crop indicated tremendous increase in organic carbon content and available nutrients in the treatments involving FYM and crop residues. A remarkable decline in soil fertility was observed in check plots. The effect of different treatments on soil fertility is the cumulative effect of 14 seasons (Fig. 12). Similar results have also been reported by Havanagi and Mann (1970), Dikshit *et. al.,* (1995), Moharram *et al* (1999) and Sharma and Namdeo (1999). Kanwar and Parihar (1992) reported that the addition of FYM significantly improved the available K status of the soil, possibly owing to the progressive incorporation of K through FYM and improved retention of bases in the soil. Application of N and P fertilizers also increased the available P_20_5 content of the soil. Similar results have also been reported earlier by Mausumi *et al.* (1998). Application of organics (FYM, crop residue, alone and in combination with $N_{20}P_{13}$) improved soil chemical properties in long run (Fig. 12).

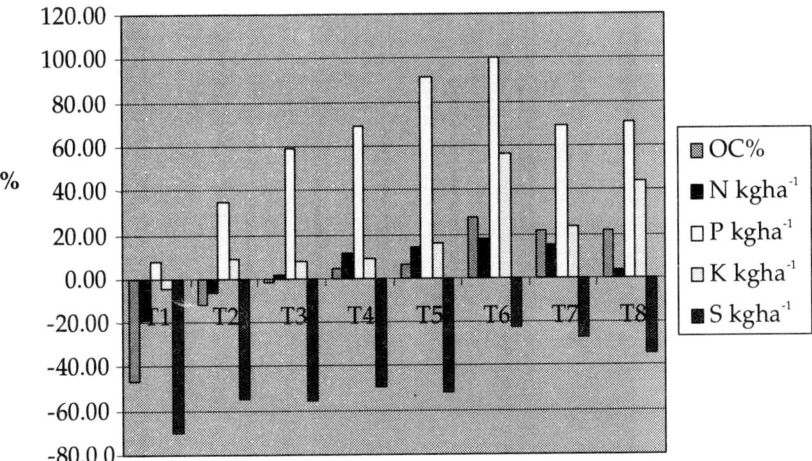

Fig. 12 : Fertility changes from the base year (1992 to 2008).

Water use and water use efficiency

The data on different components of field water balanced and profile moisture content changes during growth period of safflower. Water use and water use efficiency of safflower was worked out taking into account all the components of filed water balance namely rainfall, surface runoff, deep percolation, upwards flux of water and profile moisture content changes during the cropping season. Data presented in Table 8 revealed that water use by safflower crop varied from 105 to 152.5 mm, while water use efficiency ranged in between 5.02 to 9.34 kg ha^{-1} mm^{-1}. The highest water use efficiency of 9.34 $kgha^{-1}$ mm^{-1} was recorded from the treatment of FYM + $N_{20}P_{13}$ followed by 8.88,

7.94, 7.85, 7.83. 7.34, 7.11 and 5.02 kg ha^{-1} mm^{-1} due to treatment $N_{60}P_{35}$, $N_{30}P_{20}$, FYM alone, $N_{40}P_{26}$ Crop residue $N_{20}P_{13}$ and N_0P_0 treatments, respectively. Further evaluation of data presented in Table 12 indicated that the water use efficiency increases with increased levels of N and P. The highest water use efficiency was shown through $N_{60}P_{35}$. Application of FYM at the rate of 6 t ha^{-1} resulted in an increase of water use efficiency of safflower by 54 per cent as compared with control, which was further increased by 30per cent due to conjunctive use of FYM and $N_{20}P_{13}$.Similarly application of crop residues as surface mulch at the rate of 5 t ha^{-1} led to an increase in water use efficiency of safflower 44per cent compared with control which was further enhanced up to 24per cent when crop residues was applied in conjunction with $N_{20}P_{13}$.

Table 8 : Water use and water use efficiency of safflower as influenced by fertility treatments.

Treatments	Water use mm	Water use efficiency (kg ha^{-1}mm^{-1})
N_0P_0	105	5.02
$N_{20}P_{13}$	108.2	7.11
$N_{30}P_{20}$	105	7.94
$N_{40}P_{26}$	122	7.83
$N_{60}P_{35}$	152.5	8.88
FYM 6 t + $N_{20}P_{13}$	112.1	9.34
Residues 5 t+ $N_{20}P_{13}$	128	8.45
FYM 6 t	131.2	7.85
Residues 5 t	123.8	7.34

Thus INM strategy provides

- higher sustainability (SI 0.50),
- enhances crop yield by 10per cent as compared to RDF,
- Enhances per cent OC in soil by 76 per cent as compared to control,
- the per cent increase in N, P, K and S was 76 per cent, 39 per cent, 36 per cent, 56 per cent and 77 per cent, respectively over control,
- water use efficiency increased up to 40per cent,
- enhances nutrient uptake N (98per cent), P (108per cent), K (105per cent) and S (123per cent) by seed and straw as compared to control,
- helps to reduce the use of chemical fertilizers by 50per cent.

Effect of mulching on the productivity of crops grown in vertisols under rainfed conditions

Series of field experiments have been conducted to evaluate the comparative performance of various mulches on soybean based cropping system under rainfed conditions of Vertisols with the objective to evaluate the performance of different types of mulching on the crop productivity, water conservation and their economic feasibility. Data on seed yield of soybean and safflower as influenced by different mulch treatments are presented in Table 9. It is evident from the data that mulch treatments gave significantly higher yield of crops as compared to No mulch treatments except in 2006 where soybean yield was at par in all the treatments . In 2006 the continuous rains during the crop growth period seem to be a reason for non significant differences in soybean seed yield. But the soil moisture content was higher at the planting time of rabi crop in the mulch treatments which resulted in batter germination of safflower crop.

Table 9 : Effect of Mulch on the productivity of soybean –safflower cropping sequence.

Treatments	Soybean seed yield (kg ha^{-1})				Safflower seed yield (kg ha^{-1})				B:C Soy+ Safflower
	2006	2007	2008	Mean	2006	2007	2008	Mean	
No mulching	2379	1260	1177	1605	694	493	389	525	2.20
Gliricidia Mulch, 2 t ha^{-1}	2396	1603	1804	1934	1007	1227	678	971	3.00
Straw mulch, 4t ha^{-1}	2335	1512	1869	1905	1146	1349	729	1075	3.02
Weed biomass mulch, 5 t/ha	2587	1488	1842	1972	1059	1134	989	1061	3.03
Soil Mulching	2196	1520	1621	1779	1128	681	660	823	2.63
Polythene Mulch	2326	1687	1956	1990	1137	1146	1098	1127	2.91
CD 5%	NS	231	479	-	231	381	324	-	-

Over all results suggest that in-situ mulching with different materials for weed control and moisture conservation increased the seed yield of soybean by 11 to 24per cent as compared to no mulch treatment. The percent increase in the seed yield of subsequent *Rabi* crop i.e. safflower due to mulching was 52-110per cent. Straw mulch, weed biomass mulch and glyricidia green leaves mulch have been found effective in conserving soil moisture and enhancing productivity of soybean–safflower sequence in medium to deep vertisols under dryland conditions. Mulching has been found economically viable.

Evaluation of intercropping for vertisols

To evaluate the suitable cropping system for shallow, medium and deep Vertisols the sole soybean and Intercropping of soybean and pigeonpea were tried on shallow, medium and deep Vertisols at JNKVV, Jabalpur and at College of Agriculture, Indore. The data on soybean equivalent yield and Sustainability Index are presented in Table 10. Data clearly emphasized that intercropping of soybean and pigeonpea is more sustainable and therefore, recommended in place of sole soybean under all the three soil depths. Incase of intercropping the sustainability index ranged from 0.65 to 0.83 while in case of sole soybean it ranged from 0.37 to 0.50 only.

Table 10 : Performance of sole and intercrop on Vertisols of different depth (mean of 2001 to 2003 seasons).

Cropping System	Shallow soils		Medium Soils		Deep Soils	
	Yield*	SI	Yield*	SI	Yield*	SI
Sole soybean (JS 335)	1908	0.50	2082	0.37	2138	0.41
Soybean (JS 335) + Pigeonpea (JA 4)	2655	0.71	3668	0.73	3785	0.78

* Soybean Equivalent Yield.

Effect of land configuration on the productivity of soybean

Madhya Pradesh is endowed with well distributed rains ranging from 700 to 1200 mm. Vertisols with good moisture holding capacity can be used to grow short-duration soybean by adopting sound land management practices such as broad bed and furrow system, Ridge and furrow system to bring about 2.02 m ha accounting for 6.57per cent of the total area under double cropping of the state which were under fallowing. The adoption of proper land configuration system will help in increasing income of the farmers besides preventing land degradation due to runoff erosion.

The results of long term field experiments indicated that the small changes through land configuration in flat field conditions may help in improving the productivity of vertisols of Malwa region, which faces the problem of water logging and poor aeration during heavy rains thereby affecting crop productivity adversely. Using light machinery like bund former and *desi hal* with minor modification may improve the physical conditions and drainability. It is assumed that land treatments will help to improve soil physical conditions, root development and over all productivity in vertisols. The results of these long term studies are presented is subsequent subheads.

Raised sunken bed system

To improve the productivity of rainfed Vertisols in high rainfall areas, there is a need to provide to the resource poor farmers such a system, which will provide adequate means of surface drainage at a cheaper cost compared to sub-surface drainage and will also assure adequate root zone moisture recharge for *Kharif* crops. Jabalpur center of the AICRP on Soil Physical Constraints and their amelioration for sustainable crop production has conducted long term field experiments over more than 20 years to develop and validate Raised-Sunken Bed Technology. This technology provides: an adequate drainage during heavy rains, favourable moisture regime during dry spell in *Kharif* season to upland crops, sufficient ponding of water for paddy and channelizes the excess runoff water safely to a farm pond for recycling to provide supplemental irrigation to rainy and post rainy season crops. The technology has been demonstrated successfully at the farmer's fields in Madhya Pradesh and also found beneficial by the Parbhani (Maharashtra) center of the project. In Maharashtra approximately 0.53 m ha land experiences temporary water logging during rainy season.

Based on long-term trials at Jabalpur (M.P.) involving various crops the recommended raised-sunken bed system consists of 6 to 9 m wide and 30 to 35 m high raised bed along with the sunken bed of 6m width. The sunken beds are connected with the pipes to maintain water level at a desired height viz. 15 to 20 cm in all the sunken beds. The excess water is safely drained away and collected in a farm pond for recycling to provide supplemental irrigation to rainy season crop in case of dry spell and post rainy season crops.

Seed yield of soybean (cultivars JS-2 or JS-75-46 or JS-335) observed over a period of 16 years was in general more than two times under improved system of Raised-Sunken Bed as compared to farmers practice of flat bed sowing (Table 11).

Table 11 : Effect of land configuration on the yield (Kg/ha) of crops in vertisols of high rainfall area.

Year	Soybean seed yield		Paddy grain yield	Seasonal rainfall (mm)
	Improved system (Raised bed 6 m wide)	Farmer practice (Flat bed)	Improved system (Sunken bed 6 m wide)	
1979	1373	1000	-	472
1980	2962	2229	5100	1432
1981	1830	1050	-	700
1982	2230	1054	2030	1236
1983	2340	817	1400	1445
1985	2253	1000	2500	1380
1986	2131	647	4220	1034
1987	1840	715	1716	1216
1988	3260	1731	1877	1010
1989	2779	144	1520	627
1990	2425	581	3001	1624
1991	2523	125	2200	1215
1992	3107	581	3001	1055
1993	1733	1050	4758	1391
1994	1969	1469	3922	1090
1995	1755	1113	4473	1153

Source: Tomar *et al.* (2005).

Raised-Sunken Bed system has also been found successful for intercropping of upland crops on raised bed. Two rows of soybean or blackgram followed by one row of pigeonpea (medium duration) produced 3, 12 & 4 times higher yield for soybean sole, pigeonpea sole and soybean + pigeonpea intercrop (Table 12) under improved system as compared to farmers practice. It was observed that in most of the years pigeonpea and blackgram could not survive in flat beds (farmers practice) but performed satisfactorily under the improved system. The total productivity under improved system of soybean + pigeonpea cropping was more than 3.3 t/ha which is good yields under rainfed condition in vertisols of this region. The main advantage of the intercropping was that even if the rains ceases earlier the pigeonpea survived on residual moisture left over in the profile. The intercropping results suggest that is Raised-Sunken Bed Technology is adopted in large scale then the crop productivity can be increased in high rainfall areas of vertisols regions.

Table 12 : Effect of land configuration on the yield (Kg ha^{-1}) of soybean and pigeonpea grown as sole and intercrop in a Vertisols of high rainfall area.

Land Configuration	Cropping system	1990-1991	1991-1992	1992-1993	1993-1994	1994-1995	1995-1996
Improved system (Raised bed 6 m wide)	Soybean sole	2425	2523	3107	1733	1969	1755
	Pigeonpea sole	2012	2004	2532	1825	1626	1174
	Soybean +	1805	1977	2310	1433	1631	1344
	Pigeonpea	1805	1608	2394	1436	1426	900
Farmers practice (Flat bed)	Soybean sole	849	125	581	1050	1469	1113
	Pigeonpea sole	128	167	Failed	Failed	Failed	Failed
	Soybean +	650	95	581	1050	1469	1113
	Pigeonpea	125	151	Failed	Failed	Failed	Failed
Seasonal rainfall, mm		1631	1215	1055	1391	1090	1132

Source: Painuli *et .al.* (2002).

Performance of post rainy season crops grown on raised bed and sunken bed was evaluated against farmer's practice of flat bed sowing. As compared to farmers practice i.e. flat bed sowing under improved system the yield of each test crop was nearly two times more (Table 13). This could be attributed to the more favorable moisture regime in the seed zone under improved system at the harvest of rainy season crops resulting in better establishment and improved stand of post rainy season crop.

Table 13 : Performance of post rainy season crops on Raised and Sunken bed (yield kg ha^{-1}).

Land configuration	Crop	86-87	87-88	88-89	89-90	90-91	Geometric mean ±SE
Raised bed	Chickpea	2146	1866	1842	1722	1579	1822±84
6m width	Safflower	1412	1424	1240	1488	1348	1380±38
Sunken bed	Linseed	1170	949	1416	938	1048	1091±79
6m width	Wheat	989	1972	1218	1047	1162	1236±159

Source: Painuli *et al.* (2002).

Benefit-cost ratio based on the prices prevailing in the year 1997 for various operations and yields have revealed that the Raised-Sunken Bed System is economically viable under all the dominant cropping systems of the region (Table 14). The benefit-cost ratio showed that the land shaping cost could be recovered within 1 to 2 year. Economics viability analysis further revealed that in case of failure of post rainy season crops or even paddy during *Kharif* (due to scanty rains as was the case in 1979 and 1981) improved system gave higher returns compared to the farmers practice. Highest benefit-cost ratio was observed in soybean-pigeonpea intercropping on raised bed and paddy-chickpea in sunken bed.

Table 14 : Economic viability of raised-sunken bed technology.

Treatment combinations	Operational Cost (Rs/ha)	Yield (q/ha)	Gross Return (Rs/ha)	Net Return (Rs/ha)	B: C ratio
S: Paddy-wheat	9869	27.92 to 18.00	21147	11278	1.14:1
R: Soybean-gram		24.47 to 18.31			
S: Paddy-gram	9269	27.92 to 14.49	21894	12625	1.36:1
R: Soybean-gram		27.47 to 18.31			
S: Paddy-wheat	7359	27.92 to 18.00	15462	8103	1.10:1
R: Pigeonpea		20.93			
S: Paddy-gram	7769	27.92 to 14.49	16401	8632	1.11:1
R: Soybean fallow		24.47			
S: Paddy-wheat	8519	27.92 to 18.00	20917	12398	1.46:1
R:Pigeonpea+soybean(Intercrop)		18.81+18.11			
S: Paddy-wheat	7919	27.92 to 14.49	21664	13745	1.74:1
R:Pigeonpea+soybean(Intercrop)		18.81+18.11			
S: Paddy-gram	6759	27.92-14.49	16209	9450	1.40:1
R: Pigeonpea		20.93			

Source: Tomar *et. al.* (1996).

Conclusion

Studies carry out in vertisols of high rainfall area (>1000mm) and medium rainfall area (700 to 1000 mm) of Madhya Pradesh revealed that for rice based cropping system and soybean based production system under reduced tillage/zero tillage helps in enhancing crop productivity, improves soil quality, enhances resource use efficiency. Integrated nutrient management has proved its potential for sustaining soil and crop productivity in long run. Addition of organics helps in improving soil organic carbon, soil microbiological activities, soil quality as well as crop productivity in long run. There is a need for adoption integrated nutrient management to maintain soil health and crop quality. Rain water harvesting and its recycling will help in increasing crop productivity in Vertisols tremendously. Application of mulch helps in improving crop productivity under rainfed cultivation. On the basis of long term studies it is concluded that reduced tillage, Integrated nutrient management, adoption of improved land configuration, rain water harvesting, conservation of soil moisture, intercropping, application of deficient nutrients such as application of S and use of alternate sources of fertilizer such as rock phosphate holds the key for sustainable rainfed crop production in vertisols of Central India.

References

Dixit, P.R.; Sawarkar, N.J.; and Khatik, S.K. (1995).Comparative efficiency of oxalic acid industry waste blended rockphosphate and superphosphate on yield and nutrition of soybean in chromustert soil. *Legume Res.* **18**(3-4): 173-176.

Havengi, G.V. and N.S. Mann (1970). Effect of rotation and continuous applic. of manures and fertilizers on soil properties under dry farming conditions. *J. Indian Soc Soil Sci.*, **18** : 45-50.

Kanwar, J.S. and S.S. Prihar (1992). Effect of continuous applications of manures and fertilizers on some physical properties of Punjab Soils. *J. Indian Soc. Soil Sci.*, **10** : 243-247.

Mangala Rai (2009) Resource conservation in Indian Agriculture.Souvenir 4th world congress on conserbation agriculture, held from 4-7 Feb. 2009 at New Delhi. pp 1-6

Mausumi, R., K. Kailash, S. Raychaudhury, M. Raychaudhuri and K. Kumar (1998). Response of safflower of lime and phosphorus on an Ultisol of Manipur. *J. Indian Soc. Soil Sci.*, **46** : 628-632.

Mishra, B. (2006) Wheat: present scenario. Yojana. September, 2006 pp 8-15

Moharram, T.M., M.A. El-Mohandes and M.A. Badwani (1999). Effect of inoculation and organic manure applications on symbiotic N_2 fixation, microbial biomass and nutrients availability in sandy soils cultivated with safflower and peanut. *Ann. Agril. Sci.*, (Cairo) **44**: 27-40.

Painuli, D.K., Tomar, S.S., Tembe, G.P.and Sharma, S.K. (2002). Raised-Sunken bed technology for rainfed Vertjsols of high rainfall areas. Tech-Bulletin published by ICAR and Indian Institute of Soil Science, Bhopal, pp 1-19.

Sharma, K.N. and Namdeo, K.N. (1999). Effect of biofertilizers and phosphorous on NPK contents, uptake and grain quality of Soybean (*Glycine max.(L) Merrill*) and nutrient status of soil. *Crop Res.* **17**(2): 164-169.

Sharma, R.A. (1997). Influence of conjunctive use of organics and feritlizer nutrients on nutrient uptake and productivity of soybean-safflower cropping sequence in typic chromusters. *Crop Res.* **13**(2): 321-325.

Sharma, R.A. and B.K. Dixit (1987). Effect of nutrient aplication on rainfed safflower, *J. Indian Soc. Soil Sci.*, **35**: 452-455.

Sharma, R.A. and S.K. Sharma (2005). Evaluation of sustainability indices of improved management practices of dryland agriculture in Western Madhya Pradesh. Crop production in stress environments: Genetic and management options. Edited by D.P. Singh et.al. and Published by Agrobios (International). pp 469-476.

Subbaiah, S.V.; Kumar, R,M.; Singh, S.P.and Subba Rao, L.V.(2006). Agro techniques for increased rice production. Technical Bulletin No.18, Directorate of Rice Research, Rajendranagar, Hyderabad, A.P., India.

Tomar, S.S. (1995). Effect of soil hydrothermal regions on the performance of safflower planted on different dates. *J. Agro. Crop Science*, **175** : 141-152.

Tomar, S.S. (2008). Conservation Agriculture for Rice-wheat Cropping System. *J. Ind. Soc. Soil Sci.* **56**(4): 358-366.

Tomar, S.S. and Verma, O.P. (1985). Effect of irrigation and mulch on the performance of wheat planted on different dates. *J. Agro. Crop Sci.*, **154** : 258-266.

Tomar, S.S.; Tembe, G,P.; Sharma, S.K. and Tomar, V.S. (1996). Studies on some land management practices for increasing agricultural production in Vertisols of central India. *Agriculture water Management*, **30**: 91-106.

□□□

Chapter 15

Agri-Horti System for Sustainable Production on Degraded Lands in Semi-Arid Tropics

OM PRAKASH, R.K. DUBEY AND J.S. DESHWAL
Central Soil and Water Conservation Research and Training Institute, Research Centre, Chhalesar, Agra (Uttar Pradesh) 282 006

Integrating trees and heliophyte crops in agroforestry usually suppresses yield of latter component due to above ground shade and under ground root competition (horizontal and vertical tree root spread and tree root congestion in crop root zone specifically beyond tree cover) through altering sharing of growth resources (radiant energy, moisture, nutrients, space and air). Tree root pruning is a potential tool for managing below ground competition (Ong *et. al.*, 2008) since dominance of tree roots and shade increases as they mature (Ong *et. al.*, 2004). Trees competes with crops for light, nutrient and moisture (Ong and Huxley, 1996). They extend roots to considerably greater distances and depth than crops (Stone and Kalisz, 1991). Trees with less competitive root architecture (deep rooting and few roots in upper soil layer) may mitigate tree-crop interface in agroforestry (Rao *et. al.*, 2004). However, tree rooting also depends on other factors (site characteristics, tree age, provenance and method of propagation) (Mulalya *et al.*, 2002). Trees also exhibit rapid decline in root mass, number and length with increasing soil depth. Management of below ground tree-crop root interface is important in semi arid tropics, dry climates and infertile soils where trees and crops are grown in close proximity under scanty state of growth resources (Rao *et. al.*, 2004).

Ber is a common subtropical, drought hardy and poor man's fruit of semi arid regions. This can be grown on degraded soils of Yamuna ravines too as its potential is proven on saline and alkali soils. It does not have any evidence of allelopathy and nitrogen fixation but root mycorrhizae is reported in areas having temperature range of 7 to 50°C and annual rainfall of 125 to 2225 mm. Ber has productive potential even in seemingly infertile eco-system and produces three vital 'F's i.e. fruit, fodder and fuel which is of paramount importance in semi arid tropics facing acute shortage of fire wood and fodder. Growing crops under trees is an age old practice in arid India (Shankarnarayana *et. al.*, 1987). However, neither the ber nor root management have yet not been evaluated in most prevalent pearl millet-wheat sequence on Indo-gangetic alluvial soils of reclaimed Yamuna ravines specifically with regards to effect on pearl millet-wheat production, Ber fruit and fire wood production and economical benefits to the growers. Therefore, fruit based agri-horti system was evaluated during 1998-99 to 2001-02 in alluvial soils of reclaimed Yamuna ravines imposing pearl millet-wheat in established Ber orchard (1992) to devise a optimum ber root management practice for higher and sustained productivity from agroforestry system as a whole and to enhance monitory benefits to growers.

Material and Methods

The experiment was conducted at Research Farm of Central Soil and Water Conservation Research and Training Institute, Research Centre, Chhalesar, Agra (Uttar Pradesh) lying at 27° 10′ latitude and 78° 02′ longitude at 169 m msl. Soil of study site was well drained sandy loam (73.6% sand, 18.84% silt and 7.49% clay) having alkali pH (8.3), permanent wilting point of 5.2 per cent, field capacity of 15.8 per cent and average bulk density of 1.41 g cc^{-1} in upper 0-15 cm depth. Soil had 0.29 per cent organic carbon, 153 kg ha^{-1} available nitrogen and medium status of phosphorus and potassium. Five treatments [T_1: Tree planted in bottomless used bitumen drums of 90 cm height and 25 cm radius. T_2: Trench (40 cm deep and 30 cm wide) in two sides of tree. T_3:. Tree planted in polythene (thick black, 120 micron) lined pits (up to 90 cm) T_4: Tree without root management. T_5: Crops without Ber trees i.e. sole pearl millet-wheat] were imposed in Ber orchard (1992) planted in pits of 90 x 90 x 90 cm. The study was conducted in randomized block design with four replications in grafted Ber plantation (Cv. Ponda safeda) planted in pits of 90 cm depth and 30 cm radius at spacing of 10 m x 5 m. Ber trees were 70-80 per cent pruned every year before onset of monsoon in May. Both ber and crops (pearl millet: Cv. Pioneer 7688 and wheat: Cv. Lok 1) were raised following recommended package of practices for the area and insect pest control. Data on crop growth, yield characters and yield (grain and straw) of pearl millet and wheat were recorded. Data on tree height and volume as well as fruit and fuel wood yield (due to pruning) from ber were also recorded. Ber tree volume was worked out through using equation, canopy volume = tree height x average tree spread (E-W + N-S spread)/4. Tree height was measured in october every year when tree growth was almost complete, using straight bamboo poles and measuring tape. Average annual rainfall during the study period was 789.25 mm. Maximum annual rainfall was received in 1998 (1235.1 mm) while minimum annual rainfall was received in 2001 (446.1 mm). About 73.76, 3.10 and 23.14 per cent of mean annual rainfall was distributed in pearl millet, wheat and non crop season. Seasonal distribution of rains in pearl millet and wheat were 937.4 and 25.3 mm; 684 and 19.8 mm; 511.0 and 14.9 mm and 196.3 and 38.0 mm in 1998, 1999, 2000 and 2001, respectively. Data were statistically analyzed using randomized block design schedule.

Results and Discussion

Grain yield

Data (Table 1) revealed that T_5 (sole pearl millet-wheat) produced significantly higher mean grain yield of pearl millet (2282 kg ha^{-1}) and wheat (4243 kg ha^{-1}) which was 5.60, 12.52, 15.43 and 24.50 per cent higher in pearl millet and 6.93, 13.18 and 13.38 and 21.65 per cent higher in wheat than T_1, T_2, T_3 and T_4 respectively. Thus, tree-crop interface retarded mean crop grain which may be attributed due to variations in tree-crop root congestion; growth resources sharing by trees and crops and tree shade. T_1 produced significantly higher mean grain yield of pearl millet (2161 kg/ha) and wheat (3968 kg/ha) than other root management practices which was 6.56, 9.31 and 17.89 per cent higher in pearl millet and 6.10, 6.04 and 13.76 per cent higher in wheat than T_2, T_3 and T_4, respectively. Higher mean crop grain yield in T_1 may be attributed due to minimum tree-crop root interface on account of channelization of horizontally spreading shallow tree roots below and beyond the crop root depth. However, all tree root management options improved the mean crop grain yield over no root management (T_4) (7.28 per cent in T_3 to 13.76 per cent in T_1 in wheat and 7.86 per cent in T_3 to17.89 per cent in T_1 in pearl millet) which can be attributed due to the mitigation of tree-crop root interface, role of micro-climate in conserving soil moisture and above and underground litter input to soil. Data (Table 1) also showed that in individual years too, crop grain yield production was T_5>T_1>T_2>T_3>T_4 (except T_3 produced significantly higher pear millet grain yield than T_2 in 2000).

Table 1 : Grain yield (kg ha^{-1}) of pearl millet and wheat in agroforestry association with Ber in Yamuna ravines.

Treatments	1998	1999	2000	2001	Mean
			Pearl millet		
T_1	1562	2490	2015	2575	2161
T_2	1525	2440	1585	2560	2028
T_3	1475	2408	1725	2300	1977
T_4	1408	2144	1510	2270	1833
T_5	1680	2652	2140	2656	2282
SE+/- 5 per cent	78	87	34	27	84
CD	169	212	77	60	190
			Wheat		
T_1	3866	4116	4040	3850	3968
T_2	3558	3953	3820	3663	3740
T_3	3641	3906	3715	3706	3742
T_4	3453	3603	3450	3444	3488
T_5	4150	4409	4265	4150	4243
SE+/- 5 per cent	92	117	40	34	13
CD	201	284	90	77	30

Ber Fruit yield

Data (Table 2) revealed that various root management options produced significant variations in the Ber fruit yield over each other (T_1>T_4>T_2>T_3). Ber trees planted in bottomless drums produced the highest ber fruit yield (7347 kg ha^{-1}) which was 14.30, 18.14 and 7.62 per cent higher than T_2, T_3 and T_4, respectively. This shows that planting ber in bottomless old bitumen drum enabled the tree roots to sustained and continuous mining of nutrients and moisture from deeper soil layer below crop root zone even during soil moisture stress periods. Ber fruits yield in different root management options in individual years followed trend of mean ber fruit yield (T_1>T_4>T_2>T_3) but T_3 produced significantly higher ber fruit yield than T_2 in 2000 and 2001. T_1 also produced significantly higher ber fruit yield than other root management options evaluated in this study.

Table 2 : Fruit yield (kg ha^{-1}), canopy volume (m^3) and fire wood (q ha^{-1}) of Ber in association of pearl millet-wheat crop sequence in Yamuna ravines.

Treatments	1998	1999	2000	2001	Mean
			Fruit yield		
T_1	7000	5900	8688	7800	7347
T_2	5282	5760	7545	7100	6422
T_3	5180	4900	7616	7180	6219
T_4	6440	5240	8171	7460	6827
SE+/- 5 per cent	5.52	3.93	5.83	4.06	1.27
CD	12.48	8.88	13.20	9.18	2.87
			Canopy volume		
T_1	35.3	39.17	31.30	37.97	35.94
T_2	19.7	20.13	25.07	25.08	25.08
T_3	25.9	23.72	19.29	30.87	30.87
T_4	22.4	18.65	16.44	22.18	22.18
SE+/- 5 per cent	0.29	0.18	0.17	0.21	0.10
CD	0.65	0.41	0.39	0.48	0.22
			Fire wood		
T_1	16.1	10.7	27.0	33.9	21.8
T_2	10.7	12.9	19.7	20.8	16.0
T_3	11.5	15.3	24.3	27.4	19.6
T_4	11.1	19.5	19.6	15.5	16.4
SE+/- 5 per cent	0.24	0.22	0.24	0.44	0.11
CD	0.55	0.49	0.54	0.99	0.25

Tree volume

Data (Table 2) also revealed that various root management options produced significant variations in the tree volume (T_1>T_3>T_2>T_4). This may be attributed due to minimum tree-crop root interface in T_1 for nutrients-moisture mining from crop root zone. Tree roots in fact mined major share of nutrients and moisture from deeper soil layer below crop root zone in T_1 where relatively less scarcity of soil moisture exists during moisture stress periods.

Fire wood production

Data showed that root management options produced significant variations in the mean and individual year fire wood production or pruning yield (T_1>T_3>T_4>T_2) but variations between T_2 and T_4 and between T_3 and T_4 in 1998 was ststistically at par (Table 2). Fire wood yield depended on variations in ber tree growth that was ultimately governed by variations in availability and mining of soil moisture and nutrients by trees and crops; distribution and congestion of tree-crop roots in crop root zone and deeper soil layers and *in situ* conservation of rain water (higher in bottomless old bitumen drums). T_1 produced significantly higher fire wood (21.8 q ha^{-1}) than other tree root management options which was 36.25, 11.22 and 32.93 per cent higher than T_2, T_3 and T_4, respectively.

Conclusion

This study revealed that growing pearl millet-wheat in Ber plantations using bottomless bitumen drums is not only beneficial in producing higher and sustained crop yields due to its efficiency in mitigating tree-crop root interface and sharing of nutrients and moisture from upper and lower soil layers but it also produces significantly higher ber fruit yield, tree volume and fire wood from reclaimed Yamuna ravines. Thus, growing ber + pearl millet-wheat can lead to the highest monitory benefits to the growers on meager cost of old and used bitumen drums and can be even adopted by the resource poor farmers.

References

Mulatya, J.M., Wilson, C.K., Ong Deans, J.D. and Sprent, J.I. (2002). Root architecture of provenances, seedlings and cuttings of *Melia volkensii*; implications for crop yield in dry land agroforestry. *Agroforestry systems*. **56**: 65-72.

Ong, C.K. and Huxley, P. (1996). Tree crop interactions: a physiological approach. In CAB International and ICRAF, 385p.

Ong, C.K., Kho, R.M. and Radersma, S. (2004). Ecological interactions in multi species agro-ecosystems: Concept and rules. *In Below Ground Interactions in Tropical Agro-ecosystems: Concepts and Models with Multiple Plant Components*. Eds. M. Van Noordwijk, G. Cadisch and C.K. Ong. CABI Publishing Wallingford, U.K., 1-16.

Ong, J., Dauglas, Deans and John Okario, (2008). Tree growth and management in Ugandan agroforestry systems: effect of root pruning on tree growth and crop yield. *Tree Physiology*, **28**: 233-242.

Rao, M.R., Schroth, G., Williams, S.E., Namerembe, S., Schaller, M. and Wilson, J. (2004). Managing below ground interations in agro-ecosystems. *In Below*

Ground Interactions in Tropical Agro-ecosystems: Concepts and Models with Multiple Plant Components. Eds. M. Van Noordwijk, G. Cadisch and C.K. Ong. CABI Publishing Wallingford, U.K., 309-328.

Shankran, K.A., Harsh, L.N. and Kathju, S. (1987). Agroforestry in arid zones of India. *Agroforestry Systems,* **5**: 69-88.

Stone, E.L. and Kalisz, P.J. (1991). On the maximum strength of tree roots. For. Ecol. Manage., **46**: 59-102.

□□□

Chapter 16

Nutrient Uptake of Tomato F_1 Hybrid Avinash-2 as Affected by Irrigated Methods and Micronutrients

B. AGRAWAL[1], A. SHRIVASTAVA[2], N.K. TOORRAY[3] AND R.N. SHARMA[4]

[1] Department of Horticulture, I.G.K.V.V., Raipur, (C.G.) [2] Farm Manager (Agronomy), Krishi Vigyan Kendra, Janjgir-Champa (C.G.) [3] Subject Matter Specialist, (P.Path), Krishi Vigyan Kendra, Janjgir-Champa (C.G.) [4] Programme Coordinator (Agri. Extn.), Krishi Vigyan Kendra, Janjgir-Champa (C.G.) Department of Horticulture, I.G.K.V.V., Raipur (C.G.), Krishi Vigyan Kendra, Janjgir-Champa

Tomato is one of the important vegetable crops of Chhattisgarh and being cultivated throughout India. Tomato hybrids are very responsive for per unit agricultural inputs. Vegetable growing leads very heavy nutrient uptake because of their heavy biomass production capacity, which cause very high decline of nutrient status of the soil. In India, major nutrients are added with NPK fertilizers but micronutrient application is not in common practice where as it also declines in same pattern as major nutrients. This might be one of the major reasons of less productivity of Indian vegetable crops as compared to world average. The data about uptake of these elements are needed to standardize for specific crop and specific variety at specific yield level to increase productivity and maintaining it year after year.

Drip irrigation is the latest and most efficient method of water utilization for crop growth which leads not only appreciable saving of water but also results in achieving higher crop yield as compare to surface irrigation method. Micronutrients in water soluble form can also be applied by this system through its pipe network which are very important to the plant nutrition as major and secondary nutrients and play a significant role for maximizing production. Increase in productivity and maintaining it at same level year after year through the use of proper fertilizer application and use of other technologies at same availability or utilization of water would be a method of better utilization of available water/stored water.

Keeping the above facts in consideration the present study was carried out to standardize the uptake capacity of tomato crop variety Avinash-2.

Materials and Methods

The present study was carried out at Horticulture Research Farm, College of Agriculture, Raipur (Chhattisgarh) during the spring summer season of 1999-2000 on a well-drained sandy loam soil. The treatment comprising two main plots viz. drip and flood method of irrigation and five sub plots viz. control, Zn, B. Fe and Cu were replicated three times in split plot design. Tomato F_1 hybrid Cv. Avinash-2 was taken for the study. Drip irrigation plots were irrigated as per the water requirements of crop. The weekly water requirement of crop was calculated by using the formula referred by Indian National Committee on Irrigation and Drainage (INCID) in drip irrigation in India (Anonymous 1994). Seedlings were transplanted on 13th November; 1999 and gap filling was done at 7 DAT (days after transplanting) to maintain the desired plant density. Micronutrients were applied @ 0.5 per cent as foliar spray at 30, 40 and 50 days after transplanting. The observations were recorded on N, P, K, Zn, B, Fe, Cu uptake and yield. For uptake estimation, plant and fruit parts were oven dried and powdered and determined as per the method stated below.

(i) *Nitrogen* : Gerhardt-Kjeldotherm digestion and automatic distillation system.

(ii) *Phosphorus* : Colorimeter

(iii) *Potash* : Flame photometer

(iv) *Zn, Fe, Cu, Mn* : AAS (Atomic absorption spectrophotometer)

The fertilizer dressing to the crop were applied @ 120 kg N, 180 kg P_2O_5 and 180 kg K_2O ha^{-1}. Three split doses of nitrogen was applied as basal and remaining equal two parts at 30 and 50 days after transplanting, respectively.

Results and Discussion

Nutrient uptake

The data on nutrient uptake by tomato fruit and stover as affected by methods of irrigation and micronutrient applications is presented in Table 1 which reveals that the uptake of nitrogen (171.74 kg ha^{-1}), Phosphorus (24.33 kg ha^{-1}), Potash (163.20 kg ha^{-1}), Zinc (7.36 kg ha^{-1}), iron (7.37 kg ha^{-1}) and copper (0.054 kg ha^{-1}) was significantly superior under drip than flood irrigation. It may be due to better availability of the soil moisture as and when needed, which led to better utilization of the applied nutrients and better foliage and food supply to root. The results are in good conformity with the findings of Franco *et al.* (1999).

Maximum utilization of nitrogen (96.36 kg ha^{-1}), phosphorus (18.37 kg ha^{-1}), potash (108.80 kg ha^{-1}), zinc (4.50 kg ha^{-1}), copper (0.012 kg ha^{-1}) and iron (5.15 kg ha^{-1}) was recorded under zinc application within the different micronutrient treatments applied. Minimum was under control plots. This might be due to better foliage and vegetative growth and more photosynthesis in Zn applied plots and better supply of photosynthates to root and better utilization capacity of available nutrients and synergistic relationship between the elements.

The interaction effect of irrigation method and micronutrient applications on the uptake of nitrogen (302.72 kg ha^{-1}), phosphorus (31.07 kg ha^{-1}), potash (224.58 kg ha^{-1}), zinc (8.31 kg ha^{-1}), iron 9.53 kg ha^{-1}) and copper (0.12 kg ha^{-1}) was found to be significant. The maximum uptake was noted in Zn applied plots with drip irrigation. From these results it may be concluded that drip method of irrigation, zinc application and drip method + Zn application was effective in utilization of nutrients by tomato crop in Zn deficient soils. The results are in conformity with the findings of El Sherif *et. al.*, (1990) and Gunes *et. al.*, (1990).

Table 1 : Nutrient uptake (kg ha^{-1}) as affected by irrigation methods and micronutrient applications.

Treatments	N uptake		P uptake		K uptake		Zn uptake		Fe uptake		Cu uptake		Yield	
Irrigation methods	Fruit	Stover	Fruit	Stover	Fruit	Stover	Fruit	Stover	Fruit	Stover	Fruit	Stover	Marketable	Total
Drip	171.740	163.230	24.330	20.830	163.200	115.740	0.054	0.045	7.360	2.920	7.370	5.120	87.970	90.720
Flood	118.400	112.020	17.920	14.500	120.490	114.030	0.034	0.027	3.010	1.540	4.330	3.470	73.690	77.490
Sem ±	3.034	2.880	0.433	0.383	2.850	2.730	0.009	0.007	0.039	0.044	0.145	0.094	1.444	1.497
CD at 5% level	18.460	17.560	2.640	2.330	17.370	16.640	0.006	0.007	0.239	0.273	0.886	0.574	8.790	9.110
Micronutrient applications														
Control	96.360	91.280	18.730	15.480	108.800	104.990	0.012	0.004	4.500	1.170	5.150	3.800	68.570	71.410
Zn	242.370	233.480	25.650	21.730	203.190	191.350	0.090	0.082	6.650	3.450	7.130	5.240	102.420	106.39
Bo	172.350	163.040	22.150	18.930	161.200	151.860	0.062	0.052	5.590	2.670	5.790	4.230	86.130	89.590
Fe	109.500	103.080	19.780	16.290	119.670	114.420	0.016	0.015	4.610	1.320	5.670	4.150	74.350	77.390
Cu	104.770	97.250	19.330	15.900	116.360	111.810	0.034	0.026	4.590	1.280	5.500	4.040	72.690	75.760
Sem ±	1.9850	1.900	0.224	0.198	1.680	1.580	0.001	0.007	0.023	0.032	0.063	0.043	0.798	0.882
CD at 5% level	5.950	5.710	0.673	0.593	5.050	4.750	0.002	0.002	0.071	0.096	0.190	0.130	2.390	2.460
Interaction														
Drip+ Control	103.370	97.930	20.250	17.010	116.160	111.850	0.016	0.005	7.040	1.340	6.070	4.280	70.85	73.04
Drip+ Zinc	302.720	290.890	31.070	27.150	244.580	232.710	0.120	0.106	8.310	4.360	9.530	6.570	114.90	118.41
Drip+ Boron	215.440	205.240	26.930	23.680	195.850	185.510	0.081	0.063	7.200	3.320	7.750	5.290	97.96	100.98
Drip+ Iron	123.310	116.380	22.260	18.620	133.540	127.730	0.018	0.017	7.150	1.570	6.920	4.850	79.99	82.46
Drip+ Copper	113.840	105.700	21.170	17.710	125.850	121.190	0.034	0.033	7.100	1.490	6.550	4.590	76.18	78.69
Flood+ Control	89.370	84.630	17.220	13.950	101.450	98.130	0.009	0.004	1.950	0.990	4.230	3.330	66.26	69.77
Flood+ Zinc	182.020	176.070	20.220	16.300	161.790	149.990	0.060	0.058	4.990	2.540	4.720	3.910	89.29	94.36
Flood+ Boron	129.250	120.830	17.380	14.180	126.540	118.200	0.043	0.042	3.980	2.020	3.830	3.170	74.29	78.20
Flood+ Iron	95.700	89.780	17.310	13.960	105.790	101.410	0.014	0.014	2.070	1.070	4.420	3.460	68.20	72.31
Flood + Copper	95.700	88.810	17.490	14.100	106.870	102.440	0.34	0.018	2.070	1.070	4.450	3.480	69.20	72.83
For two M means at the same level of M														
Sem ±	2.808	2.650	0.317	0.280	2.380	2.240	0.001	0.001	0.033	0.045	0.089	0.061	1.128	1.162
CD at 5% level	8.410	8.0800	0.952	0.839	7.140	6.710	0.003	0.003	0.101	0.137	0.269	0.184	3.380	3.480
For two M means at the same different levlof M														
Sem ±	3.303	3.150	0.418	0.369	2.930	2.780	0.001	0.001	0.041	0.051	0.130	0.086	1.760	1.480
CD at 5% level	15.940	15.180	2.150	1.900	14.600	13.930	0.006	0.004	0.203	0.242	0.697	0.457	8.920	7.53

Fruit yield (marketable and total)

Yield, both marketable (87.97 t ha^{-1}) and total (90.72 t ha^{-1}) were higher with drip irrigation as compared to flood irrigation. Drip irrigation provides appropriate moisture at field capacity, better root biomass and better uptake of major as well as micronutrients that resulted luxuriant growth of the plants and ultimately higher yield. These results are in conformity with the findings of Goldberg and Shumelli (1969); Bernaar (1971); Sivanappan *et. al.,* (1974); Goel (1988); Kataria and Michael (1990) and Yohannes and Tadesse (1998).

Zn application was quite effective in increasing the marketable (102.42 t ha^{-1}) and total yield (106.39 t ha^{-1}) among the micronutrients applications. This could be attributed due to influence of Zn in increasing metabolism, biosynthesis of auxins and better nutrient uptake. Similar results were also reported by Khomchak *et. al.,* (1971), Ashour (1975), Hooda *et. al.,* (1984) and Singh and Verma (1991).

Combination of these two i.e. drip irrigation + Zn spray was quite effective in producing maximum marketable (114.90 t ha^{-1}) and total (118.41 t ha^{-1}) yield. This may be due to all the biometrics and yield attributes were positively influenced by both the factors i.e. drip irrigation and Zn application.

Conclusion

Tomato is one of the important vegetable crops of Chhattisgarh. Maximizing its production through the use of technology available is a better way for better utilization of available or stored water. The uptake of nitrogen (171.74 kg ha^{-1}), phosphorus (24.33 kg ha^{-1}), potash (163.20 kg ha^{-1}), zinc (7.36 kg ha^{-1}), iron (7.37 kg ha^{-1}) and copper (0.054 kg ha^{-1}) was found significantly superior in drip irrigation than flood irrigation. Maximum utilization of nitrogen (96.36 kg ha^{-1}), phosphorus (18.73 kg ha^{-1}), potash (108.80 kg ha^{-1}), zinc (4.50 kg ha^{-1}), copper (0.012 kg ha^{-1}) and iron (5.15 kg ha^{-1}) was recorded under zinc application within the different micronutrient treatments applied. The interaction effect of irrigation method and micronutrient applications on the uptake of nitrogen (302.72 kg ha^{-1}), phosphorus (31.07 kg ha^{-1}), potash (244.58 kg ha^{-1}), zinc (8.31 kg ha^{-1}), iron (9.53 kg ha^{-1}) and copper (0.12 kg ha^{-1}) was found to be significant. The maximum uptake was noted in Zn applied plots with drip irrigation. From these results, it may be concluded that drip irrigation, zinc application and drip method + Zn application was effective in utilization of nutrients by tomato crop in Zn deficient soils.

References

Anonymos (1994). *Drip irrigation in India.* INCID. 39-40.

Ashour, N.I. (1975). The effect of leaf sprays of 2,4-D and $ZnSO_4$ on the growth and yield of tomatoes. *Archir fiir Gartenbau* **21**(5): 411-417.

Bernaar, J.H. (1971). Comparison of drip and furrow irrigation for market tomatoes. Presented in *Agril. Plastics* at Chicago.

EL Sherif, A.F., Shata, S.M. and Youssef, R.A. (1990). Response of tomato seedlings to zinc application under different salinity levels. II. Micronutrient content. *Egyptian J. Hort.* **17**(2): 143-150.

Franco, J.A., Perez-Saura, P.J., Fernandez., J.A., Parra, M. and Garcia, A.L. (1999).

Effect of two irrigation rates on yield, incidence of blossom end rot, mineral content and free amino acid levels in tomato cultivated under drip irrigation using saline water. *J. Hort, Sci and Biotech.* **74**(4) : 430-435.

Goldbreg, D. and Schumelli, M. (1969). The trickle irrigation method for increased agricultural production under conditions of saline water and adverse soils. *Proc. of Int. Arid Lands Conference,* Africa, 150.

Goel, M.R. (1988). Tomato response to furrow, microsprinkler and drip irrigation. *Am. Soc. of Agril. Engg.* **88**:2510-13.

Gunes, A., Alpaslan, M., Cikili, Y. and Ozcan, H. (1999). Effect of zinc on the alleviation of boron toxicity in tomato. *J. Pl. Nutri.* **22**(7): 1061-1068.

Hooda, R.S., Sidhu, A.S., Pandita, M.L. and Kalloo (1984). Effect of Zinc and Boron and their methods of application on the growth and yield of tomato variety HS-110. *Haryana J. Hort. Sci.* **12**(1-2): 46-47.

Kataria, D.P. and Michael, A.M. (1990). Comparative study of drip and furrow irrigation methods. *Proc. of the XI International Conference on the Use of plastics in Agriculture,* 26th Feb-2nd March, New Delhi, India. B. 19-B. 27.

Khomchak, M.E., Shiyan, O.I. and Zaporozhan, Z.E. (1971). The effect of minor elements on tomato yield and fruit quality, *Nauchnye Trudy Ukrainskoi sel' skokhozyaaitvennoi Akademii.* **57** : 165-168.

Singh, S.S. and Verma, S.K. (1991). Influence of potassium, zinc and boron on growth and yield of tomato. *Veg. Sci.* **18**(2): 122-129.

Sivanappan, P.K., Muthukrishnan, C.R., Natarajan, P. and Thamburaj, S. (1974). *Madras Agril. J.* **61**(9) : 888-891.

Yohannes, F. and Tadesse, T. (1998). Effect of drip and furrow irrigation and plant spacing on yield of tomato at Dire. Dawa, Ethiopia. *Agril. Water Mgmt.* **35**(3) : 201-207.

□□□

Chapter 17

Impact of Long Term Fertilizer and Manure Use on Crop Productivity and Soil Fertility in Rice-Rice Cropping System in Contrast Soils

R.H. WANJARI AND MUNESHWAR SINGH
AICRP on LTFE, Indian Institute of Soil Science, Bhopal-462 038 (M.P.)

Introduction of high yielding varieties, irrigation and high analysis fertilizer accelerated the mining of nutrients other than supplied externally to soil. To sustain the productivity and soil quality it is essential to maintain the supply of nutrient in balanced way. Since large amount of nutrient has to be applied to soil in chemical form which may have impact on soil properties and soil productivity in long term. Therefore, an experiment under AICRP on "Long-term fertilizer experiments (LTFE) to study changes in soil quality, crop productivity and sustainability" was initiated by the Indian Council of Agricultural Research to monitor the changes in soil properties and yield responses and soil environment due to continuous application of plant nutrient inputs through fertilizers and organic sources.

Crop Productivity

Results indicated that grain yield of kharif rice in Inceptisol at Jagtial are larger than the yields of rabi season. Phosphorus is recorded to be growth limiting nutrients. Data presented in table 1, indicate that in general rice yields of rabi rice are larger than kharif at Bhubaneswar and Jagtial. At Jagtial also the rabi rice yields are larger in previous years but during this period due to non availability of water in canal during critical growth period less yields were recorded. The larger yield of rice during rabi season is due to greater sunshine hours due to clear weather. However, at Bhubaneswar rice did not respond to applied P during kharif but rabi rice responded. Absence of P response in kharif rice is probably due to low yield and the treatment could not express its potential.

Table 1 : Average grain yield (kg ha^{-1}) of crops during 2004-05 and 2005-06.

	Jagtial		Bhubaneswar	
	Kharif	Rabi	Kharif	Rabi
Control	3441	2746	1230	1510
100% N	4289	3560	1980	2840
100% NP	5570	5163	2090	2980
100% NPK	5778	5314	2430	2960
100% NPK+ FYM	6116	5736	2470	3120
100% NPK + Lime	-	-	2410	2870
FYM / 10 t ha^{-1}	433	3718	-	-

	Physico-chemical properties						Available nutrient status (kg ha^{-1})							
	pH		EC		OC		N		P		K		S	
	I	A	I	A	I	A	I	A	I	A	I	A	I	A
Initial	8.2	5.8	0.47	0.12	7.9	4.3	108	187	19.3	19.4	364	43.4	14	22.2
Control	8.1	5.2	0.49	0.01	6.6	4.1	153	194	12.4	9.7	295	59.8	13.4	22.4
100% N	8.1	5.2	0.51	0.08	7.9	3.9	166	227	8.60	10.4	333	61.3	13.1	39.9
100% NP	8.2	5.2	0.57	0.07	8.1	4.3	161	224	23.0	13.4	338	62.8	19.4	41.4
100% NPK	8.2	5.2	0.60	0.11	8.6	4.3	184	224	23.4	19.0	379	66.5	21.2	44.4
150% NPK	8.0	5.2	0.66	0.10	9.0	4.2	205	228	32.9	21.8	401	70.9	25.7	45.4
100% NPK+ Lime / Zn	8.1	5.3	0.59	0.09	8.4	4.2	182	208	23.6	18.9	346	69.4	20.9	43.3
100% NPK+ FYM	8.1	5.2	0.58	0.08	9.1	4.8	192	226	31.9	22.4	358	72.1	21.5	53

I = Inceptisols at Jagtial ; A =Alfisols at Bhubaneswar.

Potassium is another growth limiting nutrient. Application of K resulted increase in yield from 5570 kg ha^{-1} (NP) to 5778 kg ha^{-1} in NPK treatment. The soil of Bhubaneswar is low in K status but rice responded to K during kharif only. From analysis of water from river and the pond used for irrigation purpose during rabi has got K which is sufficient to meet its requirement. Whereas during kharif, rains are sufficient to grow rice. Application of both soil amendment like FYM (Farm Yard Manure) and lime resulted increase in yield of rice but yields were larger in FYM amended plant compared to lime.

Nutrient status of soil after 10 years

The data generated on soil analysis revealed that application of fertilizer N resulted decline in soil pH in Alfisol (Bhubaneswar) and no change has been recorded in EC. It is interesting to note that balanced application of nutrient resulted increase in soil organic carbon at both the sites compared to initial SOC recorded 9 years back. As far as nutrient status is concerned continuous application of nutrient irrespective of treatment increase in available N and S content was noted. Contrary to this, decline in available K content was recorded which could have adverse effect on productivity in long run. Continuous application of P resulted increase in soil status whereas continuous absence of P resulted decline in P status of soil. Even after continuous application of FYM @ 10 t ha^{-1} in both during kharif and rabi could not sustain the yield either any of the rice crops. Probably amount of FYM applied is low. Low productivity during rabi season at Jagtial is assigned to shortage of water due to non availability water in canal.

In Alfisol of Bhubaneswar, application of lime resulted improvement in yield of rice in only in kharif rice seasons. Incorporation of FYM over and above 100% NPK sustained the yields of rice at a higher level. But there was very little response of rice to K, P, during rabi season Probably application of river water as source of irrigation could be reason for poor response to K and P in spite of being Alfisol. The river has got catchments area of agricultural field and forest land situated at higher elevation of topography which contributes nutrients to runoff water. There is need to analyze the irrigation water which might be supplying K and P to soil.

Conclusion

Thus, integrated plant nutrient supply through NPK and FYM helped to maintain crop productivity and overall soil quality. Nutrient and Phosphorus are limiting nutrient and Potassium could be dangerous for sustainability in future. Application of P in excess to requirement accumulated in soil may cause pollution in water bodies through run all water.

□□□

Chapter 18

Edaphic Constraints vis-a-vis Conservation Practices for Increasing Productivity of Rainfed Lands in Bundelkhand Region

S.P. TIWARI, DEV NARAYAN AND H. BISWAS
Central Soil and Water Conservation Research and Training Institute, Research Centre, Datia (M.P.)

The Bundelkhand region (7.04 m ha) of central India is a semi-arid plateau that encompasses six districts of Madhya Pradesh (Chhatarpur, Datia, Damoh Panna, Sagar and Tikamgargh) and seven districts of Uttar Pradesh (Banda, Chitrakoort Hamirpur, Jalaun, Jhansi, Lalitpur, Mahoba). Planning commission of India has included these districts in the list of most backward districts of the country where 80% people depend on the agriculture and livestock rearing.

Soils of the Bundelkhand region suffer with inherent edaphic constraints (Nitant and Tiwari, 1991). These edaphic limitations of soils not only restrict their use but also adversely influence the crop

"Authorized for publication by the Director as paper No 1030/13/2009 from Central Soil & Water Conservation Research & Training Institute, Dehradun (UP) vide No 2600/Pubdatia/93-cdnIII dated14.08.09

yields in the region. Due to unfavorable physical condition of the land only 46 % of the area of the region is reported under effective cultivation, out of which about 70 % is rainfed with less than 100% cropping intensity (Narayan, 2005). The problem is further aggravated due to erratic and uneven rainfall distribution, extremes of temperatures and associated land degradation. Crop failures are common phenomenon in the region even during rainy season due to long dry spells (Bhusan *et. al.*, 1997).

Considering the severity of problem and the consequent disorders in socioeconomic conditions of the region, the ICAR established a regional research station at Datia (M.P.) in 1986 with a major objective to carry out research and develop technologies for tackling the degradation problems and enhancing the crop productivity. Based on the published research work carried out at the centre, this paper discusses the edaphic limitations of Bundelkhand soils and their management through conservation practices for efficient utilization and enhancing the productivity on rainfed lands.

Material and Methods

The information on soil resource were collected from eight pedons of the four soil series of the region (Rakar, Parwa, Mar and Kabar) at CSWCRTI, research farm Datia (Nitant and Tiwari, 1991). Max water holding capacity and saturated hydraulic conductivity of soils was calculated using textural data through hydraulic properties calculator down loaded from http://hydrolab.arsusda.gov/soilwater/Index.htm. Salient findings of the series of experiments conducted at the research centre for enhancing the productivity of rainfed lands conducted by the authors and compiled from reports have been discussed for each group of soils having different kind of edaphic limitations.

Results and Discussion

(1) Soils types and their Characteristics

Bundelkhand soils are divided into red and black groups based on their colour and topography. Each group occupies almost half of the region. Further soil four types were recognized on the basis of texture, namely Rakar, Parwa, Kabar and Mar. The Morphological, textural and hydrological properties of each soil type are presented in Table-1.

Table 1 : General characteristics of major soil types of Bundelkhand region at research farm, Datia (M P).

Particulars	Red soils		Black soils	
Taxonomy	Alfisols and Entisols		Inceptisols and Vertisols	
Morphological				
Soil types	Rakar	Parwa	Kabar	Mar
Topo-sequence	Higher	Higher-middle	Lower- middle	Lower
Soil Colour	Red	Red-Brown	Light black	Black
Soil depth	Shallow	Medium	Deep	V.deep
Surface condition	Rocky	Gravely	Mod- Fine	Very Fine
Surface Erosion	Severe	Moderate	Moderate	Slight
Stickiness	Non-sticky	Non-sticky	Slight	V. sticky
Textural				
Sand (%)	66-72	66-68	44-46	26-30
Silt (%)	18-22	19-22	28-31	22-32
Clay (%)	10-12	11-13	24-26	38-52
Texture	Gravelly, Sandy loam	Sandy loam	Loam, Silty loam	Si. clay loam, Clay loam
Hydraulic features				
Bulk density (g/cc)	1.45-1.50	1.42-1.46	1.41-1.44	1.38-1.42
Max WHC (%)	13.5 – 14.2	15.0 – 16.4	16.5 - 17.8	19.5 – 23.8
FC (%)	13-15	15-17	15-18	18-22
PWP (%)	4- 6	5-7	5-7	6-8
Saturated Hydrolic Conductivity (mm/hr)	54 - 58	43 - 47	10 – 13	1.6 -2.1
Physico-chemical				
pH	7.6-8.0	7.5-7.9	7.4-7.5	7.4-7.5
E.C. (dS m^{-1})	0.10-0.17	0.09-0.18	0.15-0.28	0.16-0.22
Organic Carbon (%)	0.26-0.28	0.32-0.35	0.36-0.38	0.43-0.55
Available N (kg ha^{-1})	160-230	235-260	201-270	260-330
Available P (kg ha^{-1})	7-10	10-15	13-18	15-19
Available K (kg ha^{-1})	200-300	290-340	450-500	630-750

Micro-Nutrients				
Available Zn (mg kg^{-1})	3.0-3.9	3.8-4.2	4.8-5.2	4.7-5.4
Available Fe (mg kg^{-1})	1.5-1.8	1.6-2.1	2.5-3.1	2.3-2.8
Available Mn (mg kg^{-1})	0.1-0.2	0.1-0.4	0.2-0.7	0.5-0.8
Available Cu (mg kg^{-1})	0.2-0.4	0.6-0.8	0.5-0.8	0.7-0.9

Edaphic limitations and scope for utilization

It is revealed from the Table-1 that all four types of soils have distinct features. Their presence at different topo-sequences, depth, texture, drainage, water holding capacity and fertility status etc pose limitations for their efficient utilization. A brief account of edaphic limitations of each group of soils is presented below.

Red soils

- The red soils (Rakar and Parwa) occur on higher elevations in topo sequence, thus are excessively drained and prone to water erosion.
- These are shallow in depth and restricts the scope for leveling and limits the feeding/moisture/ root zone for the plants and crops.
- Soils are coarse textured (sandy to sandy loam) and have very poor water retention capacity consequently often fail to support crops on residual soil moisture
- Infiltration rate/permeability is very high.
- The scope for water storage in soil profile is limited because of the presence of hard layer underlying the surface restricts the water intake in deeper layers.
- Poor organic matter and clay content in these soils leads to weak soil aggregates that collapse on exposure to rain drop impact.
- These lands have good scope for developing vegetation cover by adopting soil and water conservation practices. However development of water resource for providing irrigation is must for increasing productivity of agricultural land.
- Organic farming may be adopted to maintain the soil health on permanent basis which will improve the physical condition and

organic matter status in soils and help in conserving the natural resources.

- Water harvesting through farm ponds has limited scope due to high seepage losses

Black Soils

- Black soils (Kabar and Mar) are characterized for their occurrence in lower reaches hence, suffers with impeded drainage.
- These soils are too sticky when wet thus make these soils difficult to work during rainy season and becomes too hard to work when dry during summer season.
- Mechanical structures often crack in these soils due to swelling and shrinking properties because of high montmorillonite clay content in the soils.
- These soils are considered ideal for storage of runoff and construction of pond due to low intake rate.
- These lands are suitable for crops having high water requirement and withstand submerged condition.
- Due to higher moisture retention capacity, in general, crops grown on conserved/residual moisture in these soils sustains well.

Conservation Measures for Increasing Productivity of Rainfed Lands

(i) Effect of Cover Crops

Traditional rudimentary forms of cultivation oriented to production of cereals are common in Bundelkhand region. Since sloppy red soils (Rakar and Parwa) often fail to support long duration crops on residual moisture, farmers generally practice fallow under rainfed situation, if no water resource is available for providing supplemental/ life saving irrigation. On the other hand, initial rain period (30 days) is very crucial from erosion point of view as maximum soil loss occur during this period having higher values of crop management factor "C" for major *kharif* crops as compared to remaining period (Sharma and Santraj, 1995, and Sharma, 1999). This may be attributed due to high intensity rainfall and less canopy cover developed by the time. Keeping the land fallow during this crucial period further aggravates the situation. Studies conducted at Datia shown that runoff from 2 %

sloping cultivated fallow plot was 45 % of the monsoon rainfall and a good crop cover may bring it down to about 20 % (Tiwari and Sharma, 1994) thus, suggesting the importance of surface cover particularly during the initial period of rainy season, in reducing the erosion losses.

Since cultivated legumes develops canopy very fast and are considered as a better cover crops option. These were tested in a study and three legumes namely green gram, black gram and cowpea were identified as promising cover crops for the region. They enhanced the soil fertility, reduced erosion losses and gave sustainable grain yields of 6.0-8.0 q/ha (Annonymous, 1997). In an another study on sloping red soils, groundnut was also proyed as an efficient cover crop which recorded the minimum soil loss (3.6 t/ha) as against 8.6 t/ha under cultivated fallow plot with average yield of 7.9 t/ha (Lal *et. al.*, 2007).

(ii) Effect of green manuring

Practicing green manuring in the region has greater relevance in context of foregoing discussions then any other area having lesser surface erosion problem. Besides nutritional /fertility benefits to the land, green manuring act as an cover crop and minimize the erosion during the crucial of rainy season. On an average, green manuring increased the yield of dwarf wheat by about 35% in red soils of the region (Narayan *et. al.*, 2001). The benefits of green manuring were more pronounced in shallow and medium soil depth conditions.

Table 2 : Average yields of wheat as influenced by green manuring under different soil depths.

Soil depth	Grain yield (q/ha)		Increase over without GM (%)
Shallow	9.6	13.3	39
Medium	14.4	20.0	40
Deep	20.4	26.4	30

Source: Narayan *et. al.* (2001), GM: green manuring.

(iii) MPTs for denuded hillocks and barren land

Denuded hillocks are major source of runoff and soil erosion. Therefore, protection of available stock through closure to biotic interference is important for natural regeneration in the region.

However, stray cattle are nuisance in the region due to locally called *Anna Pratha,* in other words leaving cattle for uncontrolled grazing. Study conducted at Datia summarized that closure/protection has great impact on natural vegetation regeneration. A ten years closure/ protection against biotic interference at the research farm Datia increased the tree population having DBH >15 cm by about 6.2 fold irrespective of the species (A forest watershed, W3F) protected at the research farm from biotic interference could develop almost 100% cover with the Kardhai *(Annogises Pendula)* and Dhak (*Butea monosperma*) over a period of 10 years (Tiwari and Sharma 2000). These two forest species are very hardy in nature and generally grow in co-existence on degraded lands particularly at higher altitude. These have ablility to grow on lands having little soil mantle. Protected forest watershed of Kardhai (*Annogises Pendula)* offering almost 100% cover on steep complex slope produced 2.3 % runoff as against 12.8 % from unprotected forest watershed (Tiwari *et. al.*, 1998). This suggested that *Annogises Pendula* and *Butea monosperma* could be protected and planted on naked overgrazed hillocks for developing dense forest cover to minimize erosion and meet out the requirement of fuel and fodder in the region.

Ber is one of the important fruit trees that grows abundantly in the region. Rejuvenation of the *deshi ber* was found very effective and may be taken up extensively for increasing biomass production (fruits, fuel and fodder). Out of the three species of *deshi ber* found in the area having different growth habits, *Zyziphus mauriatiana* can be rejuvenated by patch budding technique any time during June to September with assured success of >60% whereas early budding in the month of June and July was found more successful in case of *Zizyphus numularia* and in June month in case of *Z. oinoploea* (Tiwari, 1997).

(iv) Land configuration and tillage practices

Adverse effect of shallow depth of soils on crop yields has been studied at the centre. A plot having 30 cm soil depth produced almost half of wheat grain (1.2 t/ha) than the plot having 90 cm soil depth (Narayan *et al*, 2006), suggesting the importance of soil profile depth in crop production. Land configuration and tillage practices are important for increasing profile moisture storage, pulverizing the deeper layers and creating better soil environment and ultimately minimizing the adverse effect of shallow soil depth.

Deep tillage (20 cm) practiced in sorghum resulted almost five fold increase (25.0 q/ha) in grain yield over zero tillage (4.7 q/ha). Runoff and soil loss from deep tilled sorghum plot was 35.9% and 4.98 t/ha as against 43.4% and 3.08 t/ha from zero tilled sorghum plot respectively. Growing sunhemp in between two rows of sorghum for in situ turning was further helpful in reducing runoff (30.7%) and soil loss (3.87 t/h) and increasing grain yield (27.0 q/ha) (Narayan *et. al.*, 2003). Thus, practicing zero tillage is good from conservation point of view, but deep ploughing is better choice for higher grain yields of sorghum and other cereals in the region, which may be attributed due to effective control of weeds through deep tillage, which in fact are great nuisance in rainy season.

In a another study, Nitant and Singh (1991) reported 44.9% and 36.6% increase in sorghum grain and fodder yields respectively and about 83% in pigeonpea grain yield due to deep tillage (sub-soiler) over shallow tillage using country plough. Deep ploughing with tractor mounted sub-soiler (30 cm) cracks the hard layer of shallow red soils, increased the water intake in deeper layer, thus enhance used the crop yields. Sub- soiler may be on soils having hard murram layers beneath the soil surface for improving root zone moisture regime and crop production.

(v) Potential Intercrops and Cropping-sequences

Black soils in the region suffer with drainage problem. The performance of crops/ cropping system on these soils depends on the rainfall pattern in the particular year. Studies conducted on flat (0.3%) black soils revealed that performance of the sorghum was consistent in three years out of the five years, while green gram did not perform consistently, However, yield of rabi crops (mustard, gram and safflower) after green gram were higher than fallow-rabi crops system. This finding is important in view of the fact that in black soils only rabi crops are often sown by the farmers on conserved moisture. The higher yield of rabi crops after green gram may be attributed due to additional benefit of legumes on soil fertility. Vegetative barriers of *Vetiveria ziazioides* at 30 m HI was found most effective in boosting the yield of rabi crops followed by ridge and furrow and bedding system (Sharma and Bhatt,1997)

Mono-cropping of sorghum is common on rainfed red soils in the region but many times suffers badly due to moisture stress situation. Therefore, sorghum intercropping with legume and oil seed crops has good scope in the region. A sorghum based intercropping study conducted with sesame, green gram, cluster bean and cowpea revealed that intercropping of 2 rows of sesame in between two rows of sorghum produced highest yield in terms of sorghum equivalent (41.4 q/ha) followed by one row of sesame (33.9 q/ha), two rows of green gram (30.6 q/ha) and one row of green gram (28.7 q/ha) (Dwivedi *et al*, 1994).

Gram based intercropping study conducted with mustard, taramira, lentil and linseed in black soils on conserved moisture with pre-sowing irrigation revealed that intercropping of mustard, taramira and linseed increased higher gram equivalent yield is a tune of 13.6 q/hq, 15.3q/ha and 12.6 q/ha respectively in 6:1 row ratio than 9:1 and 3:1 (Anonymous,1996).

(vi) Grasses for conservation and production

Dichanthium annulatum, Hetropogon contortus and *chrysopogon fulvus* are dominant natural fodder grasses preferably grow on uplands. Two major non palatable grasses of the region are *Vetiver* and *Saccharum munja*. While low lying fine black soils are natural habitat of *Vetiver* grass, *saccharum munja* grow abundantly on light soils near river banks / water streams. Study conducted at Datia revealed that *Dichanthium annulatum* grown on 2% slope at 50 x 50 cm spacing produced minimum average runoff (11%) and soil loss (0.3 t/ha), as compared to 50.0% and 8.8 t/ha from cultivated fallow plot, respectively. (Sharma and Santraj,1995).

Above ground biomass yield of *Vetiveria zizaneoidis, Saccharum munja and Hetropogon contortus* from 4-6 year old plantation, at plant to plant spacing of 50x50 cm was estimated to be 1.6, 1.3 and 0.8 t/ha and soil binding factor (W/r^2) 335, 284 and 192.3 respectively (Nitant *et. al.*, 1993).

(vii) Vegetative barriers/ live bunds

Vegetative barriers are closely spaced grass strips grown on contours across slope. These have been found effective in checking erosion and considered as an alternate to bund on mild slopes for soil and water conservation being relatively cheaper and pro-farmer.

A study on 2% sloping red soil revealed that vegetative barrier of *Vetiveria zizaniodes* planted at 11 m horizontal spacing in standard size runoff plots produced run off and soil loss of 32% and 3.3 t/ha as against 50.0% and 8.8 t/ha, respectively from cultivated fallow plot. (Sharma and Santraj, 1995). In an another study (1996-98), *Vetiveria zianioides* barrier planted at 11 m HI with sorghum crop on 2% sloping red soils increased grain yield by 9.9 % (21.0 q/ha) than sorghum without barrier, in spite of the fact that there was reduction in the area of sorghum by about 5.4% in Sorghum+ barrier treatment. This indicated that grass barrier has increased the yield of grain sorghum. Vegetative barrier with sorghum also produced less runoff (24.2%) and soil loss (1.74 t/ha) as compared to 47.2% and 4.6 t/ha, respectively from sole crop of sorghum (Sharma, 1999).

(viii) Water harvesting and utilization

Water harvesting has good scope in the region due to undulating topography, good annual rainfall (750-1200 mm) and presence of black soils in low lying areas. Low intake rate of black soils make them ideal for storage of runoff and construction of farm pond. Sharda and Ojasvi (2005) summarized that depending on the land use conditions runoff in the region varies from 10-70 %. In general, water yield in the region worked out to the tune of 0.2 ha m for every 5 ha catchment with flat lands and 0.3 to 0.5 ha m for every 5 ha of catchment with sloping lands. Ephemeral streams in the region offer good opportunity for storing water by constructing water harvesting structures called as stop dam.

Red soils of the region are generally not suited for water harvesting due to high seepage losses. However, experiences gained through construction of a farm pond in red uplands at research farm Datia suggested that it is possible to harvest and store rainwater in these soils for a limited period so as to provide life saving/ supplemental irrigation at very crucial stage of selective high remunerative crops. Soybean is one of the promising oilseed crops of the region under rainfed condition. However, it often suffers at pod filling stage due to moisture depletion caused by early recession of rainfall. At the same time sowing of mustard and taramira is affected badly due to low moisture content.

Results indicated that irrigation at the time of pod filling stage soybean can boost the yield by 42%. Similarly one pre-sowing irrigation followed by one irrigation at one month after sowing produced higher yield of mustard (8.8 q/ha) and toria (4.3 q/ha) as against practically fail (less then 1.0 q/ha) without any irrigation (Narayan *et. al.*, 2003) i.e. grown exclusively on conserved moisture.

Conclusion

The studies summarized that red and black soils of the Bundelkhand region are typical in nature and pose different types of edaphic limitations thus lead to unstable and poor yields particularly in rainfed areas. However various soil conservation practices could help in improving the productivity of rainfed lands. By adopting these techniques farmers can boost the yields in rainfed areas, conserve valuable natural resources and rehabilitate degraded lands for eco restoration and social upliftment.

References

Anonymous (1996) Annual Report 1995-1996, CSWCRTI, RC Datia.

Bhusan, L.S., Sharma, A.K. Tiwari, A.K., Singh B.L. and Yadav R.C. (1997). Management of Ravine watershed. Bulletin NoT-37/A-1.CSWCRTI, Research centre,Agra (UP) :1-47.

Lal ,B, Narayan, D and Katiyar V.S. (2007) Assessing crop cover influence on runoff and soil loss for red soils of Bundelkhand. Annual Report 2006-07, CSWCRTI, Dehradun: 20-23

Nitant H.C. and Tiwari, S.P. (1991) Soil resource appraisal for land use planning in Bundelkhand region. *J Soil Wat. Conserv.* India **35**(1&2):17-25

Nitant H.C. and Singh Pratap (1992) Studies on tillage practices in red soils. Annual Report 1991-92, CSWCRTI Dehradun: 134-135

Nitant H.C., Tiwari R.K. and Tiwari, S.P. (1993) Root studies of khus grass *(Vetiveria zizaneodis)* in relation to soil conservation in red and black soils of Bundelkhand. Region *J.soil wat conserve.*India.**36** (1&2) : 29-36

Narayan, D (2005) Resource conservation technologies for sustainable crop production in Bundelkhand. In (Ed Sharda *et al* Resource conserving technologies for social upliftment IASWC, Dehradun India: 256-260

Narayan,D and Lal ,B (2006) Effect of green manuring on soil properties and yield of wheat under different soil depth in alfisols of central India. Bulletin National Institute of Ecology **17**:141-146

Narayan D, Katiyar V.S. and Lal B. (2004) Tillage and surface cover management. Annual Report 2003-04, CSWCRTI, Dehradun: 22-23

Sharda VN and and Ojasvi, PR (2005) Water harvesting through watershed management in different agro-ecological regions of India. Indian *J. Agril Science* **75** (12):771-780.

Sharma A.K. and Santraj (1995) Runoff, soil and nutrients losses under different crop covers on 2% slope at Datia. Annual Report CSWCRTI, Research Centre, Datia :11-17

Sharma A.K. and Bhatt,V.K.(1998) Evaluation of farming systems for sustainability in black soils. Annual Report 1997-98 , CSWCRTI, Dehradun: 89-90

Sharma,A.K.(1999) Runoff, soil and nutrients losses under different crop covers on red soil of Datia. Annual Report CSWCRTI, Research Centre, Datia: 9-21

Tiwari, S.P. and Sharma (1994) Runoff, soil and nutrients losses under different crop covers on 2% slope at Datia. Annual Report CSWCRTI, Research Centre, Datia: 10-15

Tiwari, R.K. (1997) Studies on rejuvenation of deshi Berplants and their management. Annual Report 1996-97 CSWCRTI, Dehradun: 193-195

Tiwari, R.K. and SharmaA.K (2000) Natural regeneration of trees Annual Report CSWCRTI, Research Centre, Datia. :7-8

Tiwari AK, Sharma, AK, Tiwari, RK, Bhatt, VK and Shrivastava, MM (1999) Studies on hydrological behavior of small watershed under different land uses. Annual Report 1998-99 CSWCRTI, Dehradun. : 46-47.

□□□

Chapter 19

Sustainable Crop Production through Efficient Nutrient Management and Other Agro-Techniques — A Review

OM SINGH
Indian Veterinary Research Institute, Izatnagar, Bareilly-243 122 (U.P.), India

Nutrient-use efficiency can be expressed in agronomic, physiological and economic terms. It is primarily judged as a portion of crop uptake out of the applied nutrient. Nutrient use for agricultural production and greed for higher profits have made nutrient application in agriculture unscientific with more wastage leading to pollution of soil, air and water.

Nutrient use-efficiency expression is a ratio of crop uptake to total supply (applied + native) or incremental increase. Agronomic efficiency means the crop response to applied nutrients. Although fertilizer consumption is increasing quantitatively, the corresponding yield increase per unit of nutrient is susbstantially diminishing over the years. Nutrient use-efficiency and sustainable crop production depends upon several agronomic factor including tillage, time of sowing, appropriate crop and variety, proper crop density, sufficient water application through irrigation at right time and method of control of weeds, diseases, pests and timely application of right nutrients. These factors largely influence nutrient-use efficiency, either individually or collectively Rao (2007) found that selection of proper planting material, population density and balanced fertilization could collectively improve nutrient-use efficiency by 25 to 50per cent.

The estimated supply-demand gap would be about 1.8 million tonnes of N and 1.9 million tonnes of P by 2011-12. The entire demand for K will continue to be met through imports. The efficiency of fertilizer N is only 30-40per cent in rice and 50-60per cent, in other cereal crops. Efficiency of fertilizer P is 15-20per cent in most field crops. The efficiency of K is 65-80per cent and in case of S it's 8-12per cent only. Micro nutrients have nutrient efficiency generally < 5per cent (NAAS, 2006). In USA, fertilizer - N efficiency in maize has increased by > 30per cent over last 20 years. In Japan apparent N use efficiency for rice has resented in significant increase (Yagi and Minami, 2005). In India, large areas are rainfed have uncontrolled irrigation, imbalanced fertilization, tropical climate, traditional crop production and low investment capacity. This resulted in low fertilizer consumption (107 kg/ha), while in France, Japan, USA, U.K. highest rate of fertilizer's per unit area 300-600 kg NPK ha^{-1}. Low nutrient-recovery efficiency not only increase cost of crop production but also causes environmental pollution.

There may be strategies to improve nutrient use efficiency like (i) product strategy : enhanced efficiency of fertilization (ii) Fertilizer management strategy: enhanced efficiency of fertilization. This review is presented in the light of process and mechanism that leads to increase in use efficiency of applied nutrients, soil health, sustainable crop production and eco-friendly farming systems.

Addition of organic materials

The build-up of soil organic matter increases the potential for N mineralization. It was found that Interaction among inputs (manure, cover crops and fertilizer) and organic matter influence the rate of soil N mineralization (Horwate *et. al.*, 2006). It is also well known that N from many organic fertilizers often showes little effect on crop growth in the year of application because of slow release nature of organically bound N. Nitrogen immobilization after application of organic materials can occur, leading to enrichment of the N pool. This process increases the long term efficiency of organic fertilizers. Organic material have a major role to play in maintaining buffering capacity of soil and are important for maintaining soil physical and biological properties. A range of factors such as soil temperature, moisture and chemical composition of the organic material influence N release from organic sources in soil. Addition of green mannure alongwith NPK was promising during the first three seasons while FYM alongwith NPK as better during later three seasons, indicating long-term use of FYM for better yield in rice-rice cropping system in vertisols (Subbaiah et al, 2006). Integrated nutrient management through using green leaf manure (Melia, Neem and Subabul) with chemical N fertilizer significantly enhanced N uptake, agronomic efficiency, apparent N recovery and yield of sunflower (Panneer and Bheemaiah, 2005).

New materials of plant nutrients

Numerous products supplying plant nutrients have emerged, offering a variety of nutrient contents, physical forms and other properties. The complexities of N management exist due to its solubility, mobility and vulnerability to denitrification. The approaches have been to manipulate the granule size variations, coating with neem or coaltar, modifiers or additives to control the nutrient release rate. Controlled release fertilizers (CRFs) have higher N-use-efficiency and reduce N loss due to leading and volatilization while keeping higher yields. A slow-release and super-absorbent N Fertilizer (SSNF) can be especially useful in drought prone areas where availability of water is insufficient. The neem coated urea of NFL recoreded better shelf-life, slow dissolution as well as nitrification inhibition. Large granular forms of N and P fertilizer proved better than powered and prilled forms in increasing nutrient uptake and grain yield of rice in irrigated low lands soils (Subbaich *et. al.,* 2006).

Coating P fertilizer could limit the contact of applied P with soil resulting in early season nutrient deficiencies in crops like wheat. Silicate fertilizers increased significantly the rice yield and N efficiency.

Nano-Technologies in slow release of fertilizers

These inexpensive nanotech application to increase soil fertility and crop production could be in use in future. In China, Ferdao *et. al.* (2006) tried different nano-particles (Clay-Polyester, humus-polyester and plastic strach) for slow release of N in wheat. The increase in yield due to clay and plastic (Nano-material coating) was arround

4.5per cent over chemical fertilizer apart from saving through while leaching losses of N Fertilizer.

Use of microbial sources

Microbes harbouring in rhizosphere of crops provide benefits to crops through better nutrient availability by way of atmospheric N_2 fixation or solublising fixed mineral forms of nutrients. Rice-based cropping system offers a lot of scope for effective utilization or wide range of bio-fertilizers (Azolla, Cyanobacteria and phosphate solublizing bacteria), which help in increasing nutrient use-efficiency and yield. Phosphorus use efficiency of French bean on vertisol was improved when rock-phosphate was applied with FYM or vermicompost at 1:2 ratio alongwith phosphate solublizing bacteria (Manjunath *et. al.,* 2006).

Biological nitrification inhibition

It is discovered that some forage crops e.g. Bronchiaria grasses, have ability to regulate nitrification in soils by releasing inhibitors through root exudates. This biological mechanism has evolved to minimize N losses associated with nitrification which improve crop production and NUE (nutrient use efficiency).

Nutrient use-efficiency in cropping systems

Pigionpea substitution for rice during monsoon season in rice-wheat cropping systems improved wheat yield by 16per cent and total N and P uptake by 31 and 10per cent, respectively, due to improvement in soil physical properties. Legwuges with interminat growth are more efficient in N_2 fixation than determinate types.

Cultural practices and weed management

In general, crop yields under no till are more stable than under tilled systems with greater nutrient efficiency (Mertin, 2006). There was 24per cent increase in yield of rice-due to precision land leveling. Controlling of weeds enhances the use efficiency of applied nutrients. When weeds were removed, the rice yield increased by 10per cent under rain fed low land conditions and 17-19per cent under irrigated conditions.

Balanced use of fertilizers

Acharya and Sharma (2008) made review on INM for N-use efficiency and crop productivity. The P-use efficiency of applied fertilizer by crops ranges from 10-30per cent in the year when it was applied. The remaining 70-90per cent becomes part of the soil P pool that is released to crops was over the subsequent months and years. The best strategies for sustainable crop production and environment safety with respect to P are to produce optimum crop yields by (a) alternative P application only to winter crops (b) periodic reduction in P rates according to accumulated P, and (c) occasionally raising crops without fertilizer P to enable utilization of residual fertilizer P that had accumulated in the soil over the years (Aulakh, 2007). Application of rock phosphate alongwith green leaf manure and P solublizing fungi produced significantly higher grain yield than application of rock phosphate alone. Most Indian soils are medium to high in K status. Variable response of applied K is reported in many crops. Potassium difficiencies have been observed in crops such as cotton, maize and sorghum in vertisols (Kathryn *et. al.*, 2006).

Potessium, an essential nutrient required by plants in large quantities is present in soils in four pools (varying availability to plants), in dynamic equilibrium. The high K uptake or tissue concentration might improve the quality of produce but not the yield. Sulphur deficiency is increasing, makings as one of the limiting factors in further sustainable increase in agricultural production. Sulphur fertilizer besides enhancing yield and quality of crops, enhance nutrient uptake and fertilizer use efficiency through interaction of S with other fertilizer nutrients (Tandon and Messick, 2007). In hybrid rice in Orissa, grain yield and relative agronomic efficiency was strongly affected by reduction of P and K application as well as skipping of Ca, Zn and B.

Precision farming for efficient nutrient management

Crop modeling can provide a vast amount of real time information about fertilizer dose to achieve the target yield in different crops. Models and diagnosis indicators have been developed to meet needs of forecasts during crop cycle apart from breeding of new adopted varieties. The determination of rates and drates for N application must be more precise. Equipments that help improve N management for crops include the use of hand held active sensors, arrial photography, satellite imagery and virtually any other approach that assist in managing the variability in producer fields. Model for site-specific and balance fertilizer recommendations for wheat and maize is in practice.

Breeding of new adapted genotypes

The breeding for efficient genotypes of crops to utilize nutrients arises from balance of several process as associated with acquisition and uses. The NUE in crops has been improved through the use of combined approaches, mainly based on whole plant physiology, quantitative genetics and forward and reverse genetic approaches. It is visualized that in the 21st century, nutrient efficient plants will play a great role in increasing crop yields (Fageria, 2006). Farm yields are presently only 40-60per cent of the attainable yield potential and so an appreciable exploitable yield gap remains in production potential.

Conclusion

Soil fertility inbalances are occuring due to inadequate and imbalanced fertilization not corresponding to crop nutrient mining that is leading to development of second generation problems in nutrient management. Although fertilizer consumption is increasing quantitatively, corresponding yield increase per unit of nutrient is substantially diminishing over the years. The nutrient-use efficiency of all the major, secondary and micro nutrients continuous being low despite increasing consumption of fertilizers in India. Among all the major nutrients, nitrogen is most vulnerable and increasing its use efficiency through physical manipulation of the size of granules, coating of granules with films amendments or chemicals to prevent their fast dissolution or prevent bacterial-medicated dinitrification did not prove economically viable. Improvement in nutrient use efficiency is necessary to reduce the cost of production as well as to prevent environment polution. Split application is effective but costly. Adoption of best management practices is essential to get higher inputs-use efficiency and profitability. The recent developments in the field of biological nitrification inhibition offer good opportunity in effective nitrate management. Practicing nitrogen biological fixation through inclusion of legumes in cropping system and maintenance of higher soil organic matter will support to maintain soil-fertility and improve soil and physical and biological environment with sufficient buffering levels. Phosphorus management has been considered as proper source and the cumulative effects that can be utilized through biological process for nutrient availability and root acquisition. Identification of efficient genotypes for low nutrient situations or better response to applied nutrients through better acquisition and assimilation are encouraging. Site-specific nutrient management, adoption of conservation tillage and suitable fertilizer managment practices will certainly help in increasing nutrient-use efficiency and ultimately it will assist in sustainable crop production and good soil health as well. Practicing new agro-techniques, use of leaf colour charts, sensor-based technology, laser land leveling, good genotypes and precision farming can encourage the sustainability of soil fertility, crop productivity and ecofriendly agriculture.

References

Aulakh, M.S. (2007). Yields and nutrients use-efficiency in ground nut sunflower cropping system in Punjab, India, *Journal of sustainable Agriculture*, **31** (1):1.

Fageria, N.K. (2006). Nutrient efficient plants in improving crop yields in the 21st century. 18th world Congress of soil science, held during 9-15 July, Philadelphia, Pennsylvania, USA.

Fudao, Z. Rugang, W. Qiang, X. Yajun, W. and Jianfeng, Z. (2006). Effects of slow/controlled release fertilizer cemented and coated by nano materials on biology/plants. *Nano Science* **11** (1) : 18-26.

Horwath, W., Kabir, Z, Reed K. Kaffka, S., Miyaho, G. and Kent (2006). Long-term assessment of N use and loss in irrigated organic, low-input and conventional cropping systems. 18th World Congress of Soil Science, held during 9-15 July. Philadelphia, Pennsylvania, USA.

Kathryn, Taylor, Singh, B. and Schwenke, Graeme, (2006). Potassium dynamics in vertisols following potassium fertilization and plant uptake 18th World Congress of Soil Science, held during 9-15 July, Philadelphia, Pennsylvania, USA.

Manjunath, M.V., Patil, P.L.and Gali, S.K. (2006). Effect of organic amended rock phosphate and P solubilizer on P. use efficiency of French bean in vertisol of malaprabha right bank command of Karnataka, Karnataka. *Journal of Agricultural Sciences* **19** (1) : 36-39.

Martin, Diaz-Zortia, (2006). Innovation for improving productivity and nutrient-use efficiency : No- till grain cropping systems of South America. 18th World Congress of Soil Science, held during 9-15 July, Philadelphia, Pennsylvania, USA.

NAAS, (2006). Low and deeling crop response to fertilizers. Policy paper No. 35. National Academy of Agricultural Sciences, New Delhi.

Panneer, Selvam and Bheemaiah, G. (2005). Effect of intragrated Nutrient Management practices on productivity, nutrient uptake and economics of sunflower intercropped with Azadirachita indica and Melia azordirach trees. *Journal of oilseeds Research* **22** (1) : 231-234.

Rao, T.N. (2007). Improving nutrient use efficiency : The role of beneficial management practices. *Better crops,* India **1** (1) : 6-7.

Sublaiah, S.V. Rama Prasad, A.S.Kumar, R.M. and Surekha, K. (2006). Interaction of organic manures and granular size of inorganic NPK fertilizers on nutrient uptake and grain yield of irrigated rice invertisols. 18th World Congress of Soil Science, held during 9-15 July, Philadelphia, Pennsylvania, USA.

Tandon, H.L.S. and Messick, D.L. (2007). Practical Sulphur Guide. The Sulphur Institute, Washington, USA.

□□□

Chapter 20

Food Legumes in Cropping Systems for Resources Conservation, Food and Environment Security : Challenges & Opportunities

V.S. LATHER, R.S. PANNU, D. P. NANDAL AND ISHWAR SINGH
CCS Haryana Agricultural University, Regional Research Station Kaul, (Haryana)

Legumes have been an integral part of cropping system practiced by Indian farmers, till very recently. However, increase irrigation facilities and introduction of HYV cereals during green revolution has relegated legumes from it's traditionally areas to marginal lands due to farmers switched over to lucrative cereals and other cash crops. India has rich biodiversity in legume crops which are grown from dry temperate to humid tropics between the altitudes from sea level to snow line. Though, major emphasis for research was given on the improvement of yield of major legumes, however, their production and productivity is static during last 3-4 decades. In India, there are several known leguminous (peas, beans, grams, and forage crops) which have tremendous potential due to their efficient adaptation to low inputs conditions of moisture and fertility. Though the returns from grain legume crops were comparatively less, they had much more beneficial impact on the soil, hence played a vital role in sustaining soil fertility. Serious threat to the production and genetic diversity of these crop species found while they are important to food, nutritional and economic security. The manipulation of genetic variation in legumes is of prime importance for the successful breeding of improved cultivar with added value and desirable resistance of diseases and pest couple with much needed active state policies and supports.

Challenges to Resources Conservation, Food & Environment Security

Agricultural growth in India has always focused on growing more food grains under the burden of uncontrolled fast growing population. The green revolution was achieved through introduction of high inputs responsive dwarf HYV (high-yielding variety) of cereals (rice, wheat, maize etc.) by ignoring the ecologically adapted cereal (millets etc.), legume and oilseed crops . Though, India became self-sufficient in food grains through adoption of increase irrigation facilities, high inputs responsive HYV of cereals but that caused a serious regional imbalance in agricultural production and degradation of natural resources particularly ground water and soil health. This can be also be gauged from increase irrigation. i.e. by 1996-1997, the state of Punjab had 92.9 per cent of its area under irrigation, followed by Haryana (76.2 per cent) and Uttar Pradesh (68.7 per cent). On the other hand, only 14.4 per cent of Maharashtra and 21.9 per cent of Karnataka was irrigated.

Secondly, and more importantly, the country's cropping patterns is completely transformed in favor of mono culture crops with rice-wheat in the north, and rice-rice in the south and east, legumes in central and western India. This led to the complete marginalization of legume and minor cereals (millets) cultivation, even in their traditional areas (Table-1). Even as the area under cereal cultivation in Punjab has increased from 50.9 per cent in 1966-1967 to 73.5 per cent in 1997-1998 (rice: 0.227 million ha to 2.278 million ha; wheat: 1.4 million ha to 3.3 million ha), pulses and oilseeds grown far less widely (pulses: 13.47 per cent of cultivable area in 1966-1967 to 1.1 per cent in 1997-1998; oilseeds: 6.2 per cent to 1.8 per cent). The main cropping pattern in Indo-Gangetic plains followed by the selected sample farmers in 1996/97 indicated that rice and wheat were the major crops of the study area and occupied 81% of total cropped area. Legumes with < 10 % area still preferred in this predominantly rice-wheat system but it was much higher than the area of other crops.

Table 1 : Crops Pattern in District Karnal (Haryana), India, 1996/97

Crop group	Crop	Area (%)
Cereals (grain)	Rice	43.0
Wheat	38.0	
Others (maize)	0.2	
Cereals (fodder)	Sorghum, maize etc.)	3.4
Legumes (grain)	Pigeonpea, chickpea, lentil, mung bean, black gram	3.1
Legumes (fodder)	Berseem, lucerne	3.5
Legumes (summer)	*Sesbania* spp.	2.2
Oilseeds	Mustard, toria, and sunflower	2.8
Commercial crops	Sugarcane	3.2
Others	Others	0.6

India sprinted along an agricultural growth path guided by an imagination of more yield/ hectare of cereals only (rice, wheat etc.) using more urea, water, chemicals. However, 50 years later, now the consequences can no longer be ignored which include decline in production & productivity with serious damages to environment and natural resources :

a) Decline in Cereals Productivity despite increase use of Chemical Fertilizers

Food grain production has made a quantum jump from 50.8 mt in 1950-1951 to above 200 mt in 2000-2001 which is paralleled by an increase in fertilizer consumption, from 69,800 tonnes in 1950 to 16.7 mt in 2000-2001. However, sharp decline have been observed during 1990's in the average yield of major cereal crops. In the 1980s, increase in annual yield per ha of rice and wheat used to be 3.47 and 3.10 per cent, respectively which has now nose-dived to 0.92 per cent for rice and 2.21 per cent for wheat. Fertilizer use in India is dictated by the larger perceived need to increase food grain yield under post-independence agricultural policy. It is ironic that while India ranks third in fertilizer use worldwide, it ranks 14 and 16 respectively in the production of rice and wheat.

b) Degradation of Soils health

Mismatch between nutrients removed from the soil and those put back into it by net negative nutrient balance of 8-10 mt per annum. About 70 per cent of the total cropped area experiences nutrient depletion of more than 50 kilogram per hectare annually. The annual depletion of major nutrients (nitrogen, potassium and phosphorus) is to the tune of 5.8 mt Further, It was found a staggering 47 per cent of the soil are deficient in zinc by ICAR study and heavily skewed fertilizer policy based on 3 industrially produced nitrogenous (N), phosphatic (P), and potassic (K) fertilizers as primary nutrients and absolute inputs for crop production with major emphasis on amount of N used in form of urea(11.5 mt out of total 16.7 mt in 2000-01). The effect of skewed urea use has also unbalanced soils. While the ideal NPK ratio is 4:2:1, in 2001 the ratio stood at 28:8:1 for north India, and the national average was 7:2.7:1.

It was found in long-term fertilizer experiments conducted by ICAR that in the alluvial soils of Barrackpore, Ludhiana and New Delhi, the depletion of potassium reserves from the soil had more than doubled (366 kg per ha when only nitrogenous fertilizers were used where as 160 kg per ha where no fertilizer applied). Though, Indian soils are considered rich in this nutrient, and farmers when use adequate amounts of potassium fertilizers, the yield and quality of crops improved dramatically. Hence soils are suffering from major nutrient fatigue and are hungry for P and K. But, situation will not change unless there is coherence between the price of fertilizers and the income farmers get from produce, so as to encourage farmers to use all fertilizers.

Agriculturists in India fully realized the extent of damage micronutrient deficiency could wreak only when the fertile *terai* region in Uttar Pradesh missed a couple of paddy crops in 1966 due to zinc deficiency. Zinc is now considered the third-most limiting nutrient after nitrogen and potash. The studies by ICAR since then have shown that the practice of intensive agriculture depending on HYVs, extensive use of high analysis chemical fertilizers and reduced return of farm waste to soil is compounding nutrient deficiency in soils. Soils in many parts of Andhra Pradesh, Bihar, Karnataka, Maharashtra, Madhya Pradesh and Uttar Pradesh exhibited zinc deficiencies of over 60 per cent. Falling levels of other important micronutrients like boron, iron and molybdenum are also affecting yields (Table-2).

Table 2 : Precarious annual nutrient imbalance

State	N	P_2O_5	K_2O	Total	
Maharashtra	-252.2	-1918.7	-367.2	-2538.1	Farmers in rain fed areas
Karnataka	-92.3	-224.8	-202.6	-519.7	use very little fertilizers.
Madhya Pradesh	-17.7	-86.7	-824.7	-929.1	So nutrient mining is very high
Punjab	+497	+9	-673	–167.0	Rice and wheat corner
Haryana	-22.5	-92.0	-480.0	-594.5	over 90% fertilizers
Tamil Nadu	+78.0	+35.0	-236.0	-123.0	Somewhat balanced;
Andhra Pradesh	+207.5	-132.2	-43.7	+31.6	K levels inherently low
Gujarat	+350.7	+146.6	-365.0	+132.3	High in K levels

Source: Souvenir, Second International Agronomy Congress on Balancing Food and Environmental Security — A Continuing Challenge, New Delhi, November 26-30, 2002

C) Degradation of Environment and Natural Resources

Water over-use and increased irrigation is a third factor which has led to the current debility of food and environment security in India. To feed the rice-wheat cropping system, net irrigated area rose from 20.8 million ha in 1950 to 53.5 million ha in 1995-1996 and the agricultural area grew from a mere 20.8 per cent in 1966-1967 to 42.4 per cent of the 120 million ha of cultivated land in the country by 1997-1998. Though, the increase irrigation facilities through Canals Command Areas resulted into drastic increase in agriculture production during last 4 decades in India, however, it's over use through flood irrigation system has also caused serious degradation of soil health (increase problematic soils, water logging etc.) and environment degradation. Water logging, caused by irrigation without drainage, leads to high rates of evaporation vis-à-vis salts in the water settle on the soil which leads to salinity or alkalinity of the soil. As the salt content increases, the soil-plant system registers a reverse osmosis process which caused plant roots tend to lose water and nutrients instead of absorbing them from the soil. Similarly, water over-use through tube wells coupled with power subsidies has further degraded the ground water making major agriculture areas of Indo-Gangetic plains in dark zone for quantity and quality of water (Table-3).

Table 3 : Canal command areas display high salinity

CANAL COMMAND PROJECTS	Waterlogged area (1000 ha)	Percentage of command area	Saline area (1000 ha)	Percentage of command area
Ramganga	195	10	352	18
Gandak	562	53	400	38
Nagarjuna	114	24	69	14
Sharda	260	7	253	7
Chambal	99	20	40	8
Mahi	89	39	61	29

Source: Singh, N T, (2000) Dryland Salinity in the Indo-Pakistan subcontinent, Univ. Bonn,. Map source: Wastelands Atlas of India, ministry of rural development, 2000

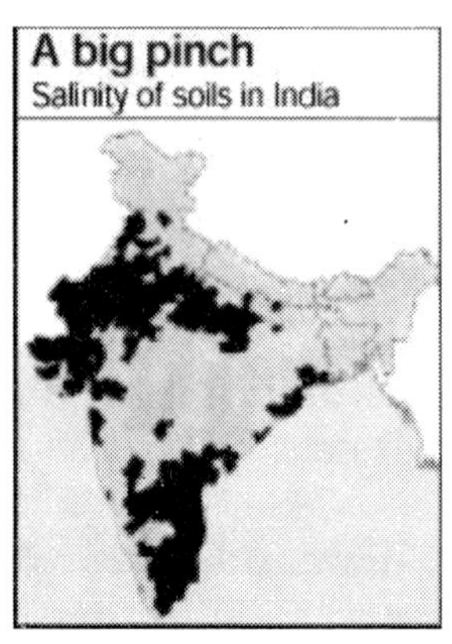

The domino effect of bad policy-making and its fallout was dwindled the nutrient quality of soils in many ways. Apart from micronutrient loss, Indian soils already weak in carbon content are today even more incapable of acting as sequesters of carbon and so reducing global warming potential. Carbon content in soil is important to soil productivity. Indian soils are relatively low in carbon content: between 0.1 per cent in desert soils and 1 per cent in soils found in the humid and sub-humid climates of Kerala and some north-eastern states. Intensive cultivation in the Indo-Gangetic plains further led to deficiency in organic carbon. Soils are a very slow renewable resource. To reclaim them requires a long-term plan and falling crops productivity give clear realization that soils cannot be blindly mined, and that humans cannot just plough through the ecology they interact with. It is time to gave up the cultivation of paddy-wheat cropping system as over-exploitation of groundwater in the Punjab state has reached 145 %, second only to Delhi," said National Rainfed Area Authority CEO Dr. Samra(2008).

Dr Barry Shapiro of the International Crops Research Institute for Semi-Arid Tropics (ICRISAT) observed that "When we compare the support given by the government to rain-fed areas as compared to irrigated areas, we see that subsidy and policy support for irrigated agriculture is much more. This is causing farmers in Punjab and Haryana to use more water and other inputs like fertilisers than they should.The growing incidence of submersible pumps against centrifugal pumps is leading to the further exploitation of its energy and groundwater resources," Dr Samra said, adding that eastern India has the potential of meeting the demands of growing paddy on a sustainable basis while crops like soyabean and legumes will result in both high returns and better resource management in NW India.

Constraints of Legumes for Resources Conservation, Food and Environmental Security

a) Importance of Legumes

The legume is the third largest flowering plants family with 650 genera and nearly 20,000 species cultivated in varied climates ranging from tropic to temperate climates and belong to family of plants that form an association with soil bacteria of the genus Rhizobium for fixing atmospheric nitrogen in soils and plants (Table-4). These plants, apart from providing protein-rich seeds for human consumption, forage for livestock, and help improve soil fertility, Resources Conservation, and Environment Security through BNF (Biological Nitrogen Fixation) and adaptation to low inputs environments (Table-4).

b) Socioeconomic Constraints to Legumes Production in India

The major problem of the rice-wheat system in the Indo-Gangetic Plain of India is the stagnating or declining yields of rice and wheat. Also, the soil nutrient status and groundwater level are deteriorating. Crop diversification through legumes can address these problems as legumes complement cereals in both production and consumption. In the production process, legumes improve soil fertility status, require less water than cereals, and their rotation with cereals helps control diseases and pests. On the consumption side, legumes are the cheapest source of protein in the vegetarian diet and supplement mineral and vitamin requirements. Despite their value in production and consumption, area under legumes in the rice-wheat system has declined after the introduction of improved technologies during the mid-1960s. Several socioeconomic factors constrains legumes production in the rice-wheat system (Table-5), and these issues were examined. (Joshi, *et al.* 2000 & Singh *et. al.* 2007).

Table 4 : List of pulses and their production (Duranti, 2006)

Main Grain Legumes	Latin name	World crop produciton, metric tons*1000
Pulses:		3819
1. Dry beans		1162
Kidney bean, haricot bean, pinto bean, navy bean	*Phaseolus vulgaris*	
Lima bean, butter bean	*Vigna lunatus*	
Adzuki bean	*Vigna angularis*	
Mung bean, golden gram, green gram	*Vigna radiata*	
Black gram, urd	*Vigna mungo*	
Scarlet ruuner bean	*Phaseolus coccineus*	
Rice bean	*Vigna umbellata*	
Moth bean	*Vigna acontifolia*	
Tepary bean	*Phaseolus acutifolius*	
2. Dry broad beans (*Vicia faba*)		
Horse bean	*Vicia faba*	255
Broad bean	*Vicia faba*	
Field bean	*Vicia faba*	
3. Dry peas (Pisum spp.)		892
Garden pea	*Pisum sativum var. sativum*	
Protein pea	*Pisum sativum var. arvense*	
4. Chickpea	*Cicer arietinum*	478
5. Dry cowpea, blackeye pea, blackeye bean	*Vigna unguiculata* ssp. *dekindtiana*	350
6. Pigeon pea, cajan pea, congo bean	*Cajanus cajan*	103
7. Lentil	*Lens culinaris*	199
8. Bambara groundnut, earth pea	*Vigna subterranea*	
9. Vetch, common vetch	*Vicia sativa*	99
10. Lupins	*Lupinus* spp.	45

11. Minor pulses:		
Lablab, hyacinth bean	*Lablab purpureus*	
Jack bean, sword bean	*Canvalia ensiformis, gladiata*	
Winged bean	*Psophocarpus teragonolobus*	
Velvet bean, cowitch	*Mucuna pruiens var. utilis*	
Yam bean	*Pachyrrizus erosus*	
Non-pulses (oil-crops):		
Soybean	*Glycine max*	6209
Peanut	*Arachis hypogaea*	1874

Nitrogen fixed by legume crops (in herbage or grain)

	No. of reports	*kg N ha^{-1} a^{-1}*
Forage Legumes Grown Alone or with Grass		
Temperate		
Various clovers (*Trifolium* spp.)	17	23-620
Lucrene and swee clover	8	164-463
Tropical		
Control	1	> 112
Stylo	1	30-196
Tick clover (*Desmodum* spp.)	1	700
Glycine wightii	1	160-450
Grain Legumes		
Temperate		
Veteh and tick beads	4	7-90
Peas	1	46
Lupins	1	128
Tropical		
Fenugreek	1	44
Lentil	2	35-77
Pigeon pea	2	41-90
Cowpea	2	73-240
Soybean	6	17-124
Guar	1	37-196
Groundnut	3	33-111
Canavalia	1	49
Chick pea	2	41-270

Main Reasons for Decline in Legumes Area

- Government's focus of support onto other cereals
- Lack of superior technology for legumes

- Biotic constraints (diseases and pests) in legumes
- Abiotic constraints (soil salinity, waterlogging and frost)
- Low productivity of cultivars

Table 5 : Adoption Constraints of Growing Pulses in India

Constraints	*(%)*	*Score Rank*
(A) Technical		
(i) Non-availability of quality seed material	98.23	II
(ii) Non-availability of sulphur based phosphatic fertilizer for balanced nutrition	94.35	III
(iii) Incidence of pests and diseases	97.52	I
(iv) Non-availability of appropriate plant protection chemicals	91.45	IV
(v) Infestation of weeds	59.47	V
(vi) Lack of proper knowledge about seed treatment	53.62	VI
(B) Socio-economic		
(i) Non-availability of inputs in time	49.32	VII
(ii) Non-adoption of proper crop rotation	41.62	VIII
(iii)Lack of knowledge about scientific cultivation practices	40.44	IX
(iv) Difficulty in assessing credit facilities	39.51	X
(v) Increasing cost of input	36.72	XI
(C) Agro-ecological		
(i) Occurrence of flood/water logging	31.33	XIV
(ii) Poor soil quality	33.47	XIII
(iii) Aberrant weather	35.25	XII

(Singh *et al.* 2007)

c) Low Profitability of Legumes

Profitability was the most important criterion for allocating area to alternative crop choices and this was largely influenced by cost of production, crop yields and output prices. Despite low cost of cultivation of legumes, the profitability of different legumes did not consistently match that of rice and wheat. However, berseem was more profitable than wheat but it was solely grown for fodder and its area expansion was restricted due to lack of market demand. Lower net profit of legumes when compared with that of rice and wheat was mainly due to their poor yield performance. However, output prices of all legumes were much higher than those of rice and wheat. Yields of legumes were so low that higher output prices could not make them more profitable than rice and wheat. The output price of pigeonpea was double that of rice whereas the yield of rice was four times higher than that of pigeonpea. Similarly, the output price of chickpea was double that of wheat, but wheat yield was 60% higher than chickpea yield.

Yield of some grain legumes (dry) 4-averages 100 kg/ha

	U.S.A	Australia	India	Ethiopia	Egypt	UK	Netherlands
Glycine	18	-	-	6	-	-	-
Arachis	23	12	7	5	21	-	-
Phaseolous	13	7	3	7	15	-	-
Pisum	17	7	8	9	14	27	33
Vicia	-	9	-	9	20	27	24
Cicer	-	-	6	6	16	-	-
Vigna	6	5	-	-	18	-	-

Historical trends in the prices of legumes, rice and wheat indicated that the minimum procurement prices of all legumes announced by the government were always kept higher than those of rice and wheat. During the past three decades, yields of legumes were always substantially lower than those of rice and wheat in the states of Haryana, Punjab and Uttar Pradesh. Yields of rice and wheat increased much faster than legumes in these states. A yield breakthrough in legumes was not realized as in rice and wheat. Although several improved cultivars of various legumes were developed, but not widely disseminated among the farmers.

d) Legumes Cultivation: A Risk to Farmers?

The yield and price risks is hindering adoption of legumes among the farmers who considered legumes production as relatively more risky than rice or wheat due to their poor stability against environmental vagaries. The coefficients of variation in yields grain legumes greater than wheat and rice in most districts in IGP of India which suggests that legumes were more prone to risk due to crop failure when compared with rice and wheat . Similarly, price fluctuations (post and pre-harvest) in grain legumes much higher than rice and wheat.

e) Profitability of Alternative Cropping Sequences

The profitability of the rice-wheat cropping sequence in relation to other alternative cropping sequences may depends under the following options:

(1) Existing prices of fertilizers and electricity charges paid by the farmers for irrigation;

(2) Without electricity subsidy for extraction of ground water for irrigation; and

(3) Without fertilizer and electricity subsidy for irrigation.

The most profitable crop sequence was rice-wheat-black gram followed by rice-berseem and rice-wheat-mungbean. The adoption of these three crop sequences was limited due to resources and market constraints. Cultivation of black gram and mungbean requires much water after the harvest of wheat, whereas berseem area expansion was restricted due to limited market determined by livestock population. Profitability of rice-wheat, the most popular cropping sequence is higher than that of rice-chickpea and pigeonpea-wheat. Even if the existing subsidies on fertilizer and electricity for irrigation are withdrawn, the rice-wheat system will still the most profitable cropping sequence. Substitution of legumes for rice or wheat means loss in earnings of farmers. To introduce legumes in the rice-wheat system, profitability of legumes needs to be raised substantially. This is possible through a substantial increase in their yield levels, which could be attained through dissemination of appropriate technologies to farmers.

f) Trade-off (% change) in Replacing Rice or Wheat with Legumes in Karnal district during 1996/97

Indicator	Pigeonpea	Chickpea	Lentil	Berseem
Profit	-49	-19	-41	+2
Food grain	-76	-64	-76	
Fixed resources	-57	-49	-61	-43
Groundwater	+95	+85	+83	-125
Soil nutrients	+65	+73	+75	+56

*In rice-wheat system, rice was substituted by pigeonpea and wheat by chickpea, lentil, or berseem

The trade-off values studies by replacing rice by pigeonpea indicated that farmers would lose 49% profit. The IGP region would need to sacrifice 76% food grain production and 57% of the fixed resources would remain unutilized. However, the region would save about 95% of the groundwater and 65% of the nitrogenous fertilizer. Similarly, trade-off was also observed when wheat was substituted by chickpea or lentil. But trade-off between wheat and berseem (a fodder legume) indicated negligible loss in profit, despite more groundwater being used for berseem than wheat. Thus substitution of wheat by this fodder legume would mean further over-exploitation of groundwater. When food grain legumes substituted rice and wheat, there was a loss in profit, food grain production, and use of fixed resources.However, there were substantial gains in conserving groundwater and nitrogenous fertilizers.

Prospects/Achievements of Grain Legumes in Resources Conservation, Food and Environment Security

The existing low yield levels of legumes will further displace them from the areas of intensive cropping system of rice-wheat rotation but their importance in resources conservation, food and environment security can not be underestimated . Therefore, more resources should be allocated for research to break yield barriers, and design innovative policies on risk and resource management. The Strategies to reduce legumes production risk may be targeted to: 1) Develop the high-yielding varieties, 2) Create assured output markets, 3) Reduce post-harvest losses and costs, 4) ?Develop appropriate crop production technologies.

Though, many improved cultivar of grain legume with increased yield and resistance against biotic and abiotic stresses has been developed recently under AICRP(Pulses) program but their dissemination and adoption among the farmer needs urgent and utmost priority. The improved grain legumes cultivars are showing promises with increase productivity (4 t/ha) with resistance against major diseases & pests along with suitability for various cropping system, mechanical harvesting (Lather,2000) and better profitability (Hedge,1992). Therefore, grain legumes areas need to be increased to harnesses their overall beneficial effects in cropping system and to alleviate the malnutrition of poor masses.

Table 6 : Production & profitiability of legumes in various cropping systems

Location	Cropping System	Av. Yield (Kg/ha) Kharif	Rabi	Net Returns (Rs/ha)
Pantnagar	Rice-Wheat	3945	4207	11218
	Rice-Lentil	4299	1506	14445
Varanasi	Rice-Wheat	3648	3419	8727
	Rice-chickpea	3838	1407	11770
Kanpur	Rice-Wheat	3614	3560	9095
	Rice-chickpea	3797	1803	13815
	Black gram-wheat	1330	3855	11861
Ludhiana	Rice-Wheat	6111	4372	17271
	Rice-Pea	6481	2064	22955

References

Hedge, D.M. (1992). Cropping systems research highlights coordinator report. In: Proc. 20th Workshop.TNAU Coimbatore, 1-4 June, 1992.

Joshi, P.K., M. Asokan, K.K. Datta and P. Kumar. (2000). Socioeconomic Constraints to Legumes Production in Rice-Wheat Cropping Systems of India. pages 176-184. *In:* Johansen, C., J.M. Duxbury, S.M. Virmani, C.L.L. Gowda, S.Pande and P.K. Joshi (eds). Legumes in Rice and Wheat Cropping Systems of the Indo-Gangetic Plain:Constraints and Opportunities. ICRISAT, Patancheru, India.

Lather V.S. (2000) Promising chickpea ideotype for higher plant density. ICPN: 7:26–27.

Samra, J.S. (2008). 'Combating land degradation for sustainable agriculture', World Day to Combat Desertification at the CII today.

Singh, S.N. Singh V.K., Singh R.K. and Singh R.K. (2007). Adoption Constraints of Pigeonpea Cultivation in Lucknow District of Central Uttar Pradesh Indian Res. J. Ext. Edu. 7 (1): January 2007.

www.knowledgebank.irri.org/.../socioeconomic/constraints/legumes/production...: Down to Earth Vol: 11 Issue: Jan 2003.

□□□

Section - 4

Livelihood Security and Watershed Management

Chapter 21

Land Use Planning and Watershed Management for Livelihood Security

J.S. BALI

Professor of Soil and Water Conservation at IIT Kharagpur and Head of Soil, Conservation Division of the Ministry of Agriculture, 215/6 Rajpur Road, Dehradun

We are concerned here with 1100 million people of India, with 800 million of them in rural areas. As many as 600 million are engaged in agriculture directly and the rest of rural population is also indirectly dependent upon agriculture. Land resources comprise 329 m ha of geographical area, with only 305 m for reportage, the rest under occupation of our "good" neighbours. We have 183 m ha of cultivable land, with net sown annual area of 142 m ha; 42 m ha under non-agricultural uses or barren ; 70 m ha under forest; and only 10 m ha of grasslands for about 500 million domestic animals.

But 145 m ha of land is degraded due to soil erosion of various types and degrees and saline, alkaline, gullied or other problems. Our average yield of 30 q of paddy per ha is half of China and one-third of Egypt. For wheat, our yield of 27 q is two-third of China, half of Egypt, one-third of France, but creditably, just about the same as in USA. We have 60 m ha under irrigation and 40 m ha under dry farming which are subject to frequent droughts. This year (2009) alone, 246 districts out of 623 have been declared drought-affected. Land resource is obviously fixed, but population is rising, and per capita land availability is shrinking.

The challenge is to adopt such practices of land use in India, which produce the maximum sustainable yields while conserving the capital resource of land. Farming is without profits. Youth want to run away from farming. From tiny land holdings, there are limits to increase of income. The challenge is to bring industry to upstream areas within integrated Bioindustrial Watersheds, with farmers themselves owning the processing industry for value addition, higher incomes and livelihood security from the small land holdings.

The challenge is to create off- farm jobs to decrease pressure on the land. The challenge is to reduce the number of cattle and improve their breeds and productivity. In short vertical increases are required from each hectare of land by knowledge, higher inputs, credit facilities and better marketing of the processed produce.

And finally, the land use planning must eradicate poverty which has made a permanent home in rural India. Presence of endemic poverty even after 62 years of independence is the biggest failure of the people and their government in India.

The urgent task of our researchers and research station should be to take up watersheds and villages and convert them into Bioindustrial Watersheds and Villages to manage natural resources and provide livelihoods to our millions in rural areas, we have been left behind in the onward march of the Indian economy.

Defective Definitions

Whatever the land capability we prescribe , based upon the soil and land characteristics, farmers are not going to change their land use, nor the forest department will yield any land for agriculture. So the emphasis or very definition of land use planning in India should shift from land use to land management. Soil and Water Conservation should be considered infrastructure of any land use. Land must not be allowed to deteriorate under intensive use. And sustainable development in India should mean permanent and stable elimination of poverty and bringing about permanent prosperity.

Planning for Profitable and Sustainable Land Use and Livelihoods

The basic pattern which will eradicate rural poverty and bring in rural prosperity would be Bioindustrial Watershed Management. India should be divided into micro watersheds of 500 ha or so and for each watershed or a cluster of them, the following components of a Bioindustrial Watershed Management Project should be implemented:

1. Micro development of surface and ground water and provision of supplemental irrigation for each hectare of agricultural land.
2. Land leveling for proper water conservation and crop stands.
3. Grassland development for maximum sustainable fodder yields, by rotational use and deferred grazing etc.
4. Contour ditching for moisture enhancement followed by afforestation of non-agricultural top lands; and better forest management related to the rural people's needs of fuelwood, fodder, and drinking water.
5. Sustainable high production from plant and animal resources by adopting better systems of cropping and animal husbandry.
6. Processing of the farm and forest produce, for value addition, through bio-industries owned by the people's group.
7. Storage and better marketing facilities, since this is where the farmer is finally beaten.
8. Credit facilities which eliminate exploitation which is at present leading farmers to suicide even in the so-called agriculturally prosperous areas of Punjab and Maharashtra.
9. Research, evaluated by poverty reduction impact, rather than the number of publications in journals of India and other advanced countries.
10. And most important of all , by providing rural infrastructure of drinking water, power, roads and communications.

Basic Field Practices for an effective Land Use Plan

1. *Agricultural Land*: Ponds and shallow tubewells or pumping from a well or a nearby perennial stream, should constitute the priority programme. Bench terraces are a must for mountain lands. Land leveling and ground water development have been mentioned above already. The repetition here emphasizes their vital importance.

2. *Grasslands* : The poor depend upon animals. Animals depend upon grazing land, which is scarce. Grazing lands are the most neglect today. There is hardly a scheme in the field for their improvement. These lands must be rehabilitated through better yielding and nutritive grass-legume mixtures as developed by the Grassland and Fodder Research Institute of Jhansi. They should not be diverted to fuel plantations of Jetropha, as at present being done in some States (specially Chhattisgarh).

3. *Forest lands* : Present forest land should be managed for providing fuel, fodder, timber and other products for the villagers. Present environment of conflict between the people and "magisterial" forest management should be replaced by benign, people-friendly technology-based forest management. People owned these forests. They were snatched away by the colonial power, without compensation.

4. *Horticulture land*: they need contour ditches and round bench terraces for water conservation and better plant growth.

5. *Plantation crop land*: Rubber plantations would need narrow (2 m-wide) bench terraces. Tea gardens need contour ditches, and water storage practices.

6. *Mining land* : Mandatory land rehabilitation should be provided in the mining acts. Mined land can be restored to agriculture where conditions are favourable. Preferably, abandoned mine lands should be put under grasses and forest. Abandoned deep holes filled with water can be used for fisheries. Most important of all is to eliminate forest mafia which is dominating the mining lands today. Forest fires, raging widely in Dhanbad-Jharia region, must be controlled. Planned explosions can starve the fire of oxygen and it gets extinguished.

7. *Roads and railways lands*: Road fills must be put under perennial grasses and bushes and suitably protected by tce retaining walls. Tree planting should be done with care, as sometimes the trees add to the unstable mass if the micro-geology is unfavourable. Idle railway lands can be leased out for farming with provisions of proper bench terraces and irrigation in the borrow-areas.

8. *Saline lands* ; Saline land needs sub-surface drainage. Cost is to be kept in mind. Practices like deep furrowing enable living with salinity sometimes. Un-controlled irrigation brings salinity. India

and Pakistan are victims of the old Irrigation Projects. Irrigation and drainage must go together.

9. *Alkali lands*; Level the land, treat with gypsum, fill clean water, and drain away after a few days to a nearby natural or planned pond or safe drainage line. Cltivate pzddy for a year or two.
10. *Acid and acid–sulphate lands*: Living with acidity has to be learnt. Liming can help, although this practice is still limited in India, though widely practiced in USA and Europe. Acidity is not yet fully recognized as a problem , though it is . Yields can go up if acidity is controlled.
11. *Ravined and gullied land*: Construct a diversion ditch to divert the runoff from table land into a drop structure and the waterway ravine. The shallow and medium gullies below the diversion ditch should be converted into bench terraces and village lands restored to original owners. Deep ravines can be afforested and grassed. Wide beds of deep ravines can also be put under bench terraced cultivation with spillways, specially towards the end of the ravine where it drains into the stream below. Successful examples of this pattern have been created, with tubewells constructed, in some ravinous districts of U.P.
12. *Land slides in the mountains*: Carefully constructed contour ditches and planting of trees, bushes and grasses, can heal the slide face. Important thing is to study the micro-geology of each slide and then provide drainage, including subsurface drainage if necessary. Operations are costly. Mere closing to grazing can also help over a time. Costs and end-value should be carefully examined before spending money. Kaulagar land slide control in Dehradun serves as a good example.
13. *Stream bank erosion* : Revetments of stones or/and bushes like *Arundodonad, Jetropha, Vitex negandu, Ipomea cornea, Ziziphus,* and others measures , can control stream bank erosion. Spurs of vegetation or stones-vegetation with sproutable wooden posts are effective. Toe of the spur is where it will erode and fail. Make it deep; also keep it under watch.
14. *Land under Brick–Kilns*: The first attack should be on the system itself. No fertile agricultural land should be allowed to go under Brick kilns. We must learn to switch to alternate construction materials like fly-ash or concrete blocks. Land is more important than bricks.

15. Myriad problems and myriad local solutions: No account can cover all field situations. Follow the principles of simple hydraulics and design the field practices needed for the particular type of land and the soil.

Some urgent concerns of Livelihood Security and Watershed Management in India are discussed below :

Land Acquisition and Rehabilitation of Affected Families

Industry needs land. Urban expansion needs land. Roads and railways need land. India has witnessed violence on the issue. Guidelines or an act needs to be prepared. Govt of India has come up with a draft, though it has not yet been presented as a Bill in the Parliament. Rehabilitation of the ousted people must be adequate. Adequate should mean better incomes and employment than before. Land for land should be given where feasible. Land must be provided with irrigation and soil conservation measures before allotment. Otherwise the rehabilitated people will abandon the new land.

Land Reforms and Aggregation of Land Holdings

Too much disintegration and scattered-ness of land holdings has occurred. Land use planning must become mandatory. Inheritance laws need to be changed so that a minimum consolidated size of land holding is assured to a family. Time has come to begin aggregation of land into bigger units with less number of farmers and more people diverted to off-farm jobs. Bioindustrial watershed management pattern can enable this.

Organization

Of late, soil conservation organizations in the States and even at the Centre, have been weakened, while land degradation problems need continuous attention. Recently, the Ministry of Agriculture took a decision to further weaken the land use planning organization, by abolishing, in one stroke, all the State Land Use Boards. Actually, they have been taken off the centrally financed pattern of projects. This amounts to abolition, since few States could perhaps find the funds to continue them. If some of them were inactive, reasons could have been found through a study and corrections made. This organization is charged with land use planning. It has been abandoned without due thought of the consequences for an agricultural country, which India still is, because more than 60 per cent of our working force continues to be dependent upon agriculture and the land.

Livelihoods in the villages should mean full, year-long employment for all workers, with net income which would lift them from poverty to prosperity.

It has to be remembered that soil and water conservation (which means effective land use management) is an **action programme,** and **not an extension programme.** Those States which have diluted the soil conservation by integration with extension , have made this vital service to the farmers, weaker.

Drought-proofing of land

Intensive micro-water development, provision of vegetative cover for non-agricultural land, construction of wells and tubewells along with recharge works, proper cropping systems and dry-land agricultural methods, can drought-proof a drought-prone area. It is time, the Government provides funds for emergency drought-proofing rather than drought relief only.

Suicides and Subsidies

Farming is an essentially profit-less or low-profit profession. But it produces food—the vital need of man and his animals. No one will go for farming when it drives one into poverty, debts and suicides. In a recent survey, it was found that 40 per cent of the rural youth in Punjab, do not want to pursue farming. Farming has to be subsidized. It is subsidized almost everywhere in the world. USA and Europe provide about 30 to 50 per cent of subsidies. France provides 80 per cent. India provides less than 10 per cent. Farming has got to be subsidized if food prices have to be kept low for the urban poor and the rural landless. There is no escape from it. Unfortunately the media in India does not still appear to understand these rural realities.

Conclusion

Finally, a word on research in land use planning and soil and water conservation. We must not blindly follow the un-applicable land capability classification of USA. Our conditions are different. Research must shift its emphasis to impact made on poverty reduction or eradication rather than publications in foreign journals. Direct benefits to India's poor farmers should be the criterion of impact assessment.

Farming is profitless. It has got to be subsidized by the people and their governments if they want food. Either raise prices or subsidize farming. Net profits, at present missing from farming, can be brought back through Bioindustrial Land Use Planning and Bioindustrial Watershed Management. Success in proper planning and effective execution of integrated schemes would work in this field of livelihoods, only when the concerned workers and departments of agriculture, rural development, forestry, water resources, micro and small industries, food processing, and panchayats learn to work not only in a coordinated way but as an integrated unit and in integrated field projects.

❑❑❑

Chapter 22

Sustainable Management of Cropping Systems in Central Indian for Food and Environmental Security

P. K. PANDA, S. K. DUBEY AND A. K. SINGH
Central Soil and Water Conservation Research and Training Institute
Research Centre, Chhalesar, Agra-282006

Central India comprising the states of UP, MP, Bihar, Jharkhand and Chhattisgarh play an important role in ensuring food and environmental security of the country. It covers the total geographical area of 85.88 million hectares which constitutes approximately 26 per cent of the country's total geographical area. This region is highly populated with a density of 452 persons per square kilometer as compared to 325 persons for the entire country. The average per capita income in this region is one of the lowest i.e. Rs 11880.00 at current prices compared to national figure of Rs 25825.00. This region is endowed with diverse physiography, climate, soil types and rainfall pattern. The cultural and socio-economic diversity add flavor and aroma to the above aspects in creating a complex inter related matrix of natural resources, human as well as animal population and their socioeconomic interactions. Under these circumstances, it is very urgent and appropriate to understand the role of profitable criopping systems for ensuring food and environmental security in central India along with natural resource conservation on a sustainable basis. Before advocating sustainable cropping system, one need to understand the soil and climate type, rainfall pattern(intensity and duration), length of growing season, food habit of the people, nature and quantity of livestock population, local resource availability, socio-economic status of the people and the market demand of the produce. In this paper a careful thought has been given to illustrate various options available to consider under different cropping systems for natural resource conservation and food security on sustainable basis.

Table 1: Demographic and socio-economic characteristics of various states in central India

S.No.	States	Area (Million hectare)	Population (Crores)	Density (Person/ Km2)	Per capita income At current price(Rs)	Literacy (%)
1	Bihar	9.41	8.29	881	5772	47
2	Chhattisgarh	13.60	2.09	154	15073	65
3	Jharkhand	7.97	2.69	338	13013	54
4	Madhya Pradesh	30.80	6.03	196	14069	64
5	Uttar Pradesh	24.09	16.61	690	11477	56
6	Central India	85.87	35.71	452	11880	57
7	India	329	102.87	324	25825	65

Source : Census of India, 2001

The term food security refers to access of safe and nutritional food for all the people at all the time to maintain healthy and active life. It has three important dimensions to fulfill. Such as

- Production
- Stability and
- Accessibility

Production

There should be sufficient production at regional, national and international level. It is the first requisite of food security.

Stability

Only sufficient production will not ensure food security to the people. There should be stability in production over the years and across the regions. It is the second requisite of food security.

Accessibility

Only sufficient production and stability cannot ensure food security to the people. There should be physical as well as economic access of the people to food. That means all the people should have purchasing power and food should be available in the open market at their door step. This is the third and important requisite of food security.

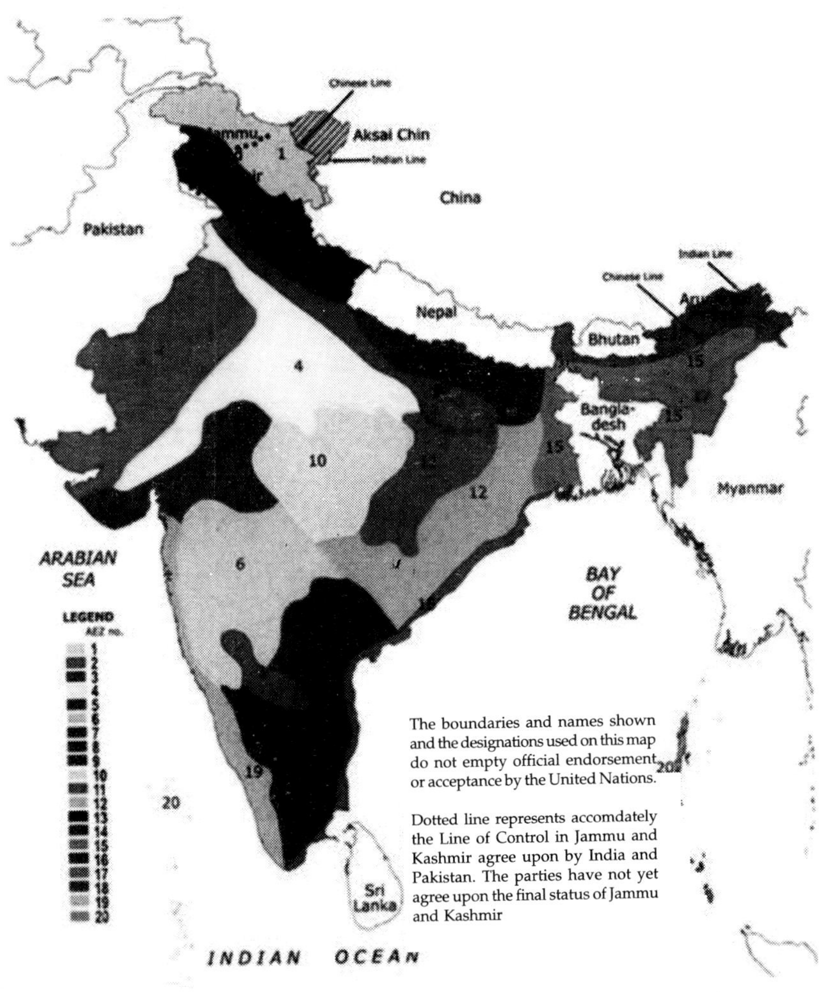

Earlier, central India comprising the states of UP, MP, Bihar, Jharkhand and Chhattisgarh have been put under four different agro-climatic zones. Such as Upper Gangetic Plain zone, Middle Gangetic Plain zone, Eastern Plateau and Hills and Western Plateau and Hills. Later on Food and Agricultural organization, United Nations developed the concept of agro-ecological regions by calculating the length of growing season, rainfall, PET and soil characteristics of a particular area. Accordingly, central India comprising the above mentioned five states can be put to ten different agro- ecological zones as follows(Anonymous,2011)

The detailed study regarding various cropping systems have been done by Yadav and Subba Rao in 2001 at Project Directorate of Cropping System Research, Meerut. State wise major and minor cropping systems practiced by the farmers of central India are listed below.

Major and minor cropping systems of Uttar Pradesh

Major cropping systems

- Rice-wheat
- Maize-wheat
- Pearl millet-wheat
- Fallow-gram
- Sugarcane-wheat

Minor cropping systems

- Pigeon pea-wheat
- Fallow-mustard
- Fallow-pea
- Fallow-lentil
- Groundnut-wheat
- Maize-mustard
- Finger millet-wheat
- Pearlmillet-mustard
- Pearlmillet-groundnut
- Rice-gram

- Rice-pea
- Rice-potato
- Rice-mustard
- Rice-sesame
- Sorghum- wheat

Major and minor cropping systems of Bihar and Jharkhand

Major cropping systems

- Rice-fallow
- Rice-wheat
- Jute-rice
- Maize-maize

Minor cropping systems

- Rice-lathyrus
- Rice-lentil
- Rice-linseed
- Rice-mustard
- Maize-wheat
- Rice-other winter pulses(paira cropping)

Major and minor cropping systems of Madhya Pradesh and Chhattisgarh

Major cropping systems

- Cotton–fallow
- Rice-wheat
- Rice-fallow
- Soyabean-wheat
- Soyabean-gram
- Rice-lathyrus
- Soyabean-fallow
- Fallow-mustard
- Fallow-gram
- Fallow-wheat

- Finger millet -fallow
- Maize-fallow
- Maize-gram
- Maize-wheat

Minor cropping systems

- Pearl millet- mustard
- Groundnut-mustard
- Sorghum-wheat
- Rice-chickpea
- Rice-linseed
- Sorghum-fallow
- Black gram-wheat
- Rice-niger

Table 2 : Productivity comparison of major crops in central India

S.No	Crop Yield (Kg/ha)	Bihar	Jharkhand	Madhya Pradesh	Chhattisgarh	Uttar Pradesh	India
1	Rice	1486	1828	824	1354	1879	2131
2	Wheat	1908	1529	1835	-	2721	2708
3	Maize	2671	1230	976	-	1335	1912
4	Jowar	-	-	1039	-	1030	844
5	Bajra	-	-	1365	-	1455	886
6	Gram	818	-	980	843	742	845
7	Pegion pea	989	645	683	--	749	650
8	Lentil	705	-	497	-	705	621
9	Soyabean	-	-	1006	-	-	1063
10	Mustard	1029	-	999	-	1057	1095
11	Groundnut	-	-	948	-	730	866

Source : Agricultural statistics at a glance, 2008

The evolution of the above cropping systems are not arbitrary one. These have been evolved through centuries based on domestic requirement, easy in marketing, peresability, use of knowledge, use of own implements and tools, low external inputs, wider choice of the crops and largely it was ecology based. As the demand for human as well as their animals increased, cropping systems were changed in the direction of high monetary return, high external input oriented, narrow choice of crops and against the spirit of the nature.

Earlier the impact of these cropping systems were invisible due to traditional and low intensive form of agriculture. However, to meet the demands of ever increasing population and their live stock, more intensive form of agriculture was adopted exhibiting serious threat to the sustainability of our natural resource base. As a result , about 57 % of the total geographical area of the country is suffering from various degradation hazards. Out of the total canal irrigated area, 3.4 million hectares are suffering from water logging and soil salinity (Singh, 1998). It has been estimated that a total of 5000 million tonnes of top soil is being eroded each year. The per capita availability of water in India will go down from 2203 m^3 in 1991 to 1340m^3 in 2025 creating a shortfall of 27.4%. By 2000, most of our soils have been found to be deficient with respect to nitrogen, phosphorus, potassium, magnesium, sulphur, iron, zinc, boron and molybdenum (Chhonkar and Rattan,2000). Low input use efficiency is another important factor in increasing input wastage and production cost. Most of the irrigation projects are running with a low efficiency of 30- 40%. The soils of the country are facing a negative plant nutrient balance of 8-10 t/ha. In addition to this we are also losing approximately 8 million tonnes of plant nutrients per year through soil erosion. The present level of nutrient use efficiency is only 30-40% leading to a huge wastage of resources. Indiscriminate exploitation of ground water without adequate recharge has resulted in depletion of ground water @ 0.5-0.8m/year in Haryana, 0.2-1.0 m/year in Punjab and 0.63- 1.13 m/year in western Uttar Pradesh. Lack of adequate recycling of organic manure like FYM, compost and crop residues have resulted in lowering of soil organic carbon to considerable extent. Due to non adoption of basic principles of crop rotations, the insect pest, disease and weed problem have emerged significantly threatening our food and environmental security. The exhaustive –exhaustive crop rotations like cereal-cereal based have not only created nutrient imbalances but also reduced the total factor productivity.

Steps need to be taken for meeting the challenges of environment and food security.

Table 3 : Agro-ecological regions of central India with their characteristics

AEZ No.	Agro-ecological region	Geographical area (million ha)	Gross cropped area (million ha)	Physiography	Precipitation (mm)	PET (mm)	Length of growing period (days)	Major crops
1.	Hot arid ecoregion with red and black soils	4.9 (1.9%)	4.18	Deccan Plateau	400–500	1800–1900	< 90	Sorghum, safflower, cotton, groundnut, sunflower, sugar cane
2.	Hot semi-arid ecoregion with alluvium-derived soils	32.2 (9.8%)	30.05	Northern Plain & Central Highlands including parts of Gujarat Plains	500–800	1400–1900	90–150	Millets, wheat, pulses, maize; irrigated cotton & sugar cane
3.	Hot semi-arid ecoregion with medium and deep black soils	17.6 (5.4%)	11.04	Central (Malwa) Highlands,Gujarat Plains & Kathiawar Peninsula	500–1000	1600–2000	90–150	Millets, wheat, pulses
4.	Hot semi-arid ecoregion with shallow and medium (dominant) black soils	31.0 (9.5%)	25.02	Deccan Plateau	600–1000	1600–1800	90–150	Millets, cotton, pulses, sugar cane under irrigation

5.	Hot subhumid (dry) ecoregion with alluvium-derived soils	12.1 (3.7%)	11.62	Northern Plain	1000–1200	1400–1800	150–180	Rice, wheat, pigeon pea, sugar cane, mustard, maize
6.	Hot subhumid ecoregion with red and black soils	22.3 (5.8%)	14.55	Central Highlands (Malwa & Bundelkhand)	1000–1500	1300–1500	150–180	Rice, wheat, sorghum, soybean, gram, pigeon pea
7.	Hot subhumid ecoregion with red and yellow soils	11.1 (4.3%)	6.47	Eastern Plateau (Chhattisgarh Region)	1200–1600	1400–1500	150–180	Rice, millets, wheat, pigeon pea, green gram, black gram
8.	Hot subhumid ecoregion with red and lateritic soils	26.8 (8.2%)	12.09	Eastern (Chhota Nagpur) Plateau and Eastern Ghats	1000–1600	1400–1700	150–180	Rice, pulses, millets
9.	Hot subhumid (moist) ecoregion with alluvium-	11.1 (3.4%)	10.95	Eastern Plains	1400–1600	1300–1500	180–210	Rice, wheat, sugar cane
10.	Hot semi-arid ecoregion with red and black soils	16.5 (5.2%)	6.19	Deccan (Telangana) Plateau & Eastern Ghats	600–1000	1600–1700	90–150	Millets, oilseeds, rice, cotton & sugar cane under irrigation

Agro – ecological zone approach

Agro-ecological approach in selection of crops and cropping system should be followed depending upon the length of growing season of a particular region.

Widen the choice of crops in our cropping systems.

Worldwide we have 30000 edible plants out of total 2,70,000 plant species available. Out of these 30000 edible plant species, only 7000 plants are used as food. Out of these 7000, only 120 plants are cultivated. Out of these 120, ninety plants contribute to only 5% of global food, where as other 21 plants contribute 20%. It is ironical to see that 9 plants contribute to 75% of total global food (FAO,1997). Earlier, we were considering millets and pulses to be the integral part of our cropping systems. But the development of irrigation facilities led to large scale cultivation of dominant crop like rice and wheat.

Incorporation of organic manure

In the waste lands, degraded lands, village community lands and in the field bunds , the fire wood plants should be planted , so that the precious cow/ buffalo dung could be made available for application in the field to maintain soil physical, chemical and bio-logical properties instead of burning them through dung cake. Similarly the potential of green manuring crops and bio-fertilizers should be tapped to the fullest extent (Singh, 2001).

Principles of crop rotation

Important principles of crop rotation should be followed for maintaining sustainable production system. The cereal -pulse crop rotation is not only required for nutrient balance point of view , but its relevance is much more felt to curb the increased menace of insect pest and diseases along with weed infestation.

Conservation technology application

The varietal selection should be based on low nutrient and water requirement with multiple resistance against biotic and abiotic competition. Feasible energy conservation technology should be adopted during mechanical manipulation of soil.

Enhancing input use efficiency

The efficiency of various agro-inputs like fertilizers, manures, pesticides and water etc. should be increased not only to reduce wastage but also for increasing farm profit.

Watershed approach in rainfed areas

The entire rainfed areas should be managed in the principles of watershed for holistic development of the village through agriculture, horticulture, forestry, animal husbandry and water resources in a sustainable manner giving adequate attention to natural resource conservation for the future generation.

Market linkage

The success of a sustainable cropping system mainly depends upon the economic return at the end it fetches to the adopters. If adequate market linkage for selling of the produce at the prevailing market price is not provided in time , farmers will not adopt those recommended crops and cropping systems in large scale.

Policy interventions

Various governmental policy intervention like minimum support price (MSP), increasing procurement price of non traditional crops like millets, pulses etc and investing in their value addition will help to provide higher return to farmers and hence it would lead to large scale adoption. Steps should be taken to remove excessive involvement of middle men in procurement and selling of farm produce so that the producers share in consumer price will be more.

After a thorough review of literature and careful considerations of agro-ecological constraints, the drawbacks of the exiting cropping systems have been analysed and the guidelines for developing sustainable cropping systems for various parts of central India have been worked out which will be not only profitable to farmers but will also be socially acceptable and environmentally sustainable in the years to come.

References

Agricultural statistics at a glance.(2008). Directorate of Economics and Statistics, Department of Agriculture and Cooperation, Ministry of Agriculture, Government of India, New Delhi.

Anonymous, (2011). Important features of agro-ecological zones of India, Water Resources System Division, National Institute of Hydrology, Jalvigyan Bhawan, Roorkee-247667.

Census of India. (2001). Directorate of Census, Ministry of Home Affairs, Government of India, New Delhi.

Chhonkar, P. K. and Rattan, R. K. (2000). Soil fertility management for sustainable agriculture. Indian Farming **49**(11) :26-31.

FAO.(1997). Food for all – success stories in the battle for food security. World Food Summit Report. Food and Agriculture Organization of the United Nations. Rome , Italy.

Singh, G. B. (1998). Natural resource management for sustainable agriculture in 21st century. Indian Farming **48**(5):7-12.

Singh, P.(2001).Sustainable crop production- limitations and strategies. Indian Farming 50(10) : 7-10

Yadav, R. L. and Subba Rao, A. V. M. (2001). Atlas of cropping systems in India. PDCSR Bulletin No. 2001-2, Project Directorate for Cropping System Research, Modipuram, Meerut-250110, pp96.

□□□

Chapter 23

Impact Evaluation of Soil and Water Conservation Measures in Tribal Area of District Satna, (M.P.)

V.P. SINGH, V.K. KANAUJIA[1] AND S.N. SINGH[2]
Krishi Vigyan Kendra (Deendayal Research Institute), Satna -485331 (M.P.) [1] Programme coordinator, Krishi Vigyan Kendra, Farrukhabad
[2] Reader, P.P.N. Degree College, Kanpur

Land and water together support plant and animal life. Conservation of these basic resources is the key of food security and food supply. Water is the prime mover in agriculture development, especially in rainfed areas. Although India receives a total precipitation of about 400 million cubic meters but much of it is lost through evaporation and run off (Singh, 1998). This is not only the water wastage, but also creates the problems of soil erosion and land degradation. Such problems are more serious in hilly and ravine areas. Watershed technology augment soil and ground water resources. It also enjoins the biophysical, social and economic inputs, which is optimally managed lead diversified high agriculture production, control of environmental degradation and provide a mechanism for the recharge of ground water aquifers. Impact evaluation of dry land technologies is essential to know an overall impact of soil and water conservation measures adopted in the particular watershed. It is also helps carrying out the project activities and to estimate the medium and long term social and economic benefits of the activities, efficiencies and impact of the project in context of its stated objectives (Gupta *et. al.,* 2000) Keeping in the view, study was undertaken to evaluate the impact of soil and water conservation measures in *Milli watershed Krishi Vigyan Kendra, Majhgawan,* Satna (M.P.).

Materials and Methods

Characterization of watershed and its problems

The study was conducted at Milli Watershed Krishi Vigyan Kendra, majhgawan, situated in the Vindhyan hill range. It extends from 24° 51′15″ to 24° 57′30″N latitude and 80° 43′30″ to 80° 54′ 15″ E longitude in Satna district (M.P.) at about 45 kms away from district place. The total area of the watershed is 12536 ha covering 18 villages. It is characterized by its backwardness, poverty, illiteracy, soil erosion and low crop productivity along with other soil economic constraints. Annual rainfall of the area varies from 800 to 1100 mm. About 90per cent of the total rainfall received during monsoon. There is no dependable source of water in the area. The maximum and minimum temperature of the area is 48°C and 13°C, respectively. The Soil in the plains is sandy loam to loamy in texture. It is shallow in depth, poor in organic matter content and other plant nutrients. The soil's moisture holding capacity is low, so it is only able to support crops of an inferior nature under rain-fed conditions. During the high rainfall months, water flows freely on the ground surface, due to poor percolation and the compact nature of the soil. Details of the land use pattern of the watershed area given in the Table 1.

Conservation measures

Various treatments planned and executed in the watershed area were broadly classified in three categories.

Mechanical measures

Mechanical measures usually involve construction of mechanical barriers across the direction of flow of rain water to retard or retain the runoff and thereby reduce the soil and water loss. These measures, as practiced in India, include contour cultivation, contour bunding, farm bunding, bench terracing, trenching, and retention or detention reservoirs. The important principles are to be kept in view at planning of mechanical control measures are:

Table 1 : Land use, soil characteristics and its problems

Land use	Area (ha)	%	Soil Texture	Problem
Forest	4913.53	39.19	Rocky, stony and murram.	Degraded, steep slope and undulated topography.
Cropped area	2038.03	16.25	Sandy loam to loamy with patches of clay.	Soil erosion, moisture scarcity, low fertility & productivity, low organic matter content gentle to moderate slopes.
Barren and wasteland	4833.72	38.56	Sandy loam to loamy with murram	Undulating, sloppy, eroded, degraded and gulllied.
Cultivated fallow	393.58	3.14	Sandy loam to loamy	Degraded, poor in fe4rtility, moisture scarcity, undulating surface.
Community & pasture land	357.14	2.86	Rocky, sandy loam to loamy	Degraded
Total	**12536.00**	-	-	-

1. Increasing the time of concentration of runoff and thereby allowing more of it to be absorbed and held by the soil.
2. Intercepting a long slope into several short ones so as to maintain less than a critical velocity for the runoff water.
3. Protection against damage due to excessive runoff.

To retard velocity of flowing water, 5078 stone check dams, 13 brushwood check dams, 104454 m contour trenches and 02 gabbion structures were constructed. To ensure water availability in the project area 36 rain water harvesting ponds and 203 *nalla* bunds were constructed at suitable sites for providing life saving irrigation and to meet domestic water needs.

Biological measures

Crop management practices such as, mulching and crop-residue management, improved varieties, seed bed preparation, seed rate and crop geometry, cropping systems, soil fertility management, pasture/ grass land development and plantations under agri-horti and agri-silvi culture systems were adopted in vast area to reduce the impact of raindrops and increase the infiltration rate.

Socio-economic measures

Various training programmes, Farmer's meating, exposure visits, demonstrations and farmers fair were organized for farmer's motivation and their active participation and creating awareness about watershed programme.

After four years, a survey was conducted for the impact study, sampling with single stage stratification technique was applied by covering different social groups of the watersheds. 170 farmers were selected from each social group. Participatory Rural Appraisal (PRA) techniques were applied to collect the informations on project accomplishments and feedback from farmers. Secondary data on land use statistics, crop yields etc., were also collected. Pre project period (1996) data were considered as the bench mark to judge the impact of the project.

Results and Discussion

Crop productivity

Implementation of watershed management programme in relation to sustain agricultural productivity was found to be very effective in changing the scenario of land capability of the area as well as to sustain the food requirements of farmers family. The productivity of major crops in watershed managed areas has increased appreciably (Table 2). The average increase (2003-2007) in the yield of paddy, shorghum, piogeonpea, chickpea, wheat, mustard was recorded 53.24, 30.28, 18.69, 58.02, 43.25 and 68.4 per cent, respectively, due to availability of water, reduced soil and water loss and adoption technologies (Kanaujia *et al.* 2000).

Table 2 : Increase in productivity of Agricultural crops.

S.No.	Crop	Production (q/ha)							
		Before treatment	After Treatment						(%) Increase
			2003	2004	2005	2006	2007	Av.	
1.	Paddy	10.30	14.20	19.60	14.60	16.40	16.50	15.78	53.24
2.	Sorghum	6.00	7.20	8.00	8.30	8.10	8.20	7.82	30.28
3.	Pigeonpea	7.40	8.60	9.50	8.50	8.50	8.70	8.78	18.69
4.	chickpea	8.10	11.00	12.50	11.30	15.70	15.60	12.80	58.02
5.	Wheat	15.80	21.00	23.50	21.60	24.20	25.30	22.63	43.25
6.	Mustard	6.30	6.50	7.80	9.40	10.02	10.5	8.42	68.40

Ground water recharge

The cumulative effect of earthen check dams, bolder dams, trenching, bunding, ponds and other treatments had caused significant improvement on recharge capacity of water in the wells and hand pipes. Prior to the project wells of the village used to become dry during the critical summer months April to June. the observations made in the wells of the village showed the positive trend of water availability (Table 3).

Table 3 : Average increase in water level of wells in Watershed Area.

Year	Annual Rainfall (mm)	Availability of water (m) In May	Increase over	In December	Increase
Pre project	905.40	0.93	-	1.80	-
2003	899.70	2.03	1.10	3.67	1.87
2004	600.50	3.09	2.16	4.36	2.56
2005	1094.85	3.85	2.92	4.85	3.05
2006	748.00	3.90	2.97	4.92	3.12
2007	635.00	3.76	2.83	4.80	3.00

There was gradual increase in the water level of wells year after year had noticed during the year 2003 and 2007 in spite of low annual rainfall in harvesting years. This is because of improvement in percolation of and down ward movement of in the wells (Kanaujia *et. al.,* 2001). It was interesting to observe that harvested rain water that automatically caused more recharge of the ground water accumulation of the water was more in the wider and deeper wells in comparison to narrow and shallow wells. It was further noticed that availability of ground water in the well had also enhanced near the perennial sources. The observation of this phenomenon further helped in planning and implementation of watershed programme in the next phases of villages.

The availability of safe drinking water in hand pipes caused significant change in social life of people of these watershed villages, as they are able to save their time for procurement of water from long distance during summer months. Further, children are less suffering from dysentery and worm problems.

Impact on irrigated area

Various water storage structures (Table 4) caused visible impact in increasing the irrigated area. This increase is due to bringing the new additional area of about 3170 ha. under irrigations through earthen embankment (2424.5 ha), ponds (409.4 ha) and natural nalla (336.1 ha).

Table 4 : Change in irrigated area.

S.No.	Sources of Irrigation	Irrigated area (ha) Pre Treatment 1996	After Treatment 2007	Increase
1.	Earthen embankment	19.5	2444	2424.5
2.	Pond	11.6	421	409.4
3.	Natural *Nalla*	34.9	371	336.1
	Total	**66**	**3236**	**3170**

Outputs and outcomes

After successful completion of watershed development the cropping intensity in the watershed area showed steady increase reaching to 136per cent and the average family income with in watershed area increased by Rs. 7800 per annum.

Change in social aspects

Concerted effort were made to Implement the watershed development programmcs for improvement of the village conditions, change in attitude of tribal families towards the education, health, environmental protection, human right, self realization for future production and co-operative life. In the beginning of the project, only 55 children were noticed to go school. But with improvement in life style of villagers, in the year 2007, 1174 students were recorded to seek education. Even most of the tribal families have started to send their children in chitrakoot and satna for better education. Awareness about environmental protection has been realized by villagers as a result more reduction in forest restoration is visible co-operative system has also developed to maintain natural resources of village as farmers are voluntarily contributing Rs. 50.00/acre as irrigation charges in watershed area.

Conclusion

This study was carried out by Milli-Watershed Krishi Vigyan Kendra, Satna (M.P.) during 2003-07 to assess the impact of watershed activities implemented during 1996-2003. Various mechanical measures viz. bunding, bench terracing, trenching, contour cultivation, construction of earthen check dams, ponds and loose bolder checks and biological measures viz. mulching, crop residue management, crop geometry, agri-horti and agri-silvi systems were adopted in addition to farmers motivation. The productivity of major crops and availability of water in the watershed managed area has increased appreciably. The average increase of 53.24 per cent was recorded is the yield of paddy, 30.28per cent in sorghum, 18.69per cent in pigeonpea, 58.02per cent in chickpea, 43.25per cent in wheat and 68.40per cent in mustard yield. The availability of ground water was also increased up to 3.12 m and tremendous increase in irrigated area (3170 ha) was also recorded. Because of watershed activities the cropping intensity, family income and educational standards were also increased.

References

Gupta, V.P., Sharma, R.P. and Goyal, R.K. (2000). Impact assessment of conservation measures in arid areas: a case study of Osian Bigmi (Jodhpur) watershed. Proc. International conf. on integrated water resource management for sustainable development, New Delhi: 1012-1019.

Kanaujia, V.K., Singh V.P., Raghuvanshi, R.P.S. and Bhagwandeen (2001). Watershed Management for Enhancing crop Production and Economic Advantages in Tribal area. In (Eds dhyani *et al.*) Resource Conservation and Watershed Management: technology options and future strategies. AISWC, Dehradun: 289-292.

Singh, P.K., Singh, J. and Mahnot, S.C. (1998). Effect of soil conservation measures on ground water recharge. Proc. Silver jubilee convention Indian society of Agril. Engg. R.A.U., Udaipur : 148-155.

□□□

Chapter 24

Assessing the Impact of Water Management in Bareilly District of Uttar Pradesh

PANKAJ SHARMA[1], R.K. SINGH[2] AND KAPIL[3]

[1] Research Associate, National Centre for Agricultural Economics & Policy Research, New Delhi-110012. [2] Head, Division of Agricultural Economics, C.S.A. University of Agril. & Tech., Kanpur-208002. [3] Senior Marketing Executive, DCM Shriram Consolidated Limited (DSCL)

More than 2 billion people worldwide live in regions facing water scarcity (Celia, 2006) and in India this is a particularly acute crisis. In 2006 between the domestic, agricultural, and industrial sectors, India used approximately 829 billion cubic meters of water every year. By 2050 demand is expected to double and consequently exceed the 1.4 trillion cubic meters of supply (Sengupta, 2006). Despite the recent rapid growth in the services and industrial production, agriculture is still an integral part of India's economy and society. Water is vital for realizing full potential of agriculture sector & country development. Agricultural growth is largely dependent of irrigation facilities (Jain & Prakash, 1993). India's agricultural sector currently uses about 90per cent of total water resource (UNICEF, 2002). Irrigated agriculture has been fundamental to economic development, but unfortunately caused groundwater depletion. Due to water pollution in rivers, India draws

80per cent of its irrigation water from groundwater (World Bank, 2005). The annual normative demand of ground water has increased drastically in all the regions of Uttar Pradesh (Tiwari 1995).

At present, irrigation through tube wells has increased threefold compared to what it was a decade ago. With over-exploitation of groundwater resources, water tables have declined sharply resulting in more energy being needed to draw water from greater depths. Development of irrigation facilities using pond water was more economical as compared to creating a new source of irrigation. Under limited water availability conditions, reservoirs and tail water recovery systems can improve profitability, reduce ground water dependence, and reduce movement of sediment, nutrients and pesticides off farm. While reservoirs may not be profitable under plentiful water conditions, cost-sharing opportunities may make them a viable means of addressing environmental concerns (Wailes, 2003).

Therefore, an in depth analysis of irrigation utilization at the micro level is essential for gaining a fuller understanding the phenomenon of under utilization. There is a need for conservation practices to be adopted to avoid the problems arising from the over exploitation of water resources. In order to conserve ground water; check its wastage through percolation and switch over the cultivation of crops having lower water requirements (Singh, 1984).

Thus, the paper examines the extent of water utilization at various levels and its management on farm households in canal system and also assessing the impact of water management under Sharda canal system.

Study Area

The area selected for the study is Sharda minor and its command. The area is located in district Bareilly in the northern part of Uttar Pradesh and considered to be under the tarai belt. As being under the tarai belt the environment is very conducive for the agriculture purpose and that is the reason that about 85 per cent of the rural population is depending on the agriculture and other allied activities. Most of the farmers in the district are the marginal farmers with small landholdings and depends on canal for their irrigation purposes. In the study area i.e. Bithri Chanpur block about 18 per cent of the total area is dependent on canal irrigation.

Methodology

As this study has been attempted at the micro-level it has been carried on the primary data. An individual farmer's schedule has been designed and administered to the selected farmers of the selected villages of Bithri Chanpur block of Bareilly district. The Bithri Chanpur block of district Bareilly was purposely selected for the study on the basis of comparatively high percentage of area irrigated by canal. A random sample of 30 farmers from the Bethri Chanpur block was selected for the study. For selection of farmers, the block was divided into three zones-head, middle and tail depending on the adequacy and dependability of water supplies. One village from each zone was selected and farmers were classified into three size groups, viz., marginal (<1 ha), small (1-2 ha), medium (2-4 ha) and big (> 4 ha) groups. The number selected from each size group was in proportion to the number of farmers in that size group. As discussed earlier the primary data related to the area, cropping pattern, area under different crops etc. for the agricultural year 2006-07 were collected by personal interviews of the sample farmers with the help of pre-tested schedules. Secondary data required about the block and canal was collected from Ziladari office, Bethri Chanpur and Irrigation department respectively. The data also constituted information regarding the holdings, irrigation methods, cropping pattern and yield of major crops. Study is primarily based on tabular analysis.

Results and Discussion

It was found that the irrigated area at different ends of the canal was mainly covered by paddy crop i.e. 19.91 per cent of the total irrigated area followed by urd/moong (18 per cent) and Wheat (16.99 per cent) whereas, crops like gram and bajra had a very little contribution in the total irrigated area i.e. 5.65 per cent and 9.96 per cent of the total irrigated area by canal system. It was also found that the area under high water demanding crops like paddy and wheat was highest in the head zone and in the case of tail zone the area was mainly covered by low water demanding crops like bajra, jowar + arhar and gram (Table 1).

Table 1 : Distribution of irrigated area (ha) by canal under different crops on the sample farms.

Crop	Different ends of canal			
	Head	Middle	Tail	Total
Paddy	6.75	6.15	5.85	18.75 (19.91)
Wheat	5.85	5.35	4.90	16.00 (16.99)
Bajra	2.85	3.01	3.52	9.38 (9.96)
Urd/Moong	7.25	5.12	4.60	16.95 (18.00)
Sugarcane	5.15	3.50	2.12	10.77 (11.44)
Jowar+Arhar	3.15	4.52	5.60	13.27 (1.09)
Gram	1.12	1.75	2.45	5.32 (5.65)
Others	1.72	1.25	0.75	3.72 (3.95)
Total	33.84	30.64	29.79	94.16 (100.00)

* Figures in paren thesis denote percentage.

Average cropping intensity on the sample farms came to 216.02 per cent. It was the highest on marginal farms followed by small and big farms. The intensity of cropping on marginal, small and big farms is worked out to 246.52 per cent, 209.60 per cent and 208.00 per cent respectively and also the average cropping intensity on the sample farms at different ends of canal came to be 216.90 per cent. There was no high variation in cropping intensity at the head and middle zone i.e. 265.41 per cent and 266.43per cent respectively, but a lower cropping intensity was found in tail zone i.e. 254.61 per cent (Table 2).

Table 2 : Cropping intensity of the sample farms at the different ends of canal.

Different ends of canal	Net cultivated area (in ha.)	Total cropped area	Cropping intensity (per cent)
Head zone	12.75	33.84	265.41
Middle zone	11.50	30.64	266.43
Tail zone	11.70	29.79	254.61
Total	35.95	94.16	261.90

The total area under cereals has increased from 17880 ha to 23309 ha for the year 1994-95 to 2004-05. But the case is opposite in the case of pulses as we can see there is a minor decrease in area of pulses i.e. from 993 ha in 1994-95 to 865 ha in 2004-05. The area under oilseeds has also decreased i.e. from 2566 ha to 1765 ha for the year 1994-95 and 2004-05 respectively. An increasing trend is observed in sugarcane an increase from 3535 ha to 3707 ha for the year 1994-95 and 2004-05 respectively.

On viewing the irrigated area under main crops we can see that the area under cereals, oilseeds and sugarcane has risen i.e. from 17119 ha to 23004 ha, 488 ha to 789 ha and 3505 ha to 3680 ha respectively for the years 1994-95 and 2004-05, although the area under pulses has decreased from 993 ha to 865 ha for the years 1994-95 and 2004-05.

There is a continuous rise in the fertilizer consumption of nitrogen, phosphorous and potash i.e. from 3195 m. ton. to 3969 million ton, 328 million ton. to 1165 million ton and 102 million ton. to 494 million ton., respectively for the year 1994-95 and 2004-05 respectively (Table 3).

Table 3 : Comparative achievements of canal irrigation at block level.

Particular	Year		
A. Condition of canal irrigation work.	**1994-95**	**1999-00**	**2004-05**
1.Canal	121 km	121 km	121 km
2. Real area irrigated by canal.	40939	43044	43915
B. Total area under main crop (ha)			
1.Cereals	17880	20839	2330
2.Pulses	993	691	865
3.Oilseeds	2566	1103	1765
4.Sugarcane	3535	3298	3707

C. Irrigated area under main crops (ha)			
1.Cereals	17119	20203	23004
2.Pulses	190	148	74
3.Oilseeds	488	663	789
4.Sugarcane	3505	3298	3680
D. Production of main crop (M.ton)			
1.Cereals	678309	918210	887081
2.Pulses	20717	13940	14367
3.Oilseeds	14187	9315	11728
4.Sugarcane	4366563	3784828	4010627
E. Fertilizer Consumption (M.ton)			
1.Nitrogen	3195	2910	3969
2.Phosphorus	328	531	1165
3.Potash	102	149	494
Total	3565	35910	5628

We took a sample of 14 marginal farmers having canal irrigation. These farmers having 8.12 ha total cultivated area increased up to 9.50 ha after adopting irrigation structure. The net irrigated area by canal also increased up to 7.25 ha from 4.68 ha due to a reliable availability of canal water.

Cropping pattern of the farmers changed according to their availability of canal water as an irrigation source. As we know the availability of canal water was low in the year 1994-95 and now it is available in ample amount. Thus, the area under main crops has increased and also the cropping intensity as well as the average yield has also risen. There is a great increase in the area of the main crops i.e. of wheat, paddy and sugarcane as these are the water demanding crops, although there is an increase in the area and production of all the crops i.e. due to using of hybrid varieties as well as many development changes have also taken place during this course of last ten years.

The area under paddy has improved to 5.65 ha in 2004-05 which was 2.68 ha in 1994-95, whereas the total cultivated area by canal irrigation has risen to 7.25 ha which was 4.68 ha for the years 2004-05 and 1994-95 respectively. Similarly, the area under wheat and sugarcane has nearly doubled from 1994-95 to 2004-05 i.e. from 2.53 ha to 4.21 ha and 0.75 ha to 1.32 ha respectively.

The average yield of main crops i.e. of paddy, wheat and sugarcane has improved from 18.18 ha to 19.19 ha, 21.35 ha to 24.56 and 621.92 ha to 625.80 ha, respectively (Table 4).

Table 4 : Comparative study of canal irrigation.

S.No.	Particulars	Years 1994-95	2004-05
1.	Total cultivated area (ha)	8.12	9.5
2.	Area (ha) irrigated by canal	4.68	7.25
3.	Cropping pattern of total cultivated area		
	a Paddy	5.11	6.25
	b Wheat	4.32	5.42
	c Gram	0.85	1.5
	d Arhar +Jowar	0.45	0.5
	e Sugarcane	0.75	1.25
	f Bajra	0.96	1.50
	g Urd/Moong	4.62	5.25
	h Other	0.75	1.75
	Total	17.81	23.42
4.	Cropping intensity (per cent)	219.33	240.20
5.	Cropping pattern of total cultivated area by canal irrigation		
	a Paddy	2.68	5.65
	b Wheat	2.53	4.21
	c Gram	0.65	0.75
	d Arhar +Jowar	0.32	0.51
	e Sugarcane	0.75	1.32
	f Bajra	0.62	0.85
	g Urd/Moong	3.12	4.62
	h Other	0.82	1.50
	Total	11.49	19.41
	Cropping intensity (per cent)	245.51	267.72
6.	Average yield of main crops per ha (q)		
	a Paddy	18.18	19.19
	b Wheat	21.35	24.56
	c Gram	8.63	9.22
	d Arhar +Jowar	7.92	9.74
	e Sugarcane	621.92	625.80
	f Bajra	9.88	11.92
	g Urd	3.85	4.34
	h Moong	3.94	2.78

The 80 per cent of farmers at the head level are using flood irrigation from the canal and only 20 per cent of farmers are using channel irrigation at the head level. In this zone there is an ample supply of water and no need of making the channels for using the canal water.

In case of middle end 80 per cent of farmers are using channel irrigation while only two farmers i.e. 20 per cent are using flood irrigation for using canal water. In this zone the ratio of water obtained is less than that of head zone and that is the reason that farmers are bound to make channels to avoid the losses of water.

The situation is similar in the tail zone to that of middle zone, although the supply of water is minimum in this zone but still the farmers are using mainly channels for harvesting the canal water as the fields are flat and only in some areas undulating fields can be observed and in that regions the farmers are also taking the help of lift canal which is only 20 per cent (Table 5).

Table 5 : Type of irrigation methods at different ends of canal.

Ends of canal	Type of irrigation method used			Total
	Flood	Channel	Lift	
Head	8	2	0	10
Middle	2	8	0	10
Tail	0	8	2	10
Total	10	18	2	30

Conclusion and Policy Implications

The following main conclusions emerge from the analyses attempted in this study.

The area under major crops i.e. paddy, wheat, sugarcane & is the highest in the head zone and has decreased in the middle and tail zone respectively, whereas the situation is opposite in the case of other crops like Bajra, Jowar + Arhar and Gram and this variation is due to the different water requirements of the different crops.

The area under main crops has increased and also the cropping intensity as well as the average yield has also risen. There is a great increase in the area of the main crops i.e. of wheat, paddy and sugarcane as these are the water demanding crops, although there is an increase in the area and production of all the crops i.e. due to using of hybrid varieties as well as many development changes have also taken place during this course of last ten years.

The average yield of main crops i.e. of paddy, wheat and sugarcane has improved from 18.18 q to 19.19 q, 21.35 q to 24.56 q and 621.92 q to 625.80 q respectively. The maximum loss occurs in the month of June, August and October, whereas the minimum loss takes place in the months of March, November and December (source: Irrigation department). Maximum numbers of farmers are using channel irrigation as a main source for harvesting the canal water, followed by flood irrigation which is limited to the head zone only and at last only two farmers are taking water by flood irrigation.

The findings that there is surplus water in canals in some periods were rather surprising from farmer's point of view. A deep probe of this problem leads to the emergence of suspicion that the farmers in the head zone make unauthorized canal cuts to irrigate their field, which might result in differences in amount of water supply reported by the canal authority and that actually available to the farmers. This shows that if the canal and field channels are maintained properly, if there are no unauthorized cuts of canal bunds and if proper leveling of land is done, and then the area irrigated can be increased substantially. Farm production and income can be increased substantially.

Policy implications

The findings of the study have a few important policy implications. First, the study has brought out that the existing crop pattern in the study is not the one which can make the most profitable use the water supplies. This means that there is need for educating and motivating farmers to change their existing cropping pattern according to return to fixed factors. Secondly, it was found that surplus water is available in canals in some periods. This shows that the farmers are either willing to use water or not able to use water because that are no properly maintained channels linking their fields to the outlets or the fields are not properly leveled. We think that the main reason for surplus water being available in the canals is the lack of channels and undulating fields, therefore the farmers should be assisted technically and financially in leveling of fields and laying out field channels, so that they could make use of surplus water.

Since we suspect that farmers in the head zone resort to unauthorized channel cuts in peak periods, steps should be taken to check these unauthorized cuts so that water is available to all farmers of the command area.

Conclusion

The study is mainly based on the primary data obtained from the 30 respondent farmers. The data was gathered by survey method with the help of prescribed questionnaire and have been analysed using simple and conventional tabular methods. The study aimed at finding out the relative efficiency of canal irrigation at its different ends, to know about the water losses at different ends of canal etc. It also examined the effect of canal irrigation on cropping pattern, yield level and fertilizer consumption. The study has revealed that the area under major crops i.e. paddy, wheat, sugarcane is highest in the head zone and has decreased in the middle and tail zone respectively, the maximum loss occurred in the month of June, August and October, whereas the minimum loss took place in the months of March, November and December, the average yield of main crops i.e. of paddy, wheat and sugarcane has improved also there is a continuous rise in the consumption of phosphorous and potash. Although the study has shown that there is ample of water in the canal but still it is not available throughout the year. Therefore, the study has suggested that the canal should be maintained by local bodies formed by the government authorities.

References

Pathak, B.D. (1983). Development of ground water resource for irrigation in India. Agricultural situation in India, **38** (5) : 287-289.

Singh, I. and Sirohi, A.S. (1977). Optimization of water resources of upper Ganga canal in Western Uttar Pradesh, India J. of Agric. Econ., **32** (1) : 42-107.

Singh, S.P. (1984). Development and management of ground water resources of Jounpur District. National Geographer. **19** (2) : 153-162.

Tiwari, L. (1995). Ground water resource utilization in the context of sustainable crop production in U.P. Unpublished M. Sc. Agri. (Rural Banking and Agricultural, Economics) Thesis, G. B. Pant University of Agri. & Tech., Pantnagar.

Wailes, Y. (2003). Assessing the Benefits of On-Farm Reservoirs and Tail-Water Recovery Systems Department of Agricultural Economics and Agribusiness, University of Arkansas, Fayetteville, AR 72701.

Celia, D. (2006). "Need for Water Could Double In 50 Years, U.N. Study Finds." *New York Times,* August 22.

Sengupta, Somini, (2006). "In Teeming India, Water Crisis Means Dry Pipes and Foul Sludge." *New York Times,* September 29.

UNICEF, (2002). India: Water Supply and Sanitation.

World Bank, (2005). India's Water Economy: Bracing for a Turbulent Future.

□□□

Chapter 25

Role of Sugarcane in Sustaining Soil Health in Rice-Wheat Sequence in Indo-Gangetic Plains

G.K. SINGH AND D.V. YADAV
Indian Institute of Sugarcane Research, Lucknow

The rice-wheat crop rotation is the most popular cropping sequence followed on about 13.5 million ha area in the indo-gangetic plains of India (10 m ha), Pakistan (2.2 m ha), Nepal (0.5 m ha) and Bangladesh (0.8). Recent reports indicate that the system is under production fatigue and productivity of this important sequence has started stagnating or even declining in certain areas, due to depletion of native nutrient reserves, emergence of multi-nutrient deficiencies and consequent decline in productivity of certain nutrients.

To compensate the effect of declining productivity, farmers have started applying greater doses of N than the recommended. Such indiscriminate use of N has further worsen the nutrient imbalance in soil-plant system, besides increasing cost of production and environmental problems. There are increasing concerns about the sustainability of soil health in a continuous rotation. This evokes the fear that future food grain production may not keep pace with demand as the population continues to increase. Therefore, improving the current levels of production and sustaining soil health is the demand of the hour to be addressed properly.

The sugarcane plant has an excellent set of genetic, anatomical and physio-biochemical attributes. It has very high potential to produce biomass and energy. In the indo-gangetic plain of India, sugarcane is cultivated in Punjab, Haryana, Uttar Pradesh, Uttrakhand, Bihar and West Bengal. In Western Uttar Pradesh, paddy-wheat- sugarcane-ratoon-moong is the most common crop rotation. In Uttar Pradesh Saharanpur, Muzaffarnagar, Bijnor, Baghpat, Meerut, Ghaziabad and Sitapur and Yamunanagar in Haryana follow mainly wheat-sugarcane system. The wheat-sugarcane system ranks second in Moradabad, Bareilly, Kheri and Hardoi districts of Uttar Pradesh, Udham Singh Nagar of Uttrakhand, Ambala, Kurukshetra, Karnal, Kaithal and Panipat districts of Haryana and Ludhiana district of Punjab.

Sugarcane and soil health

New realms of soil health

Soil productivity has been used as a indicator of soil health for quite some time ago. Recently the term soil health has been expanded to summarize the capacity of soil to function for sustaining soil and animal productivity, maintain and / or enhance water and air quality and support human/animal/plant health and habitation. This new expanded view of soil health characterizes soils based on all the functions collectively, whereas the old view characterized soils based on the prime crop productivity only.

Soil resilience and its relation to soil health

Soil resilience is related to soil health in terms of soil recovery functions. A resilient system will retain soil health which will ensure sustainability. The soil and plant community processes including physical/chemical/biological mechanisms contribute to the recovery. Recovery functions are biologically mediated which include cycling of nutrients, detoxification of pollutants and suppression of pathogenic organisms. Huge biomass of sugarcane in and above the soil surface is of great value to serve above purposes.

Enriching organic matter

The contribution of fertilizers is estimated to be the highest among various inputs of production in increasing the yield but these can't help to maintain and enhance soil organic matter status, which is the ultimate key to sustainability. Organic matter maintains porous soil structure, improves water-holding capacity and allows gaseous exchange for use by soil organisms that break down manure, crop residues and other organic matters.

Sugarcane produces 10 to 12 tones of dry leaves per ha per year. These leaves are generally called trash and contain 0.42, 0.15 and 0.57 per cent of nitrogen, phosphorus and potassium, respectively in addition to 16.8, 25.7, 236.4 and 2045.2 ppm copper, zinc, manganese and iron, respectively. Therefore, when trash is allowed to decompose *in situ,* it releases about one fifth of the nutrients required by the crop within 6 months. The rate of decomposition, however, depends on the activities of different Microbial populations that reside in soil.

Ammonia formed during decomposition, undergoes oxidation forming nitrates and nitrites. Nitrate and Nitrite react with carbohydrates and form amino acid and then proteins in the soil. Consequently there is an increase in the nitrogen status of soil. However, because of high C: N ratio in the beginning of trash decomposition, soil nitrogen gets immobilized. No nitrogen is thus left for leaching. This way nitrate leaching could also be minimized. It might also be assumed that by virtue of slow decomposition, trash also serves as a better agent for solubilising and chelating action and a long duration crop such as sugarcane is able to utilize the nutrient thus made available.

Likewise, abundant rice straw is produced every year, which is thrown away to decompose in pits or burnt in fields. However, when chopped pieces of trash, straw and legume husk were incorporated in soil, organic carbon and nitrogen contents of soil showed an improvement after one year (Yadav and Prasad, 1986). Hence it would be advisable to follow these residue incorporating practices.

Yadav *et. al.,* (1987) also found significant improvement in organic carbon content of different soils when these were incubated at their 60 per cent water-holding capacity and 27 ± 2^{0}C with cane trash chopped to 1-2 cm size added at rate equivalent to 20 t/ha, combined with 0, 75 and 150 kg nitrogen per hectare supplied through urea.

Organic carbon content of red soil increased from 0.80 per cent to 1.72 per cent with trash, 2.02 per cent with trash + 75 kg N/ha and 2.90 per cent with trash + 150 kg N/ha at the end of 120 days of incubation. The data indicated that organic carbon content of soil increased with the decomposition of trash. However with nitrogen addition, the trash decomposition was more rapid (Akinyemi and Adiwale, 1984) that further increased organic carbon content in the soil.

Sugar factory wastes such as spent wash, filter cake, bagasse etc. are other sources of organic manures. Their mixture in the ratio of 1:1:0.3 when applied @ 40 t/ha increased the availability of nitrogen, phosphorus and potassium in the soil. Thus, disposal of organic wastes generated from sugar factories in cane fields have beneficial effects on the amelioration of pollution control and plant nutrient re-cycling (Yadav, 1985).

Residue management

A sugarcane crop of 116 tonnes millable canes per hectare leaves behind in the soil, 9.6 tonnes stubbles and 4.3 tonnes roots (Singh and Srivastava, 1964). These underground portions contain 0.59 per cent nitrogen, 0.26 per cent phosphorus and 0.23 per cent potassium. Their decomposition enriches the soil by 29 kg nitrogen, 13 kg phosphorus and 11 kg potassium per hectare. Besides, they may also add 1.48 kg sodium, 7.8 kg calcium, 8.12 kg iron, 1.08 kg manganese, 0.07 kg copper and 0.10 kg zinc per hectare (Singh and Agarwal, 1984). Thus, advantages of improving physico-chemical properties of soil that accrues from organic amendments like farmyard manure might also be achieved by stubbles and roots which contain 92 per cent organic matter and 53 per cent total carbon higher than those of trash and almost equal to those of farmyard manure. However, management of stubble roots needs special care. As their C: N ratio is 106, immobilization of native nitrogen takes place immediately after the initiation of ratoon crop. This could, however, be prevented by adding a little more quantity of chemical nitrogen than the recommended dose. The mineralization of organically transformed residual nitrogen is stimulated by application of nitrogen through urea in ratoon crop as it helps in initial build up of nitrifying microbes. As a result, the availability of nitrogen to ratoon crop increases.

Mulching

Trash mulching prevents oxidation of organic matter, which is added by its decomposition *in situ* (Parthasarathy, 1961). Prasad *et. al.* (1988) observed 10.4 per cent higher cane and 9.0 per cent higher sugar yields due to mulching over unmulched crop. Mulching also upgraded the yield potential of sugarcane at every dose of nitrogen application (Yadav, 1985). Yields obtained at 50 and 100 kg nitrogen per hectare with mulching were almost equal to those obtained at 100 and 150 kg nitrogen per hectare without mulching, respectively. Thus, trash mulches are able to sustain soil health.

Australian Sugar Industry is progressively moving away from the practice of burning the crop before harvest to the system of green canes trash blanketing (GCTB), since residues that would have been lost in the fire are returned to the soil. Nutrients and organic matter may be accumulating under trash blanketing. During one year, 82 to 92% of the trash gets decomposed depending upon the rainfall, temperature and other factors.

Conversion from a burning to a GCTB management system is likely to improve fertility status of soil. The increased microbial growth under GCTB is consistent with increase in earthworms and other soil micro fauna. This eventually leads to a more resistant soil system. Soil faunal (particularly earthworms) activity has been shown to increase residual decomposition and N mineralization of residues from other crops and it seems reasonable to speculate that the rate of the composition and N release from trash may increase as soil micro faunal population.

Factory wastes

Filter cake/pressmud cake, is produced at 3-7 per cent of the cane crushed in a sugar factory. On an average, each tonne of oven dry sulphitation pressmud cake (SPMC) contains 17, 36, 14 and 23 kg of N, P, K and S, respectively besides secondary and trace elements. An economy of 50-75 kg N/ha has been reported in sugarcane cropping system by several research workers by an integrated use of SPMC with mineral fertilizers. The N-use efficiency is also increased by 4-8 per cent in sugarcane due to an integrated use of SPMC and fertilizer-N. The SPMC has been reported to increase organic carbon content and the availability of N, P, K. S, Zn and Mn in the soils. It is a good sulphur source for sugarcane cultivation.

Biofertilizers

Some of the species of *Azotobacter, Azospirillium, Acetobacter* and *Bacillus* have been found to economic biofertilizer-N by 25 per cent in sugarcane. Therefore, bio-fertilizers have a promise for N supply in sugarcane based cropping systems.

Legumes in crop sequence

Extensive evidences are available to indicate that legumes when incorporated in the cropping sequences, increased the soil N content. In the developing tropical and sub-tropical countries where costs of fertilizer are high, legumes assume considerable significance. Because of deep root system, legumes also help in mobilizing fixed phosphorus for use of the crops grown either in association or in succession.

Sugarcane produces a large quantity of biomass and remains in the field at least for three years (plant + 2 ratoons). A careful incorporation of its left over biomass by recycling in the soil may sustain soil health without compromising the productivity of cropping system.

Conclusion

There are increasing concerns about sustainability of soil health in a continuous rotation. It is noticed that farmers have started applying greater doses of N than the recommended. Such indiscriminate use of chemical fertilizers further worsen the nutrient imbalance in soil plant system. Sugarcane produces 10-12 tonnes trash, 9.6 tonnes stubbles and 4.3 tonnes roots ha^{-1}. Their decomposition enriches the soil by adding 29 kg nitrogen, 13 kg phosphorus, 11 kg potassium, 7.8 kg calcium, 8.12 kg iron, 1.08 kg manganese, 0.07 kg copper and 0.10 kg zinc per ha. Trash mulching improves soil permeability and reduces soil erosion and moisture loss, which later on incorporation adds to the soil carbon pool.

References

Akinyemi, O. and Adiwale, A. (1984). Effect of incubation temperature and different sources of N and P on decomposition of sawdust in soil. *Agricultural Wastes*, **11**: 293-306.

Parthasarathy, S.V. (1961). Value of trash in cane culture. *Indian J. Sug. Res. & Dev.* 5(2) :96-100.

Prasad, S.R., Yadav, R.L. and Singh, K. (1988). Efficient utilization of irrigation by mulching in sugarcane. *Indian J. Agron.* **33**(2): 125-130.

Singh, A. and Srivastava, S.P. (1964). Maintaining organic matter in sugarcane soils of India. Proceedings of the 5th All India Conference of Sugarcane Research and Development Workers. December 1964, ICAR, New Delhi, pp. 362-368.

Singh, M. and Agarwal, M.P. (1984). Chemical composition of sugarcane roots and its role in sugarcane ratoons. *Indian Sug.* **34**(3): 1-2.

Yadav, D.V., Singh, T. and Srivastava, A.K. (1987). Recycling of nutrients in trash with N for higher cane yield. *Biological Wastes,* **20**: 133-141.

Yadav, R.L. (1985). Enhancing recoveries of fertilizer nitrogen applied to sugarcane. *Biological Memoirs,* **11**(2): 135-142.

Yadav, R. L. and Prasad, S.R (1986). Nutrient release from soil incorporated cane trash and rice straw for sugarcane growth and yield. Indian *J. Sug. Cane Technol.* **3**: 6-10.

❑❑❑

Section - 5

Rehabilitation of Problem Lands, Salins, Alkali, Ravines, Riveine Lands, Diava Land etc.

Chapter 26

Rehabilitation of Problem Lands : Challenges and Opportunities

J.C. DAGAR[1], P.S. MINHAS[2] AND GURBACHAN SINGH[1]
[1] Central Soil Salinity Research Institute, Karnal – 132001
[2] Panjab Agriculture University, Ludhiana- 141 004

About 2 billion ha in the world is affected by various forms of human induced land degradation (Oldeman, 1991) with erosion by water being the chief contributor (1.1 billion ha) followed by wind erosion, chemical degradation and physical degradation. In India, out of 120.8 million ha degraded land, 82.6 million ha is estimated to suffer with water erosion, 24.7 million ha from chemical degradation, 12.4 million ha due to wind erosion and 1.1 million ha from physical (mainly due to stagnation of water) degradation (NBSS & LUP, 2008). Due to erosion about 5334 million tones of soil is lost (16 tones/ha) annually-highest in black soils (24-112 t/ha) followed by Shiwalik region (80 t/ha) and north-eastern region with shifting cultivation with 27-40 t/ha soil loss (Mandal *et al.*, 2008) About 11% of eroded area falls in very severe category with erosion rates of more than 40 tones/ha/year. The land degradation has both on-site and off-site impacts. Due to on-site impacts land productivity declines because of nutrient losses. In India nearly 0.8 million tones of nitrogen, 1.8 million tones of phosphorus and 26.3 million tones of potash are lost annually, amounting to 0.2% of total reserves (TERI, 1998).

Off-site effects include changes in water regime, decline in water quality, sedimentation of river bodies and reservoirs, loss of biodiversity and natural disasters like tsunamis, cyclones, floods and droughts. Besides, irrigated agriculture, especially through canal systems, has resulted in degradation due to waterlogging and salinization. It is estimated that nearly 8.4 million ha of the irrigated lands are affected by soil salinity and alkalinity; of which about 5.5 million ha is also water logged (IDNP, 2002). Long-term strategies are required to rehabilitate the degraded lands for sustainable production.

To produce 310 million tones of food grains, 190 tones of other food products, 1000 tones of fodder, 629 tones of fuel wood, and 347 tones of timber and other raw materials by 2050 from limited land resources is a big challenge (NAAS, 2009). According to DES (2007) in 140 million ha it would require a productivity of 3.3 tones/ha against an average productivity of all food grains of 1.7 tones/ha in 2006-07. Similarly, yield of 12.5 tones/ha from 120 million ha under forests and pastures would be needed to produce 1500 million tones of biomass. Forests' productivity is well below its potential of 2 tones / ha / year, mainly due to excessive grazing and over-exploitation for fuel wood (Tiwari, 2000). The loss of forest biodiversity is further endangering livelihood of millions of people who are dependent on forest land. We can not think of deviating arable lands from cultivation of food crops. To meet other requirements we have opportunities to rehabilitate the all kinds of degraded lands under perennial vegetation or any kind of agroforestry. Opportunities for alternate land uses through agroforestry systems on problematic areas including eroded, salty and water logged areas have been discussed in this paper.

Combating Land Degradation

Broadly, two-pronged strategies are needed to prevent land degradation and to bring water and wind erosions within permissible limits for sustained productivity. First is proper assessment of the degree, type, extent and severity of degradation/erosion and its effect and impact on production and nutrient losses. The second strategy is to check and reverse the process of land degradation using appropriate technologies, take conservation measures through strong research support and target the sustainable increase in production from these lands. Agroforestry is ideal option to rehabilitate all kinds of degraded lands. We have gone long way to evaluate and identify suitable plant species for site specific situations and develop required technologies for various problematic areas. We only require strategies and policies for getting these results implemented.

Why to rehabilitate degraded lands through agroforestry/alternate land uses?

New obligations emerging out of the growing population demand that, to sustain production of food and fuel, each piece of land be best utilized consistent with ecology and land use capabilities. This means that even the degraded lands including salty soils may be either reclaimed for agricultural purposes or put these lands under alternate land uses. Despite the availability of technical know-how for reclamation and management of eroded (including ravine lands), salt-affected and waterlogged lands; rehabilitation of these lands for crop production is very slow. The reasons are both social and economical. These lands are either owned by poor-resource farmers or belong to community or government agencies.

Reclamation of these lands require additional inputs such as biological or engineering barriers for checking soil erosion, application of amendments (gypsum, press mud and pyrite in case of alkali lands) or installation of drainage (for saline or saline waterlogged soils), use of fertilizers (including green manure) and infrastructure of farm operations. This is a costly preposition. Requirement of extra resources on recurring basis and lack of interest in development of common lands for practicing intensive arable agriculture have been the major bottlenecks for the rehabilitation of salty lands on an extensive scale. Further, under most of the saline soils in arid and semi-arid regions, the groundwater is saline. In the scenario of scarcity of good quality water the use of poor quality waters for agriculture is inevitable. Therefore, afforestation or agroforesty involving trees, grasses and low-water requiring crops (when using saline water for irrigation) is considered an ideal land use for reclamation and management of salty lands and utilization of degraded lands using saline water. Even the common-property lands can be brought under farm-forestry. Besides providing fuel, fodder and timber, afforestation will also lead to bio-amelioration of salty lands and will improve biodiversity and help in carbon sequestration and mitigating climate change. Afforestation of these lands will not only help in ecological and environmental considerations, but also be useful in relieving pressure on traditionally cultivated lands and forests and will generate additional employment.

It has been proved by CSSRI that establishing salt-tolerant tree plantations utilizing the saline ground water (ECiw up to 10 dS m^{-1}) provides economic use of abandoned arid lands when irrigated in furrows for initial three years of tree establishment (Tomar *et al.*, 2003b). Further, grass species could be grown successfully with dry forage production with saline irrigation (Tomar *et al.*, 2003a). Besides soil amelioration, two of the definite advantages of irrigated forages were about three to four-fold increase in grass productivity as compared with natural pastures, and extension of production period to those of conventional shortages during summer months when the most nomad populations are forced to migrate to traditionally irrigated areas. Thus, the rehabilitation of arid and degraded lands with the plantation of slat-tolerant trees or silvi-pastoral systems with saline irrigation would not only render these otherwise non-productive lands to be more productive but also ensure conservation and improvement for long-range ecological security of these lands.

Agroforestry for eroded lands

Soil erosion has socio-economic, environmental and technical dimensions. Those who suffer the most are poor farmers and landless laborers, who are least able to adopt conventional measures for its control. The low input costs of many agroforestry systems can be within the reach of poor farmers. About 39% area in the country (10 states with >50% area) has erosion of more than permissible (10 t/ha/yr) rate, resulting in reduction in agricultural production. About 11% of area fall in very severe category with erosion rate of more than 40 t/ ha / yr. Some of the states in north-east and north-west Himalayas are worst affected with more than 1/3rd of their area falling in very severe loss (40-80 t/ha/yr) category (Table 1). Protection from grazing and afforestation with suitable species are the most effective measures for checking soil erosion and consequently ravine formation. Woody species found growing in eroded habitats may find priority in afforestation program. For example, *Acacia nilotica, A. eburnean, A. leucophloea, A. catechu, Azadirachta indica, Albizzia lebbeck, Balanites roxburghii, Butea monosperma, Dalbergia sissoo, Dendrocalamus strictus, Dichrostachys cineria, Eucalyptus* spp., *Feronia limonia, Pongamia pinnata, Prosopis juliflora* and *Ziziphus mauritiana* have been found to adapt easily in the ravines of river Yamuna at Agra and Kshipra at Ujjain. Among grasses *Dichanthium annulatum, Cenchrus ciliaris, Bothriochloa pertusa, Chrysopogon fulvus, Themeda triandra, Heteropogon contortus, Sehima nervosum, Tragus biflorus, Iseilema laxum, Cynodon dactylon* and *Saccharum munja* flourish well in ravine lands.

After protecting from grazing silvi-pastoral system involving the above-mentioned tree and grass species and introducing legumes such as *Stylosanthes, Alysicarpus*, etc may be developed with great success. High value medicinal species such as *Aloe vera* and *Jatropha curcas* can easily be blended in these habitats. Integrated Watershed Management Programmes are being implemented on a massive scale since 1991, which is most sustainable multipurpose strategy. A review of more than 300 integrated watershed management projects indicated that in majority of them total crop production increased by 50 to 123 per cent (Joshi *et al.*, 2005).

Table 1 : Area of top 10 states affected by potential soil erosion (Percent of total geographical area)

State	Moderate 10-15 t/ha/yr	Moderate severe 15-20 t/ha/yr	Severe 20-40 t/ha/yr	Very severe 40-80 t/ha/yr	Extra severe >80 t/ha/yr	Total >10 t/ha/yr
Nagaland	4.09	3.80	15.96	28.48	34.94	87.27
Meghalaya	14.78	10.21	26.25	13.86	12.80	77.90
Arunachal Pradesh	5.10	5.42	23.65	27.31	11.17	72.65
Assam	4.58	18.08	14.83	28.30	-	65.79
Chhatisgarh	7.99	6.45	18.22	13.62	19.02	65.30
Jharkhand	15.55	11.44	20.95	12.20	4.63	64.77
Madhya Pradesh	12.92	9.53	18.93	9.31	8.56	59.25
Uttar Pradesh	27.58	9.95	8.26	13.44	-	59.23
Uttarakhand	7.71	7.04	9.24	34.23	-	58.22
Manipur	15.25	11.43	26.61	-	-	53.29
Country	**11.22**	**6.46**	**9.92**	**7.14**	**3.99**	**38.73**

Source: NBSS & LUP (2008)

Water-harvesting technologies resulted in 50-156% increase in irrigated area under different schemes, which increased average cropping intensity by 64% (NAAS, 2009). Apart from increasing agricultural productivity, these projects helped the stake holders in generating employment and about 47% of degraded lands have been treated for rehabilitation (Sharda *et al.*, 2008). In remaining projects also agroforestry may be incorporated as main component particularly on highly degraded areas.

Agroforestry for sand dune stabilization

The most important measures for sand dune stabilization are covering the area under trees and providing a surface cover of grasses followed by their protection against biotic interference. Besides fixing the sand dunes it is important to check the movement of loose sand by applying wind breaks and mulch. Locally available brush woods like *Leptadenia pyrotechnica, Calligonum polygonoides* (now rare due to over-exploitation), *Ziziphus nummularia and Aerva tomentosa* and grasses like *Cenchrus ciliaris, C. setigerus, Lasiurus sindicus, Panicum turgidum* and *Saccharum munja* are being used frequently. The vegetation for sand dune stabilization is highly drought tolerant with deep root system capable of extracting moisture from lower soil depths.

Trees such as *Acacia tortilis, A. jacquimontii, A. leucophloea, A. Senegal, Azadirachta indica, Balanites roxburghii, Prosopis cineraria, P.juliflora* and *Holoptelia integrifolia* in combination with grasses *Cenchrus ciliaris, C.setigerus, Dichanthium annulatum* and *Panicum antidotale* have been found most successful for sand dune stabilization. These grasses can yield 2.2 to 3.8 t/ha/yr dry fodder with no ill effect on tree growth. Silvi-pastoral system is most viable, sustainable and profitable system. If need arises one or two irrigation of saline water up to EC 10 dS/m may be applied during dry period. It will also assure intangible benefits such as amelioration of soil, climate; control of soil erosion, shelter to annual crops in vicinity and protection to wild life. Arable crops such as Moth (*Vigna mungo*), green gram (*V.radiata*) and cluster bean (*Cyamopsis tetragonoloba*), tara-mira (*Eruca sativa*) and castor (*Ricinus communis*) can successfully be cultivated when there is some rains at the time of sowing. Species suitable for shelter belts include *Acacia tortilis, A.salicina, A. aneura, A. ampleceps, A.nilotica* ssp *cupressiformis, Tamarix articulata, Parkinsonia aculeate, Prosopis juliflora* and *Eucalyptus camaldulensis*. Following suitable techniques and protecting the area by planting close-spaced wind breaks and shelter belts of suitable trees and using drip irrigation (even of saline water up to EC 10 dS/m) even fruit trees such as Ber (*Ziziphus mauritiana*), Karonda (*Carissa carandus*), pomegranate (*Punica granatum*), tamarind (*Tamarindus indica*-frost sensitive), Lasura (*Cordia rothii, dichotoma*), custard apple

(*Annona squamosa*) and date palm (*Phoenix dactylifera*) can be raised in desert environment.

Coastal dunes form a complex sequence of excessively drained ridges separated by poorly drained depressions. Along Orissa coast belts of cashew (*Anacardium occidentale*) plantations following the *Casuarina* line are quite common. Screw pine *Pandanus* is also quite frequent which may be explored commercially for its fruits yielding fragrant oil. *Casuarina equisetifolia* and *Eucalyptus tereticornis* are two very important trees along Andhra coast. Palmirah palm (*Borassus*) is most frequent in agricultural fields. Mangroves form the thick belt along protected shores and creeks. These have been denuded in many areas and are in depleted condition through out the coast. There importance was realized during tsunami and cyclones along Orissa coast. Behind mangrove belt species such as *Pongamia pinnata, Terminalia catappa, Canophyllum inophyllum, Thespesia populnea, Pandanus* spp. and *Cynomitra ramiflora* can successfully be explored for their commercial importance. Mangrove palm (*Nypa fruticons*) can successfully be cultivated along creeks for alcohol production.

Agroforestry for acid soils

In humid tropics the soils are generally acidic and low in nutrient availability. Some carry toxic levels of iron and aluminium. The low-lying coastal lands may contain acid sulphate soils derived from marine and estuarine sediments with high concentration of reduced sulphur components. Upon drainage and aeration they undergo severe acidification bringing the pH values of the soil at times below 4 in the upper 50 cm layer. Low pH adversely affects the availability of calcium, magnesium and other nutrients. Drainage results in more oxidation causing further soil degradation. Such lands are managed for rice cultivation and brackish water fish culture. Appropriate agroforestry systems may also prove useful in the management of acid soils as woody perennials can recycle nutrients, maintain soil organic matter and protect the soil from erosion and runoff. The home gardens, coffee and cacao production systems, plantation based multi-tiered dense cropping systems and alley cropping on sloping lands represent typical agroforestry systems. Aqua culture keeping mangroves intact may be ideal profitable and sustainable practice in tidal zone. Dagar (1995) and Kumar and Nair (2004) have dealt with in detail some of these systems.

Agroforestry for Salt-affected Lands

In order to rehabilitate salt affected lands, appropriate tree planting techniques and choices of tree species are very crucial for reducing mortality and consequently for improvement in the initial establishment of saplings. Since alkaline and saline soils differ from each other, methods of working the soils will also be different. For example, in alkali soils a hard *kankar* layer of calcium carbonate is generally found at a depth of about 1.25 to 1.5 m. This layer acts as a barrier for root penetration. The layer, therefore, has to be broken first to allow proper development of roots. However, saline soils do not require such preparation, as they do not have any such barriers. These require special techniques of afforestation so that salt contents in root zone are minimized.

The ideal planting method for alkali soils should provide a favorable soil environment such as by breaking the hard *kankar* layer, replacement of exchangeable sodium and additional nutrition of tree species for optimum root growth. Keeping this view in mind, pit-auger-hole technique of tree plantation has been developed by the scientists of CSSRI. In this planting method, auger-holes of 15-20 cm diameter are made to pierce the hard *kankar* layer up to 150 -180 cm deep, with the help of a tractor-mounted auger after digging pits of 35 x 35 cm. Auger holes are refilled with original soil, 3 kg gypsum, 8 kg FYM, 10 g $ZnSO_4$, and small quantity of insecticide to take care of termites. Sodicity tolerant tree saplings of 6-9 months old are planted in the refilled pit-auger holes followed by irrigation with buckets. Two to three irrigations are immediately needed for establishment of saplings. This method enables the plant roots to grow at a faster rate towards deeper soil layers where sufficient moisture and nutrients are available in alkali soils. The *kankar* layer, which creates hindrance in the development of plant roots, is broken in the process of making holes. The post-auger hole planting technique was further refined by providing sub-surface planting and furrow irrigation method (SPFIM). The pits of 45 cm in size are prepared manually and the pit-holes beyond the *kankar* layer are dug with the augers (15-20 cm diameter) and then these pits are connected by irrigation furrow. CSSRI, Karnal has been conducting several long-term experiments for developing afforestation technologies on highly saline waterlogged soils. The results suggested that furrow planting improved the survival and growth of tree species as compared to ridge planting method. Besides reducing the water application costs, it improves uniformity in water application and helps in creating a favorable zone of low salinity below the sill of the furrow through downward and lateral fluxes of water making salts move away from the furrow (root zone) especially when low salinity water is used. Creation of such niches favored the establishment of young seedlings of trees. Moreover, such a system seems to be more viable from practical viewpoint of undertaking large-scale plantations of trees.

The choice of species for alkali lands is determined by the ability of tree species to survive and withstand adverse conditions of excess sodicity. Tolerant tree species overcome high concentrations of sodicity by different regulatory mechanisms. There are very few wild plant species, which are able to grow on highly sodic soils (pH >10). On the basis of experiments conducted on highly alkali soil *Prosopis juliflora, Acacia nilotica, Casuarina equisetifolia, Tamarix articulate, Eucalyptus tereticornis* and *Parkinsonia aculeata* demonstrated a higher tolerance. The biomass of 7 years old *T. articulata, A. nilotica* and *P. juliflora* was 97.3, 69.8 and 51.3 t/ha, respectively. Singh et al (2008) reported 19.22 to 56.5 t / ha air-dry biomass from different species raised on high sodic soil (Fig.1).

More than 40 tree species of arid and semi - arid areas were evaluated for their salinity tolerance (Tomar *et al.*, 1998). The woody species like *Acacia farnesiana, Parkinsonia aculeata* and *Prosopis juliflora* were rated the most tolerant to waterlogged salinity and could be grown satisfactorily on soils with salinity levels up to 50 dS/m in their root transmission zone. Tree species like *Acacia nilotica, A. torilis, Casuarina glauca, C. obesa* and *C. equisetifolia* could grow on sites with ECe varying from 10 - 25 dS / m. It was concluded that the waterlogged saline conditions affected the survival and growth of tree species used for afforestation the most because the salt accumulation near the rooting zone was directly attributed to ground water fluctuations and the underground water was also saline. The biomass of *P. juliflora* and *C. glauca* was the highest (98 and 96 t/ha, respectively) followed by *A. nilotica* (52 - 67 t/ha) and *A. tortilis* (41 t/ha) when planted with sub surface or furrow techniques (Table 2) proving that these are the most suitable species for saline waterlogged soils (Tomar *et al.*, 1998).

Raising trees on degraded calcareous soil in dry regions irrigating with saline water

Field experiments were conducted in calcareous soils at Bir Reserved Forest, Hisar. The experimental site represented semi-arid monsoonal climate with annual rainfall of 430 mm and most of which (70-80%) occurs during July-September. Potential evapo-transpiration generally exceeds rainfall. The experimental soil was sandy loam in texture and highly calcareous in nature ($CaCO_3$ 1.8 to 15.0 %). The transplanting of 6 months old saplings of 31 tree species was done during August, 1991 in auger-holes made in the furrows. Spacing between the furrows was kept at 2.5 m and plant to plant 2.0 m. The V - shaped furrows (20 cm deep and 60 cm wide at the top) were made with the help of tractor driven furrow maker. Saplings were irrigated with saline water (ECiw 8.5-10 dS/m) during the early establishment stage. The performance rating of the evaluated species is summarized in Table-3.

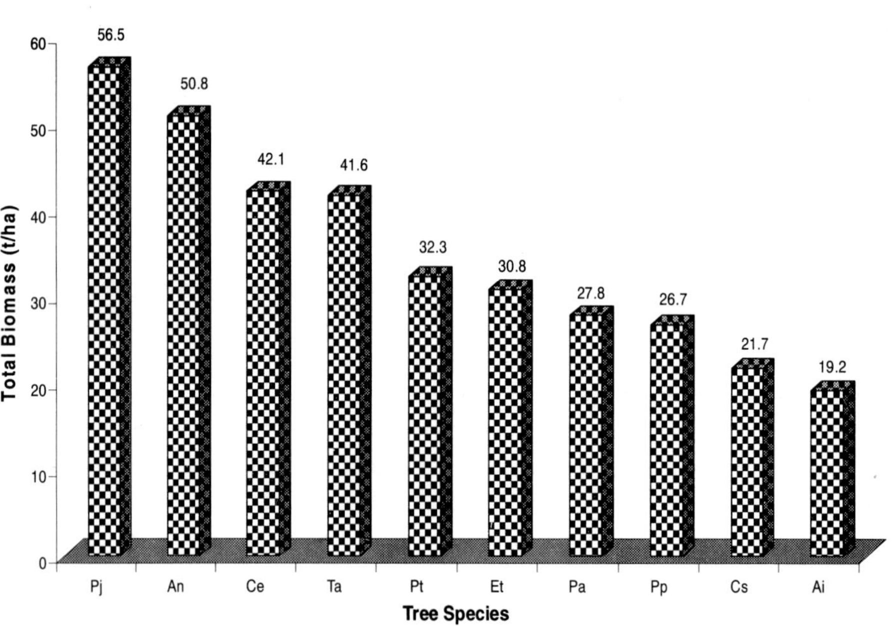

Fig. 1 : Above-ground air-dried biomass (t/ha) of different trees grown in sodic soil

Pj=*Prosopis juliflora*, An= *Acacia nilotica*, Ce= *Casuarina equisetifolia*, Ta= *Terminalia arjuna*, Pd=*Pithecellobium dulce*, Et= *Eucalyptus tereticornis*, Pa= *Prosopis alba*, Pp= *Pongamia pinnata*, Cs= *Cassia siamea*, Ai= *Azadirrechta indica*.

Table 2: Biomass estimation of trees after 9 years of planting on saline soils

Tree species	Method of planting	Range of soil salinity ECe at 0 - 120 cm depth (dS/m)	Range of water Table salinity EC (dS/m)	Estimated biomass (t/ ha)
Acacia nilotica	Subsurface	10.6 - 25.3	27 - 33	52
	Furrow	11.1 - 21.0	17 - 27	67
A. tortilis	Subsurface	6.8 - 28.1	12 -33	41
	Ridge	19.7 - 29.1	12 - 33	6
Eucalyptus camaldulensis	Furrow	10.0 - 17.9	10 - 35	28
Prosopis juliflora	Subsurface	10.3 - 24.0	32 - 36	98
	Ridge	23.5 - 57.5	32 - 36	65
Casuarina equisetifolia	Furrow	5.6 - 20.7	10 - 31	28
C. glauca	Furrow	6.5 - 33.9	12 - 19	96
C.obesa	Furrow	9.0 - 19.5	12 - 19	38
Leucaena leucocephala	Subsurface	6.9 - 23.9	10 - 25	30
Tamarix sp.	Furrow	8.2 - 21.3	10 - 32	12

Source: Tomar et al. (1998)

Table 3: Performance ratings of tree species with saline water irrigation (ECiw 10 dS/m)

Performance rating	Tree species
Very promising	*Azadirachta indica, Acacia nilotica, A. tortilis, A. farnesiana, Cassia siamea, Eucalyptus tereticornis, Feronia limonia, Prosopis juliflora, P. cineraria, Pithecellobium dulce, Salvadora persica, S. oleoides, Tamarix articulata*
Promising	*Melia azedarach, Cassia fistula, Ziziphus mauritiana*
Poor	*Acacia auriculiformis, Bauhinia variegata, Cassia glauca, C. javanica, Crescentia alata, Pongamia pinnata, Syzygium cumini, Tecomella undulata*

Evaluation and identification of promising forage grasses

A field experiment was conducted during 1993-1997 to evaluate the suitability of nine forage grasses to saline irrigation (ECiw 8.5-10.0 dS m^{-1}) and optimize its schedule. The average forage yield was found to be 0.85 t ha^{-1} in grazed condition and 2.4 t ha^{-1} when protected but when brought under judicious saline irrigation (ECiw 8-10 dS m^{-1}), the grasses raised on the same land could produce from 4.4 to 16.9 t ha^{-1} dry forage from different grasses (Table 4). *Panicum laevifolium* produced maximum forage biomass under all the treatments followed by P. maximum. Even in the lean period (when people are forced to lead nomadic life along with their herds of cattle) sufficient forage was available from all these perennial grasses. Scheduling the saline irrigation at Diw/CPE ratio of 0.4, improved the yields by about 20% while no further improvement was obtained with enhanced saline irrigation supplies.

Table 4 : Gross dry matter yield (t ha^{-1}) of different grasses (average of 3 years) irrigated with saline water of ECiw 10 dS m^{-1}

Grass species	Irrigation with saline water with Diw/CPE				
	CW	0.2	0.4	0.8	Mean
Brachiaria mutica	14.15	9.54	12.15	11.72	11.89
Cenchrus setigerus	5.61	4.64	4.57	4.38	4.80
Cynodon dactylon	13.53	8.91	9.23	10.20	10.47
Panicum antidotale	15.05	9.34	11.41	11.77	11.89
P. coloratum	15.02	6.95	10.29	8.93	10.30
P. laevifolium	22.16	13.49	16.85	16.88	17.34
P. maximum(cultivated)	19.23	10.87	13.04	12.72	13.96
P. maximum (wild)	21.59	14.00	14.72	13.72	16.01
P virgatum	17.93	9.95	12.10	11.36	12.83
Mean	16.03	9.74	11.60	11.30	12.17

Source: Tomar et al. (2003a)

Fruit based agroforestry system

One field experiment was initiated in 2002 involving three fruit tree species namely **Karonda** (*Carissa carandus*), **Aonwla** (*Emblica officinalis*) and **Bael** (*Aegle marmelos*) imposing four treatments: T_1 = traditional ring method, irrigating with available low salinity water of EC 5-8 dS m^{-1}, T_2 = planted in furrows irrigating with low salinity water (as above), T_4 = planted in furrows and irrigating with water of high salinity (EC 8-12 dS m^{-1}), and T_3 = planted in furrows irrigating alternately with above two waters. In this experiment, the plant-to-plant distance was kept 4 m (2 m in *karonda*) and row-to- row distance 5 m. One time irrigation was supplied in each dry month. The inter-spaces are being cultivated with arable crops viz. cluster bean/pearl millet as *kharif* crop(s) and barley as *rabi* crop.

During first year of establishment, on an average, 2.4t ha^{-1} grain and 10.8 t ha^{-1} air-dry fodder as straw of pearl millet and 0.0.87t ha^{-1}

grain and 2.6 t ha^{-1} straw of cluster bean could be harvested from inter-crops during *Kharif* season and 3.3t ha^{-1} grain and 3.5t ha^{-1} straw of barley was harvested during *Rabi* season (Table 5). During next three years also good crop of barley and cluster bean was harvested (Table 6). Among non-conventional crops castor (*Ricinus communis*), *Aloe vera*, Dill (*Anethum graveolens*), tara-mira (*Eruca sativa*), Isabgol (*Plantago ovata*) and lemon grass (*Cymbopogon flexuosus*) could be cultivated successfully. Agronomic practices for Isabgol and lemon grass have been developed.

Table 5 : Grain & straw yield (t ha^{-1}) of inter-crops along with fruit trees during first year

Fruit trees	Crops	T_1 Grain	T_1 Straw	T_2 Grain	T_2 Straw	T_3 Grain	T_3 Straw	T_4 Grain	T_4 Straw
Karonda	Pearl millet	2.45	11.2	2.51	11.28	2.35	11.25	2.18	10.65
	Cluster bean	0.85	2.63	0.91	2.96	0.79	2.51	0.70	2.46
Anwla	Pearl millet	2.50	10.98	2.64	11.32	2.25	11.00	1.98	10.48
	Cluster bean	0.96	3.31	1.03	3.43	0.95	3.20	0.91	2.86
Bael	Pearl millet	2.00	10.45	2.25	11.05	1.84	10.00	1.72	9.85
	Cluster bean	0.92	3.00	0.97	3.31	0.90	2.95	0.86	2.74

LSD (p=0.05)	**Grain**	**Straw**		
	Pearl millet	**Cluster bean**	**Pearl millet**	**Cluster bean**
Between fruit trees (A)	0.134	0.073	NS	0.425
Between treatments (B)	0.097	NS	NS	0.189
Interactions (A) x (B)	NS	NS	NS	NS

Table 6 : Grain and straw yield (t ha^{-1}) of barley and cluster bean with different plantations during 2006-07

Fruit tree	Treatment	Yield of barley		Yield of cluster bean	
		Grain	Straw	Grain	Straw
Karonda	T_1	4.01	4.15	1.80	2.70
	T_2	4.10	4.21	1.83	2.63
	T_3	3.81	3.89	1.78	2.37
	T_4	3.40	3.49	1.68	2.52

Aonwla	T_1	4.38	4.58	1.80	2.72
	T_2	4.23	4.32	1.85	2.77
	T_3	3.84	4.02	1.63	2.44
	T_4	3.52	3.61	1.50	2.27
Bael	T_1	3.85	3.99	1.53	2.24
	T_2	3.96	4.08	1.63	2.34
	T_3	3.62	3.73	1.48	2.00
	T_4	3.26	3.34	1.45	1.89
LSD (p = 0.05)					
Factor A (species) =		*0.13*	*0.18*	*0.16*	*0.11*
Factor B (treatment) =		0.15	0.19	0.09	0.13
Interaction (A x B)=		0.26	0.28	NS	NS

Amongst the species tested for medicinal value, the most promising was psyllium (*Plantago ovata*) with average seed yield of 1050 kg/ha and with saline water (EC 8.5 dS/m) did not show any adverse impact when compared with canal water irrigation (Tomar and Minhas 2004b, Tomar et al. 2005). Psyllium did not show any yield reduction with *Acacia* plantation even at later stages showing its suitability for partial shade tolerance. *Aloe barbadensis* was also equally tolerant and could produce 18 t/ha fresh leaves. *Ocimum sanctum* could produce 910 kg/ ha dry shoot biomass. In a separate trial dill (*Anethum graveleus*), taramira (*Eruca sativa*) and castor (*Ricinus communis*) could produce 931, 965 and 3535 kg seeds per ha, respectively when provided with three irrigation of saline water (EC 10 dS/m (Dagar et al. 2008). *Cassia senna* and *Lepidium sativum* can also be cultivated successfully irrigating with saline water up to 8 dS/m. All these high value crops can successfully be grown as inter-crops with forest or fruit trees at least during initial years of establishment (Dagar 2009, Dagar et al. 2006, 2008). It could be concluded that ornamental flowers such as *Chrysanthemum, Calendula* and *Matricaria* can successfully be cultivated irrigating with water of EC up to 5 dS/m. These species could yield 13.2, 4.7 and 3.5 t/ha, respectively fresh flowers in a season (Tomar and Minhas 2002). If good quality water is available at site, a few irrigations particularly for establishment will increase the yield of flowers. The aromatic grasses such as vetiver, lemon grass and palmarosa, when irrigated with saline water (EC 8.5 dS/m) could produce on an average 90.9, 10.4 and 24.3 t/ha dry biomass, respectively (Tomar and Minhas 2004a). Different cultivars of vetiver could produce 72.6 to 78.7 t/ha shoot biomass and 1.12 to 1.71 t/ha root biomass. The roots are used to extract aromatic oil.

Agroforestry for salty black soils (saline Vertisols) using saline irrigation

The salty soils of black soil zone are generally either contemporary or of secondary origin. The contemporary salty soils exist in the topographic situation having poor drainage conditions. However, the soils that have become sodic due to injudicious use of irrigation water can be encountered in the irrigation command area. In one long-term experiment, after 14 years of plantation it was found that *Prosopis juliflora* and *Azadirachta indica* were most successful species for these soils. Among grasses, *Aeluropus lagopoides, Leptochloa fusca, Brachiaria mutica, Chloris gayana, Dichanthium annulatum, Bothriochloa pertusa* and species of *Eragrostis, Sporobolus* and *Panicum* are most successful. Aromatic grasses such as *Vetiveria zizanioides* and *Cymbopagon martinii* can be grown easily. *Matricaria chamomile* can withstand both high pH and ESP.

In a separate fruit trial on soil of ESP 25, 40 and 60 it was found that gooseberry (*Emblica officinalis*) and Ber (*Ziziyphus mauritiana*) were the most successful plantations for these soils. Oil-yielding bush *Salvadora persica* was grown in combination with *Leptochloa fusca, Eragrostis* spp. and *Dichanthium annulatum* forage grasses on saline vertisol in Gujarat. The soil was clay loam (clay 40%, silt 31%, sand 29%) with pH ranging from 7.2 to 8.9 and ECe from 25-70 dS/m. The underground water was 0.5-2 m from surface with EC ranging from 55-60 dS/m. These grasses could produce on an average 3.7, 1.0 and 1.8 t/ha of green forage, respectively. During fourth year the seed yield of *Salvadora persica* ranged form 1.84 to 2.65 t /ha with oil contents ranging from 576-868 kg/ha at different salinity levels (Gururaja Rao *et al.* 2003).

The experiments conducted in sodic vertisols with ESP 40 growing grasses like *Leptochloa fusca, Brachiaria mutica* and *Vetiveria zizanioides,* showed that all these grasses performed well and the forage biomass increased during second year. Besides producing biomass silvi-pastoral system helped in amelioration of soil in terms of reducing soil pH, EC and ESP and increasing organic matter.

Combating waterlogging through biodrainage

In waterlogged areas near canals planting cloned *Eucalyptus* on bunds (of ~ one meter height) on farmers field (in two lines in a space of 1m x 1m) each strip separating 66m (growing crops between two strips), proved very useful, which not only controlled rise in water table (Fig 2) but also helped in revenue generation. Farmers harvested 36t/ha wood biomass (timber suitable for poles/ballies, pulp wood and fuel- wood) from 5 years and 4 months old plantation (Jeet Ram *et al.*, 2008). After harvest trees could coppice further. In saline areas lining of poly-sheets on bunds covered with thin layer of soil helped in controlling the development of salinity.

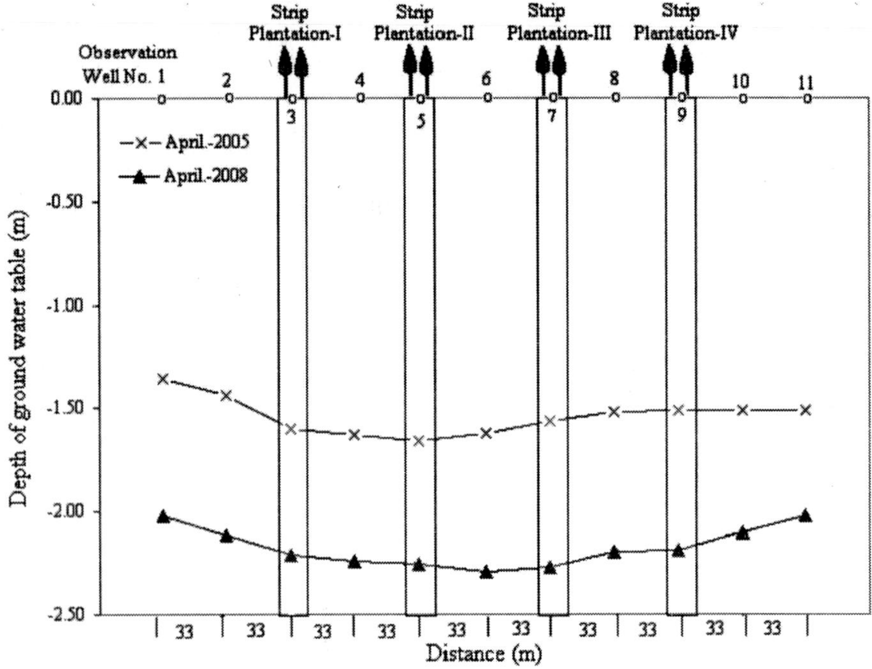

Fig. 2 : Trend of ground water table under plantations and adjacent fields

Conclusion

Agroforestry may play vital role in rehabilitation and biological reclamation of problem soils such as degraded dry lands prone to water and wind erosion including sand dunes; acid sulphate soils of humid regions characterized by low pH, toxicity of aluminium and iron and deficiency of nutrients; salt affected sodic and saline soils; and waterlogged saline soils. Profitable agriculture is possible if appropriate technologies and know how are used judiciously. For checking wind erosion, wind breaks and shelter belts involving appropriate species which require less water, possess deep root system, can be established easily with less water and grow fast will be useful tools. The trees and shrubs will play major role in improving efficiency of nutrient cycling in the system.

Biological barriers such as alley cropping and plugging in ravines, involving appropriate species play important role in checking water erosion in sloping and ravine lands. The highly alkali soils (pH>10) may be rehabilitated with tree species such as *Tamarix articulata, Prosopis juliflora*, and *Acacia nilotica*. The saline soils may be successfully brought under exotic species such as *Prosopis juliflora, Casuarina glauca, C .equisetifolia, Acacia farnesiana, A. tortilis, A. nilotica*, and *Eucalyptus camaldulensis*. These species also ameliorated the soil. The water logging in canal command areas can successfully be combating by planting trees such as Eucalyptus tereticornis on killa-line on farmers' field and in strips along canals. For arid soils irrigated with saline water, the recommended species include: *Tamarix articulata, Prosopis juliflora, Feronia limonia, Eucalyptus tereticornis, Acacia nilotica, A. tortilis, A. tortilis* (hybrid) and *Cassia siamea*. These species not only produced economic yields in terms of wood biomass but also improved soil conditions i.e. in terms of organic matter and physical properties. Grasses such as *Panicum laevigatum* and *P. maximum* can produce good forage with saline irrigation. Fruit-based agroforestry system involving Ber, Anowla, Karonda, Bael and Kaith as fruit trees and low water requiring crops such as cluster bean, pearl millet and barley as inter-crops is most suitable for calcareous soils irrigating with saline water. Non-conventional crops such as castor, *Aloe vera*, Dill, tara-mira (*Eruca sativa*), Isabgol, senna and lemon grass could be cultivated successfully.

References

Dagar JC. (1995). Agroforestry systems for Andamans & Nicobar Islands in India. *International Tree Crops Journal* **8** (2-3): 107- 128.

Dagar JC. (2009). Opportunities for alternate land uses in salty and water scarcity areas. International *Journal of Ecology & Environmental Sciences* **35** (1): 53-66.

Dagar JC, Tomar OS, Kumar Y, Bhagwan H, Yadav RK and Tyagi NK. (2006). Performance of some under-explored crops under saline irrigation in a semi-arid climate in northwest India. *Land Degradation & Development* **17**: 285-299

Dagar JC, Tomar OS, Minhas PS, Singh Gurubachan and Ram Jeet. (2008). *Dryland Biosaline Agriculture -Hisar Experience.* Techanical Bulletin 6: CSSRI, Karnal, pp28

Dagar JC, Singh G. and Singh NT (2001). Evaluation of forest and fruit trees used for rehabilitation of semiarid alkali soils in India. *Arid Land Research and Management* **15**: 115-133

DES (2007). *Agricultural Statistics at a Glance*. Directorate of Economics & Statistics (DES), Ministry of Agriculture, Govt. of India, New Delhi.

Gururaja Rao, G, Nayak AK and Chinchmalatpure AR. 2003. *Salvadora persica*: A life support species for salt affected black soils. Technical Bulletin 1/2003. CSSRI, Karnal, India, pp. 44.

IDNP. (2002). Recommondations on waterlogging and salinity control based on pilot area drainage research. *Indo-Dutch Network Project (IDNP)*, CSSRI, Karnal, India pp.100.

Jeet Ram, Dagar JC, Singh Gurubachan, Lal Khajanchi, Tanwar VS, Shoeran SS, Kaledhonkar MJ, Dar SR, and Kumar Mukesh.(2008). *Biodrainage: Eco-Friendly Technique for Combating Waterlogging & Salinity*. Techanical Bulletin 9 : CSSRI, Karnal, pp 24.

Joshi PK, Jha AK, Wani SP, Laxmi Joshi and Shiani RL. (2005). Meta-analysis to assess impact of watershed program and peoples' participation. Comprehansive Assessment Research Report 8, Comprehansive Assessment Secretariat, Colombo, Srilanka.

Kumar BM and Nair PKR. (2004). The enigma of tropical homegardens. *Agroforestry Systems* **61**: 135-152

Mandal D, Sharda VN and Tiwari A. (2008). *Soil Loss Tolerance Limits in Planning Conservation Measures in Different Districts of India*. CSWCRTI, Dehra Dun India, pp112.

NAAS. (2009). *State of Indian Agriculture.National* Academy of Agricultural Sciences (NAAS), New Delhi, pp 256.

NBSS & LUP. (2008). Harmonization of degraded lands/wastelands datasets of India-A GIS based approach. NBSS & LUP, Nagpur.

Oldeman LR. (1991). Global extent of soil degradation. Bi-ennial Report, ISRIC, Wageningen, The Netherlands.

Sharda VN, Juyal GP and Naik BS. (2008). *Watershed Development in India: status and perspective*. Allied Printers, Dehradun, India pp 219.

Singh, Gurubachan and Dagar JC. (2005). *Greening Sodic Lands: Bichhian Model*. Technical Bulletin No.2/2005 CSSRI, Karnal India.pp 51.

Singh YP, Sharma DK, Singh G, Nayak AK, Mishra VK and Singh R.(2008). *Alternate Land Use Management for Sodic Soils*. CSSRI Technical Bull No. 2/2008, CSSRI, Karnal, pp 16.

TERI. (1998). Looking Back to Think Ahead.Tata Energy Research Institute (TERI), New Delhi.

Tiwari DN. (2000). Review of Forest Management and Policies. ICLRM, Nov, 2000, SCSI, New Delhi.

Tomar OS, Gupta Raj K and Dagar JC (1998). Afforestation techniques and evaluation of different tree species for waterlogged saline soils in semiarid tropics. *Arid Soil Research and Rehabilitation* **12**: 301 - 16.

Tomar, O.S. and Minhas PS. (2002). Performance of some ornamental winter annual flowering species under saline irrigation. *Indian Journal Horticulture,* **59** (2): 201-206.

Tomar OS and Minhas PS. (2004a). Relative performance of aromatic grasses under saline irrigation. *Indian Journal of Agronomy,* **49** (3): 207-208.

Tomar OS and Minhas PS.(2004b). Performance of medicinal plant species under saline irrigation. *Indian journal of Agronomy,* **49** (3): 209-211.

Tomar OS, Minhas PS, Sharma VK and Gupta RK. (2003a). Response of nine forage grasses to saline irrigation and its schedules in a semi-arid climate of north-west India. *Journal of Arid Environments* **55**: 533-544.

Tomar OS, Minhas PS, Sharma VK, Singh YP and Gupta RK. (2003b). Performance of 31 tree species and soil condition in a plantation established with saline irrigation. *Forest Ecology and Management,* **177**: 333-346.

❑❑❑

Chapter 27

Land Degradation in Central India : The Concern of Sustainable Food Security

S.K. DUBEY AND R.K. DUBEY
Central Soil and Water Conservation Research and Training Institute, Research Centre, Chhalesar, Agra (Uttar Pradesh) 282 006 India

Land degradation refers to decline in overall quality of soil, water or vegetation conditions mostly caused due to human activities that broadly includes soil erosion, rise in water table, expression of salinity, mass movement of soil or rocks by gravity, stream bank instability and processes that results in declining soil and water quality. Vegetation Management Act (1999) states that phrase land degradation also includes a change in the ground cover to less palatable species or a change from predominantly perennial grasses to predominantly annual grasses. In terms of managing main environmental systems, India ranks at forth from below at 50 per cent sustainability index (Ratna Reddy, 2003). Therefore, effective measures to arrest degradation for sustainable development and inter generation equity on inter and intra regional basis, are needed. Land resources are linked with other resources like water, forests etc and land quality is directly affected by the agricultural practices. Land degradation affects both availability of land for agriculture and soil fertility, thus threatens the future sustainable production.

Scinerio of land degradation in Central India

Out of total degraded lands in the country (146.82 million ha), Central India accounts for 32.01per cent i.e. about 47 million ha. Madhya Pradesh and Chhatisgarh account for 26.21 million ha while undivided Uttar Pradesh for 15.32 million ha (Table 1). Cause wise water erosion accounts for major part of land degradation (63.7per cent) followed by soil acidity (10.9per cent) and water logging (9.7per cent). Extent of flood affected area vary from year to year (2.29 million ha in 1953 to 8.03 million ha in 2004 with a peak of 17.5 million ha in 1978). Bihar is worst affected by the floods in the Central India. Barren land in Central India have been estimated to be 3665 thousand ha in 1997 (Bihar: 1010, Madhya Pradesh: 1714 and Uttar Pradesh: 941). Jharkhand with considerable tribal population (22.5per cent) and an annual average rainfall of 1370 mm has severe problem of water erosion due to light textured acidic soils with low availability of phosphates, poor soil water retention and undulating topography. In Uttar Pradesh, 28.5 lakh ha of land is subject to serious soil erosion. In addition, there are degraded lands (38 lakh ha) that are sodic, ravine, water-logged or flood prone. Out of total area of 24.20 M ha in Uttar Pradesh the net cultivated area is 16.80 M ha but 2.85 M ha is subjected to serious soil erosion. Districts in Uttar Pradesh like Sonbadhra, Mirzapur, Varanasi, Chitrakoot, Rae Bareily, Kanpur Nagar, Kanpur Dehat, Kannauj, Mainpuri, Lalithpur, Jhansi and Etah have greater than 9 per cent of area under degraded land.

Table 1 : Area (thousand ha) under different kinds of land degradation in various states of Central India.

S. No.	State	Water Erosion	Wind Erosion	Water logging	Salinity/ Alkanity	Soil Acidity	Complex Problem	Degraded Area	Degraded Area (%)
1.	Bihar +Jharkhand	3024	0	2001	229	1029	0	6283	36.1
2.	Madhya Pradesh + Chhatisgarh	17883	0	359	46	6796	1126	26210	59.1
3.	Uttar Pradesh (Undivided)	11392	212	2350	1370	0	0	15324	52.0
	Total	32299	212	4710	1645	7825	1126	47817	

Causes of Land Degradation in Central India

The current trend of economic and industrial development, steady growth of human and livestock population, deforestation beyond silviculturally permissible limits (timber @ about 15 million M^3 $annum^{-1}$, FSI 1987); deforestation on soils having steep slopes or shallow depth; unsustainable fire wood and fodder extraction (260 million M^3 in 1997 as against sustainable supply of 52.6 million M^3, Pachauri and Sridharan, 1998); water and wind erosion on an area of 148.9 and 13.5 million ha, respectively (displacement erosion: 162.4 million ha, *in situ* degradation due to salinity and water logging: 21.7 million ha and nutrient depletion: 4 million ha, CSWCRTI, 1994); non adoption of adequate soil and water conservation measures; bio-physical and chemical *in situ* deterioration of soil due to indiscriminate use of fertilizers and pesticides; improper crop rotations and irrigation systems; frequent shifting cultivation with reduced fallow period from 30 to 2-3 years; encroachment in to forest lands (0.7 million ha); forest fires; over grazing; extension of agriculture to low potential lands; mining particularly the opencast mining (0.8 million ha) that disturbs physical, chemical and biological features of soil; exploitation of ground water at the rate more than the natural recharge; open access resource and poverty are the major reasons behind the grave land gradation in Central India. Increase in number of land holdings (48 to 105 million in 1960 to 1990, respectively), increase in rural population, shrinking farm size and increasing landless labors have led to non sustainable land management, a direct cause for land degradation. The current phase of economic and trade liberalization are exerting heavy pressure on India's limited land resources for competing uses in forestry, agriculture, pastures, human settlements, roads and transport and industries, a very significant factors in land degradation.

Unscientific and arbitrary land management have substantially degraded agricultural (80 out of 142 million ha) and pasture (about 11 million ha) lands. According to Waste land Atlas of India, total wasteland in the country is 63.85 million ha which accounts for 20.17per cent of total geographical area. Mining sites abandoned after excavation work follow complete leaving deep scars and traces of over burdening. In states of Jharkhand, Chhattisgarh and Madhya Pradesh, deforestation due to mining have caused severe land degradation. The mineral processing like grinding of limestone for cement industry and calcite and soapstone for ceramic industry generate huge quantity of dust in the atmosphere. It retards the process of infiltration of water into the soil after it settles down on the land. In Uttar Pradesh, over irrigation particularly in canal commands is responsible for land degradation due to high water table or water logging on account of enhanced soil salinity and alkanity. Soil erosion may decline crop productivity from 5 to 50per cent depending upon severity that ultimately depends on intensity of rains, soil erodibility and depth, steepness and length of slope, ground cover, crop management practices, resistance of crop to soil erosion and adoption of soil conservation measures. However, crop loss due to soil erosion was estimated to be 1.7 per cent by UNDP, FAO and UNEP (1993) and 4.1per cent by Bansil (1990). Crop response to erosion depends on rooting behavior, growth rate and ability of crop to provide ground cover. Response of cotton to soil erosion in black soils of Central India was excellent while ground nut showed least soil conservation potential. Annual loss in crop production due to soil erosion in India has been projected from 7.2 million tons (UNDP, FAO and UNEP, 1993) to 13.5 million tons (Bansil, 1990). Nearly 3.7 million ha area suffers from nutrient loss and/or depletion of organic matter. India loses nearly 7.4 million tons of major nutrients due to soil erosion annually (effective loss: 39 per cent), thus country loses 0.8 million tons of nitrogen, 1.8 million tons of phosphorus and 26.3 million tons of potassium every year. As per the Government of India estimate about 5.8 to 8.4 million tons of nutrients are lost every year due to soil erosion.

In India, about forty two animals graze in one ha pasture as against recommended five animals' ha^{-1} (Sahay, 2000). About one third of total fodder requirement for stall feeding is met out of the forests (MoEF, 1999). About 78per cent of forests are grazed in India. Thus, livestock population in the country is not only exerting heavy pressure on forests for fodder but also adversally affecting the natural regenration of forests. In Madhya Pradesh, over grazing is one of main reasons for land degradation. Over grazing, compaction and reduced infiltration adversely affect the natural regeneration of forests particularly of palatable tree species and accelerates the water erosion. In Uttar Pradesh, 14.5 per cent area is under ravines. Per capita forest cover in Central India is much lower (0.27 ha) than required cover of 0.47 ha $person^{-1}$. Increasing demographic pressure on land to satisfy the food, raising standards and allied needs besides industrialization, housing and other developmental projects is in order of agricultural land> grazing land and pastures > forests > barren lands > uncultivable land in Central India. Semi arid and arid tracts face severe and recurrent shortage of fuel wood. Rapid growth of population due to high fertility rates, low mortality and increasing rural-urban migration and rapid industrialization are aggravating the major ecological and socio-economic crisis that are perpetuated by land/soil degradation specifically the farming activities which contribute to soil erosion, salinity/brackishness of land and loss of nutrients. Over exploitation of land and water resources and heavy use of fertilizers and pesticides following green revolution era are gradually degrading the soil health that is being revealed in form of frequent nutrient deficiencies and nutrient related crop diseases.

Sustainable food Security comprise of availability (production + imports), access (purchasing power) and absorption (safe drinking water and environment) of food in present and future. Food security may be achieved through stabilizing population, conserving and enhancing land resources, ensuring water security, conserving and restoring forests with community participation, strengthening biodiversity, improving the atmosphere, managing common property resources, intensifying crop and animal production (Swaminathan, 2007). Though Madhya Pradesh has a strong natural resource base and may rank high in sustainability but high poverty level is a major constraint in food security. Net sown area is declining in Bihar and in the process the prime agricultural land may shift to non agricultural uses. In states like Bihar, Uttar Pradesh and Madhya Pradesh, there is an urgent need to diversify livelihood to non-crop and non- agricultural enterprises. Demand for commercial timber (64.4 million M^3 in 1996 + annual growth rate of 5 per cent) comes from industries including pulp and paper, plywood, packaging, housing, matchwood, sports goods, furniture, agricultural implements and railway coaches (FSI, 1995 and 1987).

Uttar Pradesh has nearly 7.4 million ha of different kinds of degraded lands and poor soil cover (5.86 per cent). Poor soil cover, high annual rainfall (for instance 1200-1400 mm in easten region), lack of follow up of soil and water conservation measures etc are main reasons behind land degradation in the state. Soil cover ranges from a minimum of 0.1per cent in Sant Ravidash Nagar to a maximum of 20.36 per cent in Chaudauli. More than 45 districts in Uttar Pradesh have effective forest less than 5.0per cent. Increasing demand for timber (3.65 M M^3 in 1988 and 5.81 M M^3 in 2011) as well as increasing livestock pressure (63.04 million in 1982 to 74.89 million in 1993). In Sharda Sahayak Cannal Command Area, between Ghaghra and

Ganga in Uttar Pradesh, satellite studies have shown that 15.19 per cent of the command area has become surface water logged and rendered infertile for the winter crops (Shah and Mathew, 1989). Inefficient and excessive irrigation in areas with poor soil permeability and drainage cause water logging and salinization of the soil that adversely affects crop yields in canal-irrigated regions of Uttar Pradesh. Industrialization and urbanization exert pressure on available agricultural lands and leads to land conversion from agriculture and forests to commercial uses that is inevitable in a growing economy but appropriate land use planning is essential to safeguard agriculture and ecological degradation.

Jharkhand has 92-93 per cent of total coal mined in the country and receives maximum royalty among coal producing states in India. Mining has changed large tracts of forests in to wastelands in Damodar valley and Singhbhum. This led to loss of 0.4 million ha forest during 1997 to 1999 (Forest Survey of India's State Forest Report). Mining waste pollutes rivers, nearby water courses and ground water as well when untreated effluents, slimes or tailing are released. The most polluted Damodar receives about 130 million liters of industrial effluent, 65 million liters of untreated water, 45 t of fine coal along with toxic metals like arsenic, mercury, chromium, nickel etc. Industrial effluents settle and form hard layer which adversely affect soil infiltration and permeability and ground water recharge. Industries also leave mining dust and noxious fumes in the air which cause respiratory and eye distress. Roro asbestos dumped in open in West Singhbhum district is causing health hazard to human being and cause low back pain, blindness or severely reduced vision, chest pain and respiratory distress.

Effects of land degradation

Land degradation leads to desertification (loss of good soil and nutrients), droughts (due to low water holding potential) and loss of bio-diversity in eco-systems. This may lead to disappearance of animals and plants from many habitates. Over exploitation and mis management of soil and water resources is aggravating land degradation. According to experts, "the amount of crops, animals and other biomatter that we extract from the earth each year exceeds what the planet can replace by an estimated 20 per cent, meaning it takes 14.4 months to replanish what we use in twelve, deficit spending of the worst kind". Long-term agricultural yield growth is declining in Central India for instance in Western Uttar Pradesh during 1990 to 1995 growth in rice yields was only one per cent compared to five per cent during 1980 to 1990. Similarly, growth in wheat yields was 1.6per cent compared to 2.4 per cent during same period. Sedimentation and pollution of water bodies, low nutrient and water use efficiency, decline in level of agricultural/ eco-system productivity, food and livelihood insecurity, decline in ground water recharge, change in general vegetation in habitates to less preferred tree and grass species etc are diverse effects of land degradation.

Measures to Control Land Degradation

The cost of land reclamation can be measured in many ways like yield loss, enhanced use of farm inputs or as the cost of rehabilitating the degraded land including placement of lost nutrients. A recent World Bank assisted Uttar Pradesh Sodic Lands Reclamation Project (UPSLRP) spent about US$194.1 million to reclaim 150,000 hectares of sodic land in Uttar Pradesh. It is further estimated that an equivalent of UA$ 15.528 billion is required to rehabilitate the entire sodic lands in the state (World Bank, 1998). During 1997-2000 Uttar Pradesh has successfully reclaimed more than 10per cent of the total sodic lands with external assistance. A two-pronged strategy is called for rehabilitating the degraded lands in Central India. Limited and inelastic land has to be protected first followed by conservation of diverse variety of natural resources found on it, in order to solve myriad environmental, economical and social problems and poverty. World over, the impacts of land scarcity in both quantitative and qualitative terms are being experienced acutely barring a few countries. Though nearly one-fourth of the earth's land is covered with soil only about a little over one tenth is naturally suited for agriculture. Per capita availability of arable land is falling all over productive lands from degradation and recovering already degraded lands (has been slow) for production not only in the states of the Central India but throughout different countries on earth planet. This recall for scientific land use planning with specific consideration to maintain and improve the interrelationship between land and water cycles to ensure generation and security of sustainable livelihood system on a long-term basis. Common methods of arresting land degradation comprise of afforestation, shelter belts, controlling over grazing, management of waste lands and controlled mining activities. However, plantations have poor survival due to factors of high summer temperature, droughts of variable duration and intensity, hard sub soils, gravely soils having dominance of stone part and other several reasons. Higher growth and regeneration of natural vegetation as well as higher survival of plantations of trees and grasses particularly in rainfed production system is only possible when there is maximization of *in situ* conserved rain water on watershed basis using relevant mechanical and vegetative measures as well as integration of these two. Watershed mode of development for arresting water erosion and *in situ* rain water conservation using suitable mechanical and vegetative measures is prime option for sustaining diversified production systems including agro-forestry in different regions of the Central India. To satisfy timber demand without affecting our forests, region specific pilot plantation schemes are needed to be launched on waste and barren lands as well as through developing appropriate agroforestry systems under different agro-ecological regions strictally adopting scientifically recommended package of practices and soil and water conservation measures.

Government Programs to Combat Land Degradation

1. Uttar Pradesh Land Reclamation Project with World Bank Assistance, launched in 1993 has treated 1.90 lakh ha area up to 1999 with an expenditure of 1092.5 crores.
2. Scheme for reclamation of alkali soils in Uttar Pradesh aims to improve physical condition and productivity of alkali soils through providing assured irrigation water and on farm development works like land leveling, bunding and plowing, community drainage system and application of soil amendments and organic manures.

Conclusion

Central India accounts for 32.01per cent (47 million ha) of total degraded lands in India (146.02 million ha) of which undivided Madhya Pradesh, Uttar Pradesh and Bihar account for 26.21, 15.32 and 6.28 million ha, respectively. Per capita land availability is declining in Central India due to increasing population, land degradation and land conversions leading to insistent rise in pressure on agricultural, forest and pasture lands for food, fodder, fuel wood, fiber etc. This is also leading to increase in number of land holdings, shrinking farm size, increasing landless labors and poverty resulting in non sustainable land management, a direct cause for land degradation. Land degradation actually refers to decline in overall quality of soil specifically its potential to grow plants that broadly covers soil erosion, rising water tables, expressions of salinity, alkalinity, acidity, floods, mass movement of soil by gravity, stream bank instability etc. Land degradation affects both the availability of land for agriculture and soil fertility, thus threatens sustainable future production. Current economic and industrial development trend; steady growth of human and livestock population; deforestation beyond silviculturally permissible limits; deforestation on soils having steep slopes or shallow depth; unsustainable fuel wood and fodder extraction; unscientific and arbitrary land management; water logging due to unlined canals; faulty irrigation systems; soil salinity, alkalinity and acidity; water and wind erosion; nutrient depletion; non adoption of adequate soil and water conservation measures; bio-physical and chemical *in situ* deterioration of soil due to indiscriminate use of fertilizers and pesticides; improper crop rotations; frequent shifting cultivation with reduced fallow cycles; encroachment in to forest lands; forest fires; over grazing; extension of agriculture to low potential lands; mining particularly the opencast mining; dust pollutants of industries like lime stone, ceramics etc; exploitation of ground water at the rate more than the natural recharge; open access resource and poverty stand as the major reasons behind the grave land degradation in Central India. The crop, soil and nutrient losses in finance terms apart from its collective ecological implications caused by land degradation vary according to cause. Therefore, this paper reviews the status and causes of land degradation in five states of Central India (Uttar Pradesh, Bihar, Jharkhand, Madhya Pradesh and Chhattisgarh) with specific reference to various options of rehabilitation of degradation lands and future sustainable food security.

References

Bansil, P.C. (1990). *Agricultural Statistical Compendium.* New Delhi: Techno-economic Research Institute. **1**: 581 pp.

CSWCRTI (1994). *Annual report (1994/95).* DehraDun: Central Soil and Water Conservation Research and Training Institute. 195 pp.

FSI (1987). *The State of Forest Report.* Dehra Dun: Forest Survey of India, Ministry of Environment and Forests, GoI, New Delhi.

FSI (1995). *The State of Forest Report.* Dehra Dun: Forest Survey of India, Ministry of Environment and Forests, GoI, New Delhi.

MoEF (1999). *National Forestry Action Programme.*Ministry of Environment and Forests, Govt. of India, New Delhi, Vol. **I**, pp. 79.

Pachauri, R.K.; Sridharan, P.V. (1998). Looking Backto Think Ahead: growth with resource enhancement of environment and nature. New Delhi: Tata Energy Research Institute. 346 pp.

Ratna Reddy, V (2003). Land degradation in India: Extent, costs and determinants. Economic and Political Weekly, November, 01, 2003.

Sahay, K.B. (2000) 'Problem of livestock population'. Delhi: *The Tribune, 11 April* 2000.

Shah, P.N. and Mathew, A.C. (1989). Assessment of the status of water logging in Sarda Sahayak Command Area of Uttar Pradesh, India using IRS-1A, LISS-1 data. GIS development.net. AARS. 10th Asian Conference on Remote Sensing. Poster Session 2.

Swaminathan, M.S. (2007). Atlas of the sustainability of food security in India. M.S. Swaminathan Research Foundation, Chennai.

UNDP; FAO; UNEP. (1993). Land degradation inSouth Asia: its severity, causes, and effects upon thepeople [World soils research report no. 78]. Rome: FAO, 100 pp.

□□□

Chapter 28

Soybean Based Intercropping System for Sustainable Production under Rainfed Condition in Bundelkhand Region

DEV NARAYAN AND H. BISWAS
Central Soil and Water Conservation Research and Training Institute, Research Centre, Datia-475 661 (M.P.)

Madhya Pradesh is the largest soybean producing state in India but the yield levels of soybean in Bundelkhand region are very low. Rainfed agriculture is a common practice during rainy season as about 70per cent area comes under rainfed agriculture and the crop intensity is about 135per cent in the region. The overall trend of monsoon rains is very erratic and uncertain in the region which may be virtually responsible for creating scarcity of moisture and crop failures under rainfed conditions. Out of 5 years, 2 years experienced normal rainfall years, 2 years as drought and one year as excessive rainfall year. Long dry spell during *kharif* season are the common features. Intercropping can play a significant role in enhancing productivity and profitability per unit area and time through more efficient use of land, water and solar energy besides assuring insurance against crop failure or failure of one or the other crop on account of vagaries of weather or disease/ pest epidemics in rainfed agriculture. Under limited soil moisture conditions intercropping coupled with improved tillage can further play an important role in higher intake of rain water into the soil profile and moisture conservation (Narayan *et. al.*, 2007). Information on this aspect is meager particularly for Bundelkhand region, therefore, the present investigation was undertaken to find out the most productive and remunerative intercropping combination under rainfed conditions of Bundelkhand region.

Materials and Methods

A field experiment was conducted during the rainy seasons of 2006 to 2007 at experimental farm, Central Soil & Water Conservation Research and Training Institute, Research Centre, Datia, Madhya Pradesh. The experimental red soil was sandy loam in texture with pH 7.3, organic carbon 0.31per cent, EC 0.12 dS m^{-1} and available N, P and K 396.7, 20.5 and 230.7 kg ha^{-1}, respectively, however, experimental black soil was sandy clay loam in texture with pH 7.2, organic carbon 0.36per cent, EC 0.11 dS m^{-1} and available N, P and K 482.5,29.0 and 277.8 kg ha^{-1}, respectively. The total rainfall received during the crop growth period was 340 and 268 mm during 2006 and 2007, respectively. The treatments were 2 tillage depths (i. Farmer's practice - two pass harrow, one pass cultivator and planking and ii. Improved tillage- M.B. plough once, cultivator and planking) and , 6 sole/ intercropping systems [1. Sole soybean ('Cv. PK 1042'), 2. Soybean + sorghum ('Cv. CHS 14') 2:2, 3. Soybean + clusterbean ('Cv. Naveen') 2:2, 4. Soybean + sesame ('Cv. Gujrat 2') 2:2, 5. Soybean + pigeon pea ('Cv. IPCL 87') 2:1 and 6. Soybean + castor ('Cv. Kranti') 2:1 proportion. The treatments were fitted in split plot design keeping tillage levels in main plots and cropping systems in sub-plots, replicating 3 times. Recommended dose of nutrients (30 kg N and 60 kg $P_2 O_5$/ha) was applied as basal to the sole soybean and soybean based intercropping treatments. No additional nutrients were given to the intercrops. Sole soybean and crops in intercropping systems were sown at 40 cm row spacing. Standard package of practices were followed to raise good crop. The soybean equivalent yield was calculated by converting the yield of intercrops into the yield of soybean on the basis of existing market price of individual crop during 2006-07.

Results and Discussion

The year 2006 and 2007 were drought years as only 340 and 268 mm rainfall was received during crop growth period in respective years consequently low crop yields were recorded. The results revealed that improved tillage and intercropping system was beneficial as compared to farmers practice and sole cropping of soybean, on both red and black soils (Table 1 & 2).

Table 1 : Soybean equivalent yield (kg/ha) as influenced by different tillage and intercropping treatments in red soils of Bundelkhand during *kharif* 2006 and 2007.

Treatment	Farmer' practice (conventional tillage)		Improved tillage practice		Mean	
	2006	2007	2006	2007	2006	2007
Soybean (Sole)	19	9	40	34	30	21
Soybean + sorghum (2:2)	215	340	327	433	271	387
Soybean + pigeon pea (2:1)	0	0	0	0	0	0
Soybean + clusterbean (2:2)	356	395	465	587	411	491
Soybean + sesame (2:2)	124	282	200	435	162	358
Soybean + castor (2:1)	333	417	496	705	414	561
Mean	175	241	255	366	-	-
C. D. (for intercropping)					83	113

C. D. (for tillage practices): Non-significant in 2006 and 2007.

Effect of tillage practices and intercropping systems on soybean equivalent yield

Higher soybean equivalent yield was recorded under improved tillage over conventional tillage in red and black soils. In red soils, there was 46 and 52 per cent increase in soybean equivalent yield during 2006 and 2007, respectively over conventional tillage. However, in black soils there was 59 and 88 per cent increase taken place because of improved tillage in 2006 and 2007, respectively over farmer's practice.

All intercropping systems recorded significantly higher soybean equivalent yield over sole cropping of soybean on both the soils except soybean + pigeon pea which failed during 2007 on black soils and both the years under red soils. Under red soils, soybean + castor intercropping system recorded the highest soybean equivalent yield (414 and 561 kg/ha in 2006 and 2007, respectively). In black soils, soybean + sesame intercropping system registered the highest soybean equivalent yield (600 and 855 kg/ha in 2006 and 2007, respectively) followed by soybean + clusterbean (500 kg/ha in 2006 and 756 kg/ha in 2007) and soybean + sorghum (494 kg/ha in 2006 and 774 kg/ha in 2007).

Table 2 : Soybean equivalent yield (kg/ha) as influenced by different tillage and intercropping treatments in black soils during *kharif* 2006 and 2007.

Treatment	Farmer' practice (conventional tillage)		Improved tillage practice		Mean	
	2006	2007	2006	2007	2006	2007
Soybean (Sole)	41	21	135	36	88	28
Soybean + sorghum (2:2)	400	687	588	861	494	774
Soybean + pigeon pea (2:1)	425	0	665	0	545	0
Soybean + clusterbean (2:2)	356	516	643	995	500	756
Soybean + sesame (2:2)	502	492	698	1218	600	855
Soybean + castor (2:1)	275	284	444	638	359	461
Mean	333	333	529	625	-	-
C. D. (intercropping)					197	115

C. D. (tillage practices): Non-significant for 2006 and 2007.

Economics of tillage practices and intercropping systems

Improved tillage was more remunerative than conventional tillage in producing higher additional returns in red and black soils (Table 3 & 4). In red soils, improved tillage recorded 40 and 45 per cent higher additional return in 2006 and 2007, respectively over conventional tillage. However, in black soils, improved tillage recorded 35 and 88 per cent higher additional returns in 2006 and 2007, respectively over conventional tillage. In general, the intercropping systems were more remunerative than sole cropping of soybean on red as well as in black soils. In red soils, soybean + castor intercropping system recorded the highest additional returns (Rs 5773 and 8096 in 2006 and 2007, respectively) followed by soybean + clusterbean (Rs 5718 and 7049/ha in 2006 and 2007, respectively). In black soils, soybean + sesame recorded the highest additional returns (Rs 7686 and 12403/ha in 2006 and 2007, respectively) followed by soybean + sorghum (Rs 6098 and 11183 in 2006 and 2007, respectively) than sole cropping of soybean.

Table 3 : Additional returns (Rs/ha) over sole cropping of soybean in red soils as influenced by different treatments during *kharif* 2006 and 2007.

Treatment	Farmer' practice (conventional tillage)		Improved tillage practice		Mean	
	2006	2007	2006	2007	2006	2007
Soybean (Sole)	-	-	-	-	-	-
Soybean + sorghum (2:2)	2936	4974	4307	5991	3621	5482
Soybean + pigeon pea (2:1)	0	0	0	0	0	0
Soybean + clusterbean (2:2)	5055	5802	6361	8296	5718	7049
Soybean + sesame (2:2)	1580	4102	2407	6017	1993	5060
Soybean + castor (2:1)	4708	6120	6838	10072	5773	8096
Mean	2856	4199	3987	6075		

Table 4 : Additional returns (Rs/ha) over sole cropping of soybean in black soils as influenced by different treatments during *kharif* 2006 and 2007.

Treatment	Farmer' practice (conventional tillage)		Improved tillage practice		Mean	
	2006	2007	2006	2007	2006	2007
Soybean (Sole)	-	-	-	-	-	-
Soybean + sorghum (2:2)	5387	9993	6809	12373	6098	11183
Soybean + pigeon pea (2:1)	5760	0	7958	0	6859	0
Soybean + clusterbean (2:2)	4732	7436	7633	14385	6182	10911
Soybean + sesame (2:2)	6919	7071	8453	17735	7686	12403
Soybean + castor (2:1)	3510	3942	4642	9030	4076	6486
Mean	5262	5689	7099	10705		

The higher yield advantage under intercropping systems over sole cropping of soybean was mainly resulted due to efficient utilization of available resources (land, water and solar light) and improved tillage was further helpful in conserving higher rainwater into deeper soil layers which was advantageous to the crops for a longer duration during reproductive phase after cessation of monsoon rains. The results corroborates the findings of Narayan *et al.* (1999), Prakash *et al.*, (2005), Jadav *et al.*, (2007), Patel *et al.*, (2007) and Narayan *et al.*, (2007).

Conclusion

This study recommends that soybean should be intercropped with suitable crops like sorghum, sesame or castor than growing under sole cropping coupled with improved tillage for obtaining higher yield and returns in red and black soils of Bundelkhand region.

References

Jadav, K.V. , Solanki, R.M., Vekariya, K.D. , Bhalu, V.B. and Savalia, R.L. (2007). Productivity improvement of groundnut-castor intercropping system through row proportion and castor sowing time. *Annals of Arid Zone* **46** (1): 37-41.

Patel, D.K., Patel, P.G. and Patel, M.M. (2007). Intercropping studies on rainfed castor under north Gujarat conditions. *Annals of Arid Zone* **46** (1): 99-101.

Prakash, V., Kundu, S. and Srivastava, A.K. (2005). Production and monetary returns of pigeon pea (*Cajanus cajan*) + groundnut (*Arachis hypogaea*) intercropping under rainfed conditions in mid-hills of North-Western Himalayas. *Indian Journal of Agricultural Sciences* **78**(12) : 797-9.

Narayan, D., Om Prakash & Bhushan, L.S. (1999). Intercropping of Indian mustard (*Brassica juncea*) with Barley (*Hordeum vulgare*), rocket salad (*Eruca sativa*) and linseed (*Linum usitatissium*) under limited irrigation conditions in reclaimed ravine land of river Yamuna. *Indian Journal of Agricultural Sciences* **69** (5) : 15-18.

Narayan, D., Tiwari, A.K., Lal, B., & Katiyar, V.S. (2007). Influence of tillage and sunnhemp mulching on runoff, soil loss and sorghum yield. *Indian Journal of Agricultural Sciences* **77** (6) : 387-389.

❑❑❑

Chapter 29

Watershed Based Reclamation and Control of Chambal Ravines

G.P. VERMA[1], Y.P. SINGH AND S.K. DUBEY[2]
1 Ex-Director Research Services, (J.N.K.V.V.), 108, Balwant Nagar, Gwalior- 474001 (M.P.) Lead paper presented at FESCO-2009, held at Agra on September 16-18, 2009.
2 Head, CSWCRT Research Center, Agra (U.P.)
R V S Krishi Viswa Vidyalaya, KVK-Zonal Agricultural Research Station, A.B. Road, Morena, Madhya Pradesh 476001

Land (comprising soil and associated flora and fauna) and water are the most precious natural resources of a nation, and how well these natural resources are managed decides the prosperity of the nation. Rain-water is a boon to man-kind but if not handled properly, it causes severe soil-erosion and ultimately land-degradation. Chambal ravines represent the most degraded from of once cultivated land. It is one of the several types of wasteland rather wasted land. Such a land has been so severely eroded by run off water that it is almost completely unsuitable for agricultural production. Consequently, it is classed as wasteland. Other types of wasteland are: saline soil, degraded forest land, mined area land, waterlogged/marshy land, inundated land, and so on. At least two thirds of arable land in our country suffers from degradation of one kind or another and one third part is almost unsuitable for cultivation. It results not only in loss of non-renewable land and soil resources but also leads to other processes destructive to national economy, such as floods in rivers, siltation of water reservoirs and consequent loss in their storage capacity, damage to railway lines, roads and other public properties. Ravines also encroach upon inhabited villages which have to be shifted to new sites to avoid loss of lives. Besides these economic losses, an important socio-economic problem created by ravines, particularly in Madhya Pradesh, Utter Pradesh and Rajasthan states, is shelter to dacoits and consequently making the area dacoit-infested. Thus, management of ravinous land so as to control ravine formation, and to reclaim the ravines already formed, is necessary for the survival of present and future generations.

Extent of Ravinous Land

Of 2 billion ha of once productive agricultural land in the world but presently out of production, 35per cent is due to soil erosion. In India, according to Bali (1974), 69 billion ha land is in critical stage i.e. liable to degradation in 10-15 years. No recent and accurate data on the extent of ravines are available. However, the severity of the problem of formation of ravinous land and its spread can be judged from the fact that more than 4 million ha of land is already covered by it presently and 0.5 per cent of the net cultivated land gets converted into ravines every year. The state-wise data on area under ravinous land as reported by a working group (Anon, 1972) are presented in table 1.

Table 1 : State-wise Area under Ravines (Anonymous, 1972).

S.No.	State	Ravine Land (Lakh ha)
1.	U.P.	12.30
2.	M.P.	6.83
3.	Bihar	6.00
4.	Rajasthan	4.52
5.	Gujrat	4.00
6.	Punjab	1.20
7.	West Bangal	1.04
8.	Tamilnadu	0.60
9.	Maharashtra	0.20
10.	Other States	3.27
	Total	39.97

District-wise data on area under ravines in M. P. are presented in Table 2.

Table 2 : Area under ravines in different districts of M.P.

S.No.	District	Ravine Land (Lakh ha)
1.	Morena	1.92
2.	Bhind	1.19
3.	Gwalior	1.08
4.	Guna	0.97
5.	Chhattarpur	0.40
6.	Datia	0.26
7.	Shivpuri	0.26
8.	Sagar	0.25
9.	Damoh	0.25
10.	Panna	0.20
11.	Tikamgarh	0.05
	Total	6.83

Chambal is an inter-state river as part of its catchment is in Rajasthan and Uttar Pradesh. It emerges from the northern slopes of Vindhyan range near Mhow town of Indore district. It flows northward in M.P. for 320 km and in north-east direction for 286 km. through Rajasthan, forms boundary between Rajasthan and M.P. for 216 km, boundary between M.P. and U.P. for 112 km. and finally flows through U.P. for 64 km. before joining the river Yamuna. Therefore, Chambal valley extends through M.P., Rajasthan and U.P. states and the river Chambal with its tributaries forms the infamous ravines-Chambal ravines, not only degrading the land but also providing a shelter to dacoits. Their area in 3 states is shown in Table 3.

Table 3 : Area under Chambal Ravines in M.P. Rajasthan and U.P. (River-wise in lakh ha).

M.P.		Rajasthan		U.P.	
River	Ravine area	River	Ravine area	River	Ravine area
Chambal	0.950	Chambal	1.303	Chambal	0.790
Asan	0.182	Kali Sindh	0.400		
Sank	0.023	Parwati	0.400		
Seep	0.136				
Parwati	0.572				
Kuno	0.071				
Total	2.260	Total	2.103	Total	0.790

A study by CADA, Gwalior, revealed that in chambal division, ravines covered 2.283 lakh ha in 1943-44 and 3.107 lakh ha in 1975-76-an increase of 36per cent in 32 year. Singh and Rao (1996) used remote sensing techniques and reported that the area under gullies and ravines in 66 villages of Morena district increased to 218.695 km^2 in 1994 from 130.15 km^2 in 1969, i.e. an increase of 88.545 km^2 in 25 year. This amounts to an increase of 3.5418 km^2/year or 2.72 percent of the fertile land per annum from 1969.

Earlier Efforts on Ravine Reclmation

The seriousness of the problem of ravines-formation and their spread was realized not only by the state Govt. and the National Govt. but also by the International Agencies such as World Bank, European Economic Communities (EEC) now called European Union (EU) and so on. Consequently, many efforts at various levels were made to reclaim the ravines and put them to agricultural use.

As early as 1919, the ruler of the erstwhile Gwalior State appointed a commission with fallowing terms of reference:

1. To suggest ways and means to arrest further extension of ravines.
2. To suggest means for making available fodder, fuel and timber for agricultural purposed to such villagers who had no access to forest nearby.

The Govt. of M.P. executed several ravine reclamation projects for Chambal ravines viz. Chhonda project (1955-56) Bagchini project (1955-56), Nayakpura Project (1956-57), Deori Hingona (1959-65), Jawasa Project (1962-70) and Dimini- Chandpur Project (1967-68). Under these programme, an area of 9080 ha was treated from first five year plan to 1970-71, comprising 3100 ha of ravine reclamation and 5980 ha of ravine afforestation at a cost of Rs. 44.23 lakhs.

Later on, Chambal Multipurpose Hydel Project for Chambal Valley was started as a joint venture of the Govt. of the then Madhya Bharat (now part of M.P.) and Rajasthan to exploit the irrigation potential of the river Chambal. Accordingly, besides construction of Dams and network of canals to irrigate 2.83 lakh ha in Chambal Division, rectangular fields were crated by large scale land-leveling using bull-dozers on the part of command area under ravines. But soon there were problems particularly with rectangular fields and a World Bank supported Command Area Development Project was taken up in 1975 with a financial credit of 24 million. But the results were again not worth the investment (Anonymous, 1997).

During 1988-92, ravine reclamation was undertaken in M. P. and U. P. as per Govt. of India instructions under Dacoity Prone Area Development Programme. Construction of 476 km of peripheral bund along with table land treatment of 5229 ha and shallow ravine treatment of 2776 ha was undertaken at a cost of Rs 362.44 lakhs, in M. P. (Table 4).

Table 4 : Earlier Ravine Reclamation work (Action Aid, India, 1996).

S. No.	Scheme	Sector	Period	Area covered (ha)	Expediture (lakh)
1.	Ravine reclamation	State	1955-73	1240	30.19
2.	Ravine reclamation	Central	1970-84	1748	102.15
3.	Ravine erosion control (CADA)	State & World Bank	1976-87	13500	201.20
4.	Dacoity Prone Area Development Prog. (U.P.& M.P.)	Central	1988-92	8005	362.44

During 1987-94, European Union (EU) funded a project entitled "Integrated Watershed Management Programme in ravinous areas of Chambal and Yamuna catchments" on 48000 ha land in Agra and Etawa districts at a cost of about Rs. 80 crores. Very good work on land-development (leveling/bench terracing), introduction of improved and high yielding varieties of crops, development of road and introduction of tube-well irrigation, was undertaken and immediate results were startling. But the approach was very expensive leveling of land to create rectangular fields with pipe-outlets for run-off from field to field create rectangular fields with pipe-outlets for run-off from field to field worked very well initially but gave up after some years due to lack of maintenance the requirement of which was rather. Alternatively, if an effort was made to manage the run-off rain water on watershed basis by converting main gullies into grassed-water ways allowing rain-water to run from field to field over grassed field bunds/appropriate cheap out lets and finally into stabilized waterway through appropriate waste-weir, the rain-water surplussing arrangement would have been much cheaper yet more stable and needing less maintenance as observed by Monitoring and Evaluation Team of WAPCOS (Verma, 1999).

Why were Earlier Efforts Not Fruitful

While the projects on ravine reclamation undertaken in M.P. were of a pioneering nature in the country, the benefits derived from these were not commensurate with the investment. In all these projects, either the priorities for multisectoral development were not provided, or even if provided, these were predominantly devoted to development in one or two selected sectors. The scope of work and the nature of funding was so limited as to have no multisectoral impact in any case. Invariably, all these projects were treated as purely government schemes because the State Government was the main player and executioner and people's participation was more of a rhetoric rather than reality. Other reasons for the failure of these projects were:

1. Based on borrowed technology, too much emphasis was laid on dozing for leveling, and bunding for making rectangular fields for cultivation. There were breaches in field-bund as no appropriate provision was made for surplus run-off to get out of field safely (without causing erosion).
2. Appropriate and adequate size structures were not made due to high cost and lack of basic information on rates of run-off, rainfall intensity, etc. In absence of such information, over-designing was necessary but not followed perhaps because of cost factors. The well built masonary structures are still there but surplus run-off did not pass through them. Instead, the bund breached on either side of the structure and made a way for the run-off. At some places, run-off tunnelled through under the masonary crest-well, leaving the structure intact but redundant. The probable reasons for such failures could be (i) under-size of structures, (ii) inappropriate location of these, and (iii) lack of integration of vegetative cover and mechanical structures.
3. Ravine afforestation (including aerial seeding) was not always undertaken in the same watershed as the ravine reclamation area.
4. Watershed approach for simultaneous treatment of tableland, peripheral land and the ravine land was completely lacking in most projects, and

5. Emphasis was given only on reclamation of ravines and not on prevention of formation of ravines by managing the run-off on water-shed basis.

Watershed Management Technology for Ravinous Land

An attempt has been made to formulate the appropriate technology of ravine reclamation and control by putting together various components of the technology based on (i) lessons learnt from failures of soil conservation schemes implemented earlier, (ii) experiences of gully-control on watershed basis gained from Indo-UK Dry Farming Project, Indore (Anon, 1979), (iii) results of research work conducted at Central Soil & Water Conservation Research and Training Institute, Research Centre, Chhalesar, Agra, and at Central Grassland and Forest Research Institute, Jhansi, (iv) expertise gained from Integrated Watershed Management Programme on Ravinous Land of Chambal and Yamuna catchments" in Agra and Etawa Districts of U. P. (Verma, 1999), and (v) experience of "Watershed based development of Ravine land" at Noorabad, District Morena (Verma,2003). Based on these experiences, the main components of the technology are as follows:

1. Resource Inventory

An inventory of natural resources of the entire watershed including ravinous area has to be made through

(i) Reconnaissance soil survey

(ii) Contour survey/use of aerial photographs or satellite imagery (to get drainage lines)

(iii) Available records on rainfall, potential evaporation, wind velocity, maximum and minimum temperatures, underground water, mineral resources, vegetation,

(iv) Agro-economic survey (existing land-use pattern, distribution of holding sizes, socio-economic status of residents of the watershed, source of power, labour, present crop-yield, extent of adoption of improved practices, roads/other infra-structure, marketing facilities etc.) and meetings with farmers are to be held to know their problems and the remedies they suggest.

2. Land and Water Management Works

Land and water management works have to be planned for the entire watershed starting from the ridge and including the table land which usually exists just below the ridge and above the ravinous land. Again the ravinous land maybe classified into 3 categories depending upon the depth of ravines viz. (i) shallow ravines (less than 1 m deep), (ii) medium ravines (1-3 m deep), and (iii) deep ravines (more than 3 m deep) and appropriate works have to be planned for different depths of the ravines.

Table land

Usually it is flat or gradually sloping and thus run-off water may not cause serious erosion. However to increase the productivity of such land through conserved water, field bunding (in alluvial soils) is necessary. If the average rainfall of the area is less and infiltration rate of the soil high, field outlets may not be necessary as most of the rain-water will seep into ground water quickly. In case of high intensity rain, the run off may flow over these table (grassed) bunds (15-20 cms high). If bunds are constructed afresh, compaction of bunds and planting of grass may be done or natural grass maybe allowed to take over to create a vegetative cover on the bunds. Bunding will prevent the sheet and rill erosion by interfering with free flow of run-off.

Peripheral bund

Between the table-land and the ravinous land a peripheral or water division bund must be constructed to lead the run off from table land to a safe place (water reservoir, farm pond or a stable grassed water way depending upon the location). Peripheral bund is essentialy a graded bund the cross-section of which will depend upon the run off to be tackled and the type of soil. The bund should be covered with vegetation like grasses, agave, munj, kans, etc. on the upper side (the side along which water will flow) and the utility trees on the outer side (side facing ravine land). Suitable deed rooted tree species like babul (*Acacia species*), shisham (*Delbergia sissoo*) or fodder-cum-fuel species like subabool (*Luchaena leucocephela*) or fruits tree like ber, guava, aonla, jamun, etc. can be planted.

Table land

As soil under ravinous land is very susceptible to water-erosion emphasis must be laid on raising kharif crop on sloppy land at least, as a good crop-cover (like that of soybean) is the cheapest soil conservation measure. On sandy loam soil dominant ravinous land, groundnut and pigeonpea are much more remunerative than pearl millet, moong, urd and sorghum. There is an immense potential to increase their yields by choice of proper varieties, timely sowing and adopting other improved practices of crop production. Growing a crop during rainy season results in not only making the best use of rain-water but also serves as the cheapest and effective soil conservation measure. As a matter of fact, if the land has a slope less than 1 per cent, the soil erosion due to run-off water on such land can be fully controlled by raising a well nourished crop, planted against the land-slope on a grade of 0.2 to 0.3 per cent. If improved practices of growing this crop are adopted, the crop cover provided by it will be much better and more effective. An experiment at Powarkheda during late fifties to assess the effect of cover-crop on surface run-off and soil-loss revealed that silt wash and loss of water was highest from the cultivated fallow followed by cotton and lowest from sunhemp and urid plot. Thus cropping in kharif, adopting high yielding varieties of crops and improved package of practices to grow these will not only result in high yield and consequent high income to the farmer but will also protect his land and ensure stable farming system.

Shallow ravines

Agri-horticulture system is most suitable for management of ravinous land with shallow ravines. Fruit plants such as guava, grafted plum, aonla (*Emblica officinalis*), karonda (*Carissa carandas*) and kareel can be planted along with kharif crops such as urid, moong, pigeonpea (short duration variety) depending upon topography of the land. Mahpatra and Jha (2009) found that Aloevera planting in the inter-tree spaces of plum trees gave a net return of Rs. 75000/ha.

Medium ravines

On land having medium ravines, silvripastrol systems can be developed growing grasses in washes and low level valleys and trees on higher spots (banks of gullies, monds of soil, etc.) suitable trees for this are subabool (*Lucaena lucocephala*), shevia (*Sesbania aegyptica*), agasthi (*S. grandiflora*), sirus (*Albizzia lebek*), babool (*Acacia nilotica*) and nilgiree (*Eucalyptus hybrida*). Various grasses for the system are dinanathgrass (*Pennisetum pedicellatum*), anjan (*Cenchrus ciliaris*), peela anjan (C. setigerus), marvel (*Dicanthium annulatum*), mushel (*Iseleiema laxum*), ponia (*Sehima nervosum*), tor (*Thalipogon aligans*), broom grass (*Thysanolaena maxima*) and lemon grass (*Cymbopogon flaxosus*) with legumes such as stilo (*Stylosanthes hamat, S. scabra*), siratro (*Phaseolus atropurpureus*), pilli pasara (*Phaseolus trilobus*), chipkani (*Desmodium contortum*) and shaora (*Alysicarpus rugosus*).

Deep ravines

After initial stabilization and reclamation work (a series of water diversion/graded bunds with suitable surplussing disposal systems), agri-horticulture and silvi-pastoral systems as above can be tried. Alternatively, afforestation with fuel and fodder trees can be undertaken. Even bamboo plantation can be tried. Afforestation of deep ravines is often recommended, but afforestation by aerial seeding tried in Chambal ravines failed terribly. Planting of fruit trees with constant care by the owner is as good as or better than afforestation. With run-off collection suitably, beautiful orchards of mango and other fruit trees can be established. Afforestation as a single measure is no absolute cure for ravines as serious gullying was noticed even in some reserved forests. If table land at the top yields excessive amount of run-off, its concentration and consequent erosive power can not be checked by trees alone. That is why mechanical structures are important for reclamation of deep ravines.

Run-off collection and Recycling

Run off collection and its recycling and its use for recharge of underground-water is a very important component of ravine reclamation and control technology. Run off collection can be done in several ways depending upon the site in watershed. Runoff will get stored temporarily even against peripheral bund, contour/graded bunds and submersible check dams, the construction of which will be necessary for the control of the ravine formation. Also, at suitable sites farm ponds or tanks can be constructed to store runoff. If at any site, construction of a peripheral bund is not feasible, construction of a bund along the boundary of two farmers both having ravinous land along the boundary, will be very appropriate to hold silt and water upstream the bund. The bund has to be of an appropriate cross-section according to the catchments above bund. Also, it should be compacted and protected by planting munj, agave, grass, etc. Further, it must be provided with a waste-weir of appropriate type and size and at an appropriate point on the bund, to dispose off excess of run-off. This bund will serve double purpose of holding water (for recycling/charging) and reclaim and control ravine-formation. Formation of gullies which was taking place earlier will stop because as soon as run-off water from table-land meets the stored-water, it loses its velocity/abrasive power.

At certain sites just below the table land, the rain-water causes small land-slides forming deep-pits. It will be quite appropriate to construct a bund such that a number of pits can be converted into a tank. The tank will not only store run-off but will also stabilize the table-land. On land with deeps ravines if it is not possible to afforest, suitable dam can be constructed against a number of deep and U-shaped ravines to create a good water-reservoir fit for fish culture. Needless to say that the dam will have to be ably designed and constructed.

Establishment of agro-based cottage industries

Agro-based cottage industry should be set up using locally available raw-material e.g. (a) preparation of milk products from surplus milk, (b) promotion of self help groups for making tenti pickle, aonla murabba, mango pickle, etc. can be done depending upon raw material locally available. Cottage industries for (c) sausage making, (d) straw board making, (e) bamboo basket making, (f) spinning and weaving, and (g) darrie and galicha making, can also be established.

Practices often recommended but not practical

Certain practices though often recommended, are not practical under farmers' conditions, e.g.

i. *Closure of the area to grazing and other biotic interference* : This is not practical because grazing on fallow lands is accepted as a right of grazers. Consequently, the structures have to be modified so as not to impound too deep water against the bund and the gabions. It means the height of gabions may be kept less than the calculated, in the first year. The desired height will be achieved by raising it year after year.

ii. *Grassed ramps in the bunds for safe disposal of excess run-off* : This is the cheapest measure but it can not be established well during pre-monsoon period to be fully effective during monsoon season.

Recent efforts of Ravines Reclamation

Being conscious of non-availability of land for various types of plantations being promoted by the Government, efforts are still on to reclaim ravine-land. Most of the ravinous land, which earlier belonged to farmers inhabiting the area, has become either revenue (Govt.) land or forest land. That is, due to wrong "Land Policy" of the Govt., the original land-owners (farmers) have no right of ownership on once their own land. Can there be a greater injustice to poor farmers deprived of their means of livelihood? Crores and crores of Rupees have been invested on ravine reclamation in the name of farmers' welfare without any benefit at all to farmers. Although the failure of earlier efforts to reclaim revines by land-leveling was well established, even recently efforts are being made to reclaim ravines by leveling the land and bringing it under plantations.

Some individuals/organizations have got large areas of ravinous land allotted to them on lease by M.P. Govt. They leveled the land, raised nursery of Jatropha seedlings and planted these seedlings on leveled land. While there is good growth of Jatropha plants the canopy provided by them against the soil-erosion is not adequate to control ravines-formation. Consequently, gully formation through development of rills has already started and no surprise if the land reverts back to its original form in course of time. The effort would have met a grand success if following precautions/steps were taken:

1. The area selected was a micro/milli watershed.
2. The main gully/gullies would have been stabilized by construction of appropriate gabion/masonary structures and planting of grass, munj, etc. and allowing natural vegetation to take over.
3. Conversion of gullies into grassed water-ways
4. Construction of contour bunds across the slope to guide run-off water into stabilized waterways and planting grass, munj, agave, etc. on the bunds to protected these.
5. Smoothening of land surface between the contour bunds instead of flat levelling which is very expensive yet not long lasting.

References

Action Aid India (1996). Chambal Ravines-A glimpse of the scenario. Paper presented at the "Workshop on Chambal Ravines" held at Gwalior on Feb. 6-7 1996.

Anonymous (1972). Report of the working group on Ravine Reclamation in Dacoity infested areas of Uttar Pradesh, Madhya Pradesh and Rajasthan. Govt. of India, Ministry of Home Affairs, New Delhi, Oct. 1972.

Anonymous (1997). Chambal Command Area Development Agency, Ninth FYP (1997-2002) scheme, Chairman, CADA, Gwalior, M.P.

Bali, J. S. (1974). Soil degradation and conservation problems in India, UN-FAO consultation of experts on soil degradation, Food and Agriculture Organization, Rome.

Mahapatra, K. P. and Jha, P. (2009). Aloevera cultivation in interspaces for supplementing productivity of ber orchards in reclaimed Yamuna ravines. Central Soil and Water Conservation Research and Training Institute, Research Centre, Chhalesar, Agra, U. P.

Singh, S. and Rao, K.K. (1996). Land use pattern changes due to ravines using remote sensing techniques in Morena Distt. M.P. Paper presented at the "Workshop on Chambal Ravines " held at Gwalior on Feb. 6-7 1996.

Verma, G.P. (1982). Land and Water Management in semi-arid Rainfed Areas. Review of Soil Research in India, Indian Society of Soil Science, New Delhi.

Verma, G.P. (1983). Watershed Management Practices for the Control of Erosion and Water-logging under Rainfed Condition. Lecture delivered at the "Training on Management of Black soils under Rainfed conditions" organized under the auspices of NARP at JNKVV Campus Indore, M.P. from Sept. 1 to 10, 1983.

Verma, G. P. (1999). Monitoring and evaluation Report of Central Water and Power Commission Evaluation Team for Project on "Integrated Watershed Management Programme on Ravinous Land of Chambal and Yamuna Catchments". Central Water and Power Commission, New Delhi.

Verma G.P. (1999). Ravine soils and their Management. Paper presented at the workshop on Land Research of M.P. held at Bhopal on Aug. 17, 1999 under the auspices of Department of Agriculture, M.P.Govt. and N. B.S.S.& L.U.P.Nagpur.

Verma G.P. (2003). Final Report submitted to Deptt. of Sci.& Tech. GOI, New Delhi on the project "Watershed-based Development of Ravine land" Centre of Sustainable Agri.& Rural Dev., C. P. Colony, Morar, Gwalior.

❑❑❑

Chapter 30

Deriving Homogeneous Land Unit through GIS by Integrating Biophysical and Socioeconomic Aspects of Karnal District

D. BARMAN[1*], R.N. SAHOO, N. KALRA, K. KAMBLE D. CHAKRABORTY

Division of Agricultural Physics, IARI, New Delhi-110012, India

[1] Corresponding Author E-mail : danbar@rediffmail.com, CSWCRTI, RC, Sunabeda, Post Box No. 12, Dist. Koraput, Orissa-763002

To reduce the vulnerability of food system to climate change, to conserve the land degradation and to recover the soil fertility and productivity; improved resources and inputs management options for food production, choice of crops and cropping systems to enhance the production, food distribution and economic access will be important to ensure that they contribute to sustainability. An outcome of this concept is precision farming, the site-specific socio-economically viable knowledge-driven management system to increase farm profits by using the modern tools viz. remote sensing, Geographic Information System (GIS), Global Positioning System (GPS), and simulation models. Adaptation options need to be assessed with regard to their socioeconomic and environmental efficacy, but a greater understanding of the interactions of food systems with Global

Environmental Change (GEC) is needed to be able to do with this confidence (Aggarwal *et al.*, 2004).

Soil information is important inputs in determining crop suitability of different soils Tamgadge *et al.*, (1999). National Bureau of Soil Survey and Land Use Planning (NBSS & LUP), Nagpur has delineated India into 20 agro-ecological regions (AER) in 1992 by superimposing the dominant bioclimate cum length of growing period (LGP) combination over dominant soil-scapes. In order to have realistic idea about yield contributing soil characteristics/qualities, data on level of crop management and the socio-economic data are required. Sys *et al.*, (1993) have categorically stated that deriving soil suitability zones is a 'Multi criteria Decision Making' problem.

The information on spatial variability in soil fertility status and crop condition is a pre-requisite for adoption of precision farming; one can improve the productivity or reduce the cost of production and diminish the chance of environmental degradation caused by excess use of inputs (Pierce and Nowak, 1999). Thus, mapping and analysis of within field variability is an essential input for precision crop management. In this paper, we used GIS tool to delineate agricultural land units by including various biophysical and socioeconomic aspects.

Methodology

The study area was Karnal district of Haryana extending from 29.41° - 29.95° N latitude 76.48° - 77.2° longitude and 246 m altitude from mean sea level. Various themes on major physico-chemical properties of soil such as texture, pH, electrical conductivity (EC) and organic carbon (OC) were transformed in spatial format where all the layers were geo-referenced and digitized in ArcGIS 8.3. The point data of these parameters have been transformed into surface models by using inverse distance weightage interpolation technique.

Deriving Homogeneous Units through Clustering/Overlaying

Historical data regarding the area, production, productivity, fertilizer, pesticides and irrigation for various major crops grown in Karnal district were collected from office of Government of Haryana. Historical weather data were available from IMD were used to create Theissen polygons to cover the regions of agricultural produce. Soil attributes such as sand, silt, clay, pH and EC and ground water pH,

EC and depth of water table were obtained from district soil survey office of Karnal. All the soil properties were determined by standard methods.

Land Units

By using the raster calculation module of ArcGIS 8.3, soil units were generated by crossing/overlaying those thematic layers of soil properties like soil texture, organic carbon content, degree and extent of soil salinity and sodicity. The overlay of the rainfall map with the soil unit map produced the homogeneous agro-ecological units. Further overlaying the irrigated area and district map of Karnal with the agro-ecological units produced the land units. But due to the cent per cent irrigated area of the Karnal district, there is no demarcation between the agro-ecological and land units. These digitized land units have been linked up with the biophysical and socioeconomic resources in blocks of Karnal district to precise the agricultural management practices.

Linking Socioeconomic Databases with Biophysical Resources

Homogeneous land units were generated by including biophysical resources such as (i) soils: particle size distribution, organic carbon content, extent and degree of salinity and sodicity; (ii) climate: distribution of magnitude of rainfall, and (iii) land: irrigated area, district boundaries. These homogeneous units were linked with tahsilwise socioeconomic resources assigned by some indices.

Results and Discussion

Delineation of Homogeneous Land Units

There was a significant spatial and temporal variation in the availability and use of biophysical and socioeconomic production resources in the region. Therefore it is important to demarcate homogeneous areas for better production strategies. Therefore, the homogeneous land units were generated in this study by integrating biophysical and socioeconomic aspects.

Soil units

Based on the soil particle size distribution, organic carbon content, salinity and sodicity, soil units were generated. The numbers of samples fall under each category of suitability classification of soils.

Particle size distribution : The maximum probable distribution in case of sand, silt and clay were 40-70, 15-30 and 10-30 per cent, respectively. The soils of the district mainly belong to most fertile Indo-Gangetic Alluvium type having light textured soils.On the basis of relative proportions of the sand, silt and clay fractions, soils of Karnal can be broadly classified into two main textural classes such as sandy loam. Three blocks viz. Karnal, Indri and Gharaunda in Eastern part of the district have the soils of sandy loam type covering 51.95 per cent area and the western part constituted by Nilokheri, Nissing and Assandh are of loamy type covering 48.05 per cent area of Karnal district.

Sodicity and salinity : For sodicity assessment, indicator pH was used to classify the soils of the region into mildly alkaline (pH, 7.5-7.8) and moderately alkaline (pH, 7.9-8.6) categories. Mildly alkaline covers 93.54 per cent area and only 6.46 per cent area was under moderately alkaline category.

To characterize salinity, soils of Karnal were clustered into two groups (normal and medium) on the basis of electrical conductivity (EC, dSm^{-1}). The maximum area, 97.40 per cent is under normal soils having EC ranging from 0.8-1.0 dSm^{-1} and only 2.6 per cent area was of medium category having EC range 1.0-1.5 dSm^{-1}. The result revealed that there was no salinity area found in Karnal district due to long term remedial cultivation and management practices. Another reason may be the extensive cultivation of rice and wheat crops for which leaching loss of salts due to intensive irrigation practices followed for the crops in the district.

Soil Organic carbon : Soil organic carbon content is an important indicator of soil fertility. By using the inverse square distance interpolation technique, the representative point data of organic carbon content were used to make suface map of organic carbon. Two classes were found such as very low (0.2 to 0.3 per cent) and low (0.31 to 0.5 per cent) content. Very low area accounts 61.45 per cent and low area accounts 38.55 per cent of the district. In general, organic carbon content of Karnal is in low category because of extensive agricultural practices in this region.

Thematic maps of each physical and physico-chemical parameter of soils were overlaid in GIS environment resulted in 13 different soil units. Describes each soil unit in terms of soil characteristics and area occupied. The soil unit number 1, 2, 7 and 8 constituted the maximum area (91.44 per cent) in Karnal district.

Agro-ecological units

Although Karnal is a small district, it has a large variation in climatic features, especially rainfall due to its geographical location along the monsoon track. The unimodal rainfall varies from 400 mm to 900 mm per annum. Almost 80 per cent rainfall comes from the southwest monsoon and received in three months *i.e.* July to September. There is an increasing gradient of rainfall from west to east direction of the district. Therefore, the district is divided into four rainfall zones by generating rainfall map on the basis of total annual rainfall. Among these, the extreme southwestern part receives lowest rainfall amount (400-500 mm) which occupied 1.56 per cent of the total geographical area of Karnal and then 500-600 mm rainfall in a very small area (3.13 per cent) towards the east. The central part of the district receives 600 to 700 mm rainfall accounts maximum of 71.20 per cent area whereas eastern part receives highest rainfall (700-900 mm) pertains 24.11 per cent area. Temperatures are generally very high during summer (reaching a maximum of 45°C during day time) to very low during winter (reaching a minimum of 6°C).

The overlay of rainfall map (four zones) with the soil map (13 units) resulted in 27 homogeneous agro-ecologocal units. This district is covered by the agro-ecological region 4 and 9 and agro-eco subregions 4.1 and 9.1 AESR-4.1 covers 59.95 per cent area and AESR-9.1 covers 40.05 per cent area in the district.

Land units

Overlaying the district map and irrigated area map with that of 27 agro-ecological units resulted land units. Karnal being fully irrigated area, total land units are same as the agro-ecological units. Frequency distribution indicated the major five land units are 1, 5, 7, 14 and 18 covering 79 per cent area and rest is covered by 22 land units.

Socioeconomic Resources of Karnal

The economic wealth that meets the needs of human being for completing their life cycle is called socioeconomic resources. Some of them are analyzed and discussed as below:

Land holdings

The total number of holdings in the district is 103735. Of these about 70.4 per cent of the holdings are marginal (<1 ha) and small (1-2 ha) size occupying 22.4 per cent of the total area. The semi-medium (2-4 ha) and medium (4-10 ha) holdings of 26.7 per cent occupy 51.8 per cent of the area in the district. The large (>10 ha) size holdings of 2.9 per cent occupy 25.8 per cent of the area. As such the average holding size of the Karnal district is 2.12 ha.

Socio-economic indicators

Socio-economic resources have been presented by some indices for comparative study of Karnal with Haryana and India. Within the Karnal district, tahsil wise spatial distribution of the socioeconomic resources were identified. It was clear that population density of Karnal district (505.6 per sq. km) was much higher than Haryana (478 per sq. km) and India (324 per sq. km), but decennial growth rate of population was somewhat lower in Karnal (23.06 per cent) as compared to Haryana (28.43 per cent) and little higher than India (21.54 per cent). It also showed that the density of population per hectare of gross cropped area (GCA) was only 3.3 which resemble Haryana but it was lower than India (5.39 persons). The similar trend was observed with regard to the density of population per hectare of net cropped area (NCA). This signified that the population pressure on land in Karnal district and Haryana was almost 60 per cent as compared to India. The literacy rate of Karnal (67.74 per cent) resembled the Haryana (67.91 per cent) but slightly higher than India's average which indicated the human resource development in Karnal and Haryana was at per with India. Per hectare consumption of fertilizer in Karnal (416.8 kg $ha^{-1}yr^{-1}$) was much higher as compared to the state average consumption (175.7 kg $ha^{-1}yr^{-1}$) and India's average (106.7 kg $ha^{-1}yr^{-1}$). Proportion of gross irrigated area to gross cropped area 99.84 per cent signified almost all the crops cultivated in Karnal were irrigated which showed the tremendous irrigation potential of Karnal district. Irrigation and cropping intensity in Karnal district (198.5 and 198 per cent respectively) and in Haryana (184.6 and 182 per cent) were much higher as compared to India (139.4 and 135.3 per cent). The net sown area was 79.67 per cent of total geographical area which was at par with state average but much higher than India's average (46.06 per cent). This showed that the maximum area of Karnal and Haryana were under plow as the soils of these regions are very fertile. Among all the cultivated crops cereals were predominant in the district and state and cereals productivity was leading in the country due to higher subsidies in irrigation and more potential of farmers in capital investment.

Conclusion

Thirteen homogeneous soil units and twenty seven land unit have been identified for better agricultural management system in the Karnal district. Karnal is divided into 2 halves of agro-ecological sub-regions, the north-eastern portion is constituted of hot sub-humid (dry/moist) climate, 120-150 days of LGP and south-western portion is constituted of hot semi arid (dry) climate and 90-120 days of LGP. Inherent fertile soils of Karnal are productive for most of the crops for its favorable pH and EC, although the organic carbon content is of low category. There is a need to link the relational layers with the decision tools for appropriate land use types for better profits and also safeguarding the health of eco-system. The dissemination mechanism for appropriate knowledge to farmers is the major concern, and the present study aims in this direction to prepare the knowledge base for subsequent techno-transfer to the farming community.

References

Aggarwal, P.K., Joshi, P.K., Ingram, J.S.I. and Gupta, R.K. (2004). Adapting food systems of the Indo-Gangetic plains to global environmental change: key information needs to improve policy formulation. *Environ. Sci. Policy*, **7**: 487–498.

Aggarwal, P.K., Kalra, N., Singh, A.K. and Sinha, S.K. (1994). Analyzing the limitations set by climatic factors, genotype, water and nitrogen availability on productivity of wheat. The model documentation, parameterization and validation. *Field Crop Research*, **38**: 73-91.

Ladha J.K., Fischer K.S., Hossain M., Hobbs P.R. and Hardy B. (2000). Improving the productivity and sustainability of rice-wheat systems of the Indo-Gangetic Plains: a synthesis of NARS-IRRI partnership research. Discussion Paper no. 40. International Rice Research Institute, Philippines, p. 31.

Pierce, F.J., and Nowak, P. (1999). Aspects of precision agriculture. *Adv. in Agron*, **67**: 1-85.

Rabbinge, R. (1995). Eco-regional approaches, why, what and how. In: Eds J. Bouma, A. Kuyvenhoven, B.A.M. Bouman, J.C. Luyten and H.G. Zandstra, Eco-regional approaches for sustainable land use and food production. Kluwer Academic Publisher, Dordrecht, The Netherlands, pp. 3-11.

Sachdev, C.B., Tarsem Lal, K.P.C. Rana and J. Sehgal, (1995). Soils of Haryana for optimizing land use. NBSS & LUP, Nagpur, India, pp. 59 +2 maps.

Sys, IR C, Van Rast, E., Debaveye, Ir J. and Boernaert, F. (1993). Land evaluation. Part- III Crop requirements. Agricultural Publication No. &, General Administration for Development Cooperation Place du Champ de Mars 5 bte 57 1050, Brussels, Belgium, 115p.

Tamgadge, D.B., Giakwad, S.T.and Gajbhiye, K.S. (1999). Soils of Madhya Pradesh-II: Land use capability, cropping systems and soil degradation. *J. Indian Soc. Soil Sci*, **47**: 114-118.

□□□

Chapter 31

Evaluation of Integrated Resource Management and Planting Time on Yield and Water Expense Efficiency (WEE) of rice (*Oryza sativa*)

ABDHESH K. SINGH[1] & N.K. VERMA[2]

[1] Programme Co-ordinator, K.V.K. Manpur, Gaya, 823003

[2] Deputy Project Director, ATMA, Gaya, 823001

Tillage, irrigation and nitrogen are three basic and costly inputs for successful crop production. In North Bihar, rice transplanting is generally delayed due to non availability of water and paucity of labour at proper time. The present field experiment has been carried out to optimize the doses of basic inputs as well as to maximise their efficiencies under timely and late sown conditions.

Materials & Methods

The field experiments of 27 (twenty seven) treatment combinations of tillage, irrigation and nitrogen were conducted at calcareous belt of North Bihar having tillage in main plot while irrigation and nitrogen were placed under sub and sub-sub plot, respectively. Each treatment had three levels viz. wedge plough twice + planting (T_1), wedge plough four times + planting (T_2) and mould board plough + cultivator twice

+ planting (T_3), irrigation at initial tillering & flowering (I_1), I_1 + panicle initiation (I_2), I_2+ maximum tillering and milk stage (I_3) and O (N_1), 60 (N_2) and 120 (N_3) kg N/ha^{-1} were the levels of nitrogen. The paddy variety used was Saket-4 at the planting distance of 15 x 15 cm. The dates of transplanting were 24th July (timely) and 16th August (late). A basal dose of 60 kg P_2O_5, 40 kg K_2O and 25 kg $ZnSO_4$ per hectare and half of the total doses of nitrogen as per treatment was applied after imposing tillage treatment. WEE of rice was estimated by dividing the grain yield by the total water expense including rainfall, irrigation water and profile water depletion (Prihar *et al.*, 1976).

Results

The observation on grain yield of rice (Table 1 & 2) reveals that tillage treatment increased grain yield significantly under both the situations. This is also supported by Mehta *et al.*, 1992 and Sharma *et. al.*, 1988. The maximum grain yield recorded under T_3 was 28.31 and 24.12 q ha^{-1} and minimum under T_1 was 22.31 and 19.14 q ha^{-1} in timely sown and late sown conditions, respectively. The yield under T_2 was statistically at par with T_3 but significantly superior to T_1 during both timely and late sown conditions, respectively.

Irrigation at I_3 fetched the maximum yield which was 16.3 and 8.50 per cent higher than I_1 but the impact of I_2 on grain yield was statistically at par with I_1 and significantly inferior to I_3 by producing 10.4 and 6.50 per cent less under timely and late transplanting, respectively. The result further indicate that each increasing level of nitrogen produce significant enhancement in yield that is supported by Bali *et al.*, 1995. The maximum yield at N_3 was 31.19 and 25.61 q ha^{-1} which were 75.53 and 15.00 per cent more than N_1 under timely and late sown conditions, respectively. Interaction effect of tillage and nitrogen (TxN) were significant under timely transplanting only.

The computed data on WEE on (Table 3 & 4) showed the maximum value with $T_3I_3N_3$ which were 43.92 and 34.67 kg ha^{-1}cm^{-1} under timely and late transplanting respectively. As far as the influence of individual treatment was concerned tillage and nitrogen treatment affected WEE significantly due to their increasing level under both the situations without any significant interaction. Among the irrigation treatments, the higher value of WEE was obtained at lower level of irrigation due to the fact that increase in grain yield was comparatively lower than water expense which is also confirued by Singh *et al.*, 1987 and Bharambe *et al.*, 1997.

Table 1 : Effect of tillage, irrigation and nitrogen on rice grain yield (q ha^{-1}) : Timely Transplanted (24th July).

	T_1				T_2				T_3				Average over tillage			
	I_1	I_2	I_3	Mean	I_1	I_2	I_3	Mean	I_1	I_2	I_3	Mean	I_1	I_2	I_3	Mean
N_1	14.27	16.27	18.05	16.19	18.83	18.17	19.05	18.68	20.07	19.75	20.18	20.00	17.72	18.06	19.09	18.29
N_2	21.38	23.40	25.32	23.36	25.67	26.57	30.63	27.62	27.97	28.59	34.72	30.72	25.00	26.18	30.22	27.14
N_3	25.55	27.85	28.78	27.39	29.03	30.47	35.47	31.65	32.37	33.21	38.00	34.52	28.98	30.50	34.08	31.19
Mean	20.40	22.50	24.05	22.31	24.51	25.06	28.38	25.98	26.80	27.18	30.96	28.31	23.90	24.92	27.80	25.54

Source	T	I	N	TxI	TxN	IxN	TxIxN
SEm ±	0.6782	0.5483	0.2999	0.9496	0.5194	0.5194	0.896
CD at 5%	2.66	1.69	0.86	NS	1.49	1.49	NS

Table 2 : Late Transplanted (16th August).

	T_1				T_2				T_3				Average over tillage			
	I_1	I_2	I_3	Mean	I_1	I_2	I_3	Mean	I_1	I_2	I_3	Mean	I_1	I_2	I_3	Mean
N_1	13.28	13.73	14.55	13.85	16.32	16.38	17.82	16.84	17.73	18.19	18.27	18.06	15.78	16.10	16.88	16.25
N_2	19.90	20.23	21.67	20.60	22.97	24.07	25.45	24.16	2597	25.15	27.40	26.17	22.95	23.15	24.84	23.64
N_3	22.05	21.98	24.90	22.98	24.73	25.50	26.95	25.73	27.20	27.77	29.39	28.12	24.66	25.08	27.08	25.61
Mean	18.41	18.65	20.37	19.14	21.34	21.98	23.41	22.24	23.63	23.70	25.02	24.12	21.13	21.44	22.93	21.83

Source	T	I	N	TxI	TxN	IxN	TxIxN
SEm ±	0.5383	0.4013	0.2126	0.6950	0.3682	0.3682	0.6377
CD at 5%	2.11	1.24	0.61	NS	NS	NS	NS

Table 3 : Timely Transplanted (24th July).

	T_1				T_2				T_3				Average over tillage			
	I_1	I_2	I_3	Mean	I_1	I_2	I_3	Mean	I_1	I_2	I_3	Mean	I_1	I_2	I_3	Mean
N_1	22.02	22.80	21.22	22.01	28.33	25.05	22.76	25.38	30.38	26.87	27.46	28.24	26.91	24.91	23.81	25.21
N_2	32.22	32.10	29.48	31.26	37.26	35.48	36.08	36.27	41.00	38.25	41.50	40.25	36.82	35.28	35.69	35.93
N_3	38.01	37.76	33.06	36.27	41.53	40.33	41.77	41.01	47.13	44.15	43.92	45.07	42.22	40.75	39.38	40.78
Mean	30.75	30.89	27.92	29.85	35.70	33.62	33.33	34.22	39.50	36.42	37.63	37.85	35.32	33.64	32.96	33.97

Source	T	I	N	TxI	TxN	IxN	TxIxN
SEm $\pm$	0.9101	0.8171	0.3931	1.4153	0.6809	0.6809	1.1794
CD at 5%	3.58	NS	1.13	NS	NS	NS	NS

Table 4 : Late Transplanted (16th August).

	T_1				T_2				T_3				Average over tillage			
	I_1	I_2	I_3	Mean	I_1	I_2	I_3	Mean	I_1	I_2	I_3	Mean	I_1	I_2	I_3	Mean
N_1	18.79	19.33	17.57	18.56	23.65	23.88	22.00	23.18	25.82	26.25	22.34	24.80	22.75	23.16	20.64	22.18
N_2	27.78	28.24	25.54	27.20	32.77	34.63	31.11	32.84	37.27	35.91	25.89	35.36	32.61	32.93	29.85	31.80
N_3	30.52	30.46	29.06	30.01	34.61	36.40	32.50	34.50	38.67	39.14	34.67	37.50	34.60	35.33	32.08	34.00
Mean	25.70	26.02	24.06	25.26	30.34	31.64	28.54	30.17	33.92	33.77	29.97	32.55	29.99	30.47	27.52	29.33

Source	T	I	N	TxI	TxN	IxN	TxIxN
SEm $\pm$	0.7389	0.5407	0.2930	0.9365	0.5075	0.5075	0.8789
CD at 5%	2.90	1.66	0.84	NS	NS	NS	NS

Conclusion

Among the 27 treatments combination maximum grain yield of 38.00 and 29.29 q h^{-1} were recorded under $T_3I_3N_3$ combination at the water use efficiency (WEE) of 43.92 and 34.67 kg/$ha^{-1}cm^{-1}$ under timely and late transplanting situations. Significant interactions of tillage and nitrogen (TxN) and between irrigation and nitrogen (IxN) on grain yield of rice were observed under timely sown crop only.

References

Bali, A.; Siduque, M.; Ganai, B.A.; Khan, H.U.; Singh, K.N. and Bali, A.S. (1995). Response of rice genotype to nitrogen level under transplanted condition in Kashmir valley. *Indian J. Agron.* **40** (1): 35-37.

Bharambe, P.R.; Bhalerao, J.B. and Oza, S.R. (1997). Effect of nitrogen and soil water regimes on soil plant water relationship, yield and water use efficiency of summer sunflower. *J. Indian Soc. Soil Sci.* **45** (4) : 701-705.

Mehta, R.K.; Singh, G. and Singh, O.P. (1992). Response of deep water rice to tillage practices.*Annals Agril. Res.* **13** (4): 340-343.

Prihar, S. S.; Khera, K.L.; Sandhu, K.S. and Sandhu B.S. (1976). Comparison of irrigation schedule based on pan evaporation and growth stages in winter wheat. *Argon J.* **68** : 650-653.

Sharma, P.K.; Dutta De, S.K. and Redulla, C.A. (1988). Tillage effects on soil physical properties and wet land rice yield. *Agron. J.* **80**:34-39.

Singh, P.N.; Joshi, B.P. and Singh, Gurumal (1987). Water use and yield response of wheat to irrigation and nitrogen on an alluvial soil in North India. *Agril. Water Management.* **12** (4): 311-321.

□□□

Chapter 32

On-Farm Participatory Trials on Deep Tillage and Soil Amendments to Assess Their Effect on Soil Properties and Yield of Pearlmillet-Wheat Grown on Sodic Soil

Y.P. SINGH AND SUDHIR SINGH*
RVS Krishi Viswa Vidyalaya, KVK-Zonal Agricultural Research Station, Morena, M.P. 476001, India

Salt affected soils are an important ecological entity in India. Around 8.6 million ha area (Tyagi, 1998) is affected with menace of salinity in different agro climatic regions. Of this, 1.122 million ha area (Singh, 1992) is in Rajasthan alone where soil reclamation programme is less effective due to water scarcity as ground-water is either the major or the only source of irrigation. As per survey reports, marginal and poor quality ground waters constitute a greater part of the ground water resources in Rajasthan state. As regards the distribution of water quality in Rajasthan state, only 16% is good, 16% is marginal and 68% is poor quality underground water. Further, under poor quality water category, distribution of saline, sodic and saline sodic waters are about 16, 35 and 49%, respectively (Yadav and Kumar, 1995). Due to unavailability of good quality water, use of poor quality under-ground water for irrigation has deteriorated the soil properties and reduced crop yields. Rajasthan state has developed several water projects for irrigation and drinking purposes. Galwa, Moti and Mansi dams for rain water conservation were developed in Tonk district. These dams are used to supply irrigation water through canals at critical stages of crops. Expansion of irrigation facilities through inter-basin transfer of water has created problems in different command areas through out the country, either in the form of water-logging or a combination of water-logging and salinity in different irrigation commands. Moreover, sodicity problem in command areas is created due to indiscriminate irrigation with sodic under-ground water, as several times canals fail to supply irrigation water when needed. Further, undulated topography, low rainfall and high evaporation rate have also contributed to salt accumulation resulting in deterioration of soil condition and consequent low crop-yields. It calls for management of such soils to improve their physico-chemical condition for attaining higher sustainable crop production in command areas. The present study of farm participatory research was accordingly conceived to study the effect of different levels of gypsum (as per gypsum requirement), deep tillage in summer, green manuring and application of farm yard manure on yields of pearlmillet-wheat crop sequence and on physico-chemical properties of sodic soils in command area.

Materials and Method

Study area and characteristics of the farming system

The study was carried out in two villages of Tonk district situated in south-eastern part of Rajasthan state. The study area is located at 23^0 3′- 30^0 12′ N latitude and 69^0 30′- 78^0 17′ E longitude in north-western part of India. The district lies at an altitude of 264.32 meters above mean sea level and receives an annual rainfall of 500 mm (average of last 22 years). Monsoon in this region normally commence during last week of June or 1st week of July and withdraw by 15th September and maximum intensity of rain was during the month of July and August. The total rainfall received during the experimentation period in different years was 396, 473 and 361 mm during 2004, 2005, and 2006, respectively. Pan evaporation was generally higher than rainfall with the exception of the months of July and August. Maximum pan evaporation occurs in May and June when the fields remain fallow. Ten representative farmers were selected from two villages for on farm participatory trials. The selected soil of farmer's field in two villages was sandy loam to clay loam in texture and classified as Ustifluvents.

The soils were deficient in organic carbon, available N, P and Zn, but medium in available K. Available Cu, Fe and Mn contents were above the critical limit at both the sites. The average farm size of villagers varied from 1.5 to 8.3 ha of owned land. As per survey report, only 18% area was under Kharif crops, viz. Jowar (*Sorghum vulgare* P.), pearl millet (*Pennisetum typhoides* B.), sesame (*Sesamum indicum* L.) and blackgram (*Phaseolus mungo* R.). However in Rabi season, 100% area was covered by crops like wheat (*Triticum aestivum* L.) followed by mustard (*Brassica campestris* L.), barley (*Hordeum vulgare* L.) and chickpea (*Cicer arietinum* L.). In summer (Zaid) season, the land is usually kept fallow except very little area under vegetables. Farmers of this region, indiscriminately use canal/underground poor quality water for irrigation purposes.

Methodology

On-farm trails were initiated in May, 2004 in Galwa command area of Uniara panchayat samiti of Tonk district of Rajasthan, India. Uneconomic farming in the villages due to soil sodicity led to migration of villagers to cities for earning their livelihood by working as laborers. During the initial phase of the project, a survey was conducted on sixty four farmers of two selected villages to see their response to sodic soil reclamation and identify the constraints in the adoption of reclamation technology. Simultaneously, soil samples were also collected and analyzed for physico-chemical properties (Table 3). The initial and after harvest of wheat crop, soil samples were taken from 0 - 0.30 m soil layer with an auger 4 cm in diameter. The soil samples were air dried and ground to pass through a 2 mm sieve. Soil pH, electrical conductivity (1:2 soil water ratio) and exchangeable sodium of soil samples were determined using pH meter, conductivity bridge and flame photometer, respectively. The exchangeable sodium percentage was determined as per U.S.D.A. Hand book 60 (Richards, 1954). Organic carbon, available N, P, K and micronutrients were determined by Walkley and Black (1934), Kjeltec-II auto analyzer, Olsen P, NH_4OAc-extractable K, and DTPA extraction using atomic adsorption spectrophotometer (Jackson, 1973), respectively.

Experimentation

The experimental fields were divided in to seven sub plots, each having 500 m^2 areas with 0.30 m high bunds for rain water conservation. Before making plots, field-leveling and deep tillage up to 24 cm was done by using disc plough during last week of May to first week of June (summer season). The trails were conducted from 2004 to 2007. The trails were laid out in randomized block design with five replications. The selected interventions for soil reclamation were as follows: T_1: farmer's practice (control), T_2: gypsum requirement (GR) 25% + green manuring (GM), T_3: GR-50% + GM, T_4: GR-25% + GM + farm yard manure (FYM) @ 10 t ha^{-1}, T_5: GR-50% + GM + FYM @ 10 t ha^{-1}, T_6: deep tillage (DP) + GR-25% + GM + FYM @ 10 t ha^{-1} and T_7: GR-50% + DP + GM + FYM @ 10 t ha^{-1}. These seven interventions were applied only in first year from 28th June to 2nd July in five trails on sandy loam and five on clay loam textured soil. The sesbania seed was sown after application of gypsum as per gypsum requirement. The gypsum requirement varied from 3.75 to 5.46 t ha^{-1} as per sodicity level and soil texture. The recommended package of practices was followed as given in Table 5. The sesbania crop for green manuring was incorporated in to soil by harrowing after 50 days of sowing and pearl millet crop was grown as fully rain-fed. During Rabi wheat crop was grown in winter (Rabi) season during 2004-05, 2005-06, 2006-07 and pearlmillet crop was grown in rainy (Kharif) season during 2005, 2006, respectively. The pearlmillet crop was raised fully as rain-fed. During Rabi irrigations to wheat crop were applied at crown root initiation (25 days after sowing), tillering (45 days after sowing), booting (75 days after sowing) and milking (105 days after sowing) growth stages. As per availability of irrigation water from canal, pre irrigation, crown root, tillering and booting stage irrigations were given by canal (good quality) water and only at milk stage, irrigation were given by under-ground (sodic) water during Ist year and during IInd and IIIrd year wheat crop irrigated by canal (good quality) water, respectively. The composition of irrigation water used is given in Table 4. Such sodic waters are typical of the ground waters having high RSC and are found in south-eastern part of Rajasthan state of India.

Results and Discussion

Knowledge and adoption of sodic soil reclamation practices by farmers

Primary survey was conducted about knowledge and adoption of sodic soil reclamation practices by 64 farmer's families of two selected villages of experimental sites. The recommended practices of sodic soil reclamation were leveling of field, bunding for rain water conservation to facilitate leaching of salts, gypsum application @ 50% of gypsum requirement and green manuring. Out of 64 farmers in surveyed villages, about 14.1% farmers used FYM for sodic soil reclamation,

9.4% used only bunding for rain water conservation, 4.7% used only gypsum without knowing gypsum requirement and 3.1% used gypsum along with FYM and bunding for rain water conservation (Table 1). Out of 64 farmers, 44 farmers had not adopted any practice of sodic soil reclamation. Main reason of non-adoption of land reclamation as stated by 59.1% farmers was lack of availability of good quality irrigation water and poor quality of underground water (Table 2). Other factors responsible for poor adoption were lack of risk bearing capacity, undulated topography, high cost of gypsum, fragmented holdings, lack of investment power and knowledge etc.

Table 1 : Prevailing practices of sodic soil reclamation adopted by farmers in the selected villages (sample size N=64).

S. No.	Practices	No.	Percent	Rank
1.	Application of farm yard manure	9	14.1	I
2.	Bunding of rain water conservation	6	9.4	II
3.	Applications of gypsum	3	4.7	III
4.	Application of gypsum along with FYM and bunding for rain water conservation	2	3.1	IV
5.	Did not follow any practice	44	-	-

Table 2 : Reasons for non-adoption of sodic soil reclamation practices by farmers in the selected villages (sample size N=44).

S. No.	Component	No.	Percent	Rank
1.	Lack of availability of good quality water	26	59.1	I
2.	Lack of risk bearing capacity	18	40.9	II
3.	Undulated topography	17	38.6	III
4.	High cost of gypsum	16	36.3	IV
5.	Fragment holdings	12	27.3	V
6.	Lac of investment power	11	25.0	VI
7.	Lack of knowledge of soil reclamation	08	18.2	VII
8.	Limited availability of gypsum	07	15.9	VIII

Table 3 : Initial properties (0 to 0.30 m depth) of soil selected for experimentation (mean values of 5 farmers' field from each site).

Parameter	Site-1 (Sonthra village)	Site-2 (Khelania village)
EC_2	0.86 (± 0.16)	0.66 (± 0.30)
pH_2	9.14 (± 0.24)	9.26 (± 0.25)
ESP	32.51 (± 5.5)	37.52 (± 6.5)
Bulk density (Mg m^{-3})	1.47 (± 0.03)* 1.52 (± 0.04)**	1.43 (± 0.02)* 1.45 (± 0.02)**
Organic carbon (g kg^{-1})	1.5 (± 1.1)	1.7 (±1.6)
Available N (kg ha^{-1})	137 (± 10.5)	151 (± 10.0)
Available P (kg ha^{-1})	10.39 (± 1.65)	7.2 (± 0.91)
Available K (kg ha^{-1})	237 (± 15.5)	206 (± 9.0)
Available Cu (mg kg^{-1})	0.29 (± 0.02)	0.24 (± 0.10)
Available Fe (mg kg^{-1})	5.4 (± 0.3)	5.1 (± 0.5)
Available Zn (mg kg^{-1})	0.44(± 0.15)	0.57 (± 0.114)
Available Mn (mg kg^{-1})	2.3 (± 0.3)	2.5 (± 0.6)
Sand (%)	73 (± 2)	45 (±3)
Silt (%)	16 (±1)	29 (±4)
Clay (%)	11 (±1)	26 (±2)
Soil texture	Sandy loam	Clay loam

*values of 0-0.15 m depth and **values of 0.15 to 0.30 m depth.

Table 4 : Composition of irrigation waters used in participatory trials.

Characteristic	Canal water (good water)	Under ground water (sodic water)
Electrical conductivity (dS m^{-1})	0.25	1.07* (± 0.31)
Ion concentration ($mmol_c$ L^{-1})		
Ca^{2+}	0.87	2.65 (± 1.46)
Mg^{2+}	0.49	0.73 (± 0.63)
Na^+	0.64	16.84 (± 5.81)
$CO_3^{2-} + HCO_3^-$	0.57	8.91 (± 2.36)
Cl^-	0.25	2.67 (± 1.26)
RSC (residual sodium carbonate)	0.8	9.0 (± 2.3)
SAR (sodium adsorption ratio)	0.92	14.7 (± 3.0)

*mean of 10 farmers field values of under ground water.
RSC = $(CO_3^{2-} + HCO_3^-) - (Ca^{2+} + Mg^{2+})$ (all ions expressed in $mmol_c$ L^{-1}).
SAR= Na+/ $[(Ca^{2+} + Mg^{2+})/2]^{1/2}$ (all ions expressed in $mmol_c$ L^{-1}).

Table 5 : Details of experimental operations.

Operation	Sesbania	Pearlmillet	Wheat
Date of sowing	28th June to 2nd July	5-7th July	15-22 November
Variety	Local	HHB-67	Raj-3765
Line to line spacing (cm)	15	20	15
Seed rate (kg ha^{-1})	40	20	100
N application (kg ha^{-1}) through urea	Nil	60	90
P_2O_5 application (kg ha^{-1}) through diammonium phosphate	Nil	60	60
Zn application (kg ha^{-1}) through zinc sulphate	Nil	-	05
Pre-sowing irrigation (70 mm)	First monsoonal rain	First monsoonal rain	1
Post-sowing irrigation (50 mm)	Rain fed	Rain fed	4
Date of harvest	24-28th August*	5-9th October	9-12th April

* ploughing for green manuring by harrowing.

Effect on yield

For sodic soil reclamation, application of gypsum as per GR @ 25% and 50% alone or in combination with green manuring, FYM @ 10 t ha^{-1} and deep tillage in summer were very effective and significantly improved the grain and straw yields of pearlmillet and wheat as compared to control (Table 6). Pooled yield indicated that the treatment of deep tillage in summer + GR 50% + green manuring and FYM @ 10 t ha^{-1} (T_7) was the best treatment as the highest mean yields of 2.54 and 11.44 t ha^{-1} of grain and straw, respectively, of pearlmillet and 4.29 and 4.91 t ha^{-1} of grain and straw, respectively, of wheat was recorded. The lowest mean yield of 1.06 and 4.80 t ha^{-1} of grain and straw, respectively, of pearlmillet and 1.33 and 1.83 t ha^{-1} of grain and straw, respectively, of wheat was recorded with farmer's practice (T_1). The application of gypsum enhanced the availability of soluble calcium directly and indirectly through dissolution of native $CaCO_3$. The calcium thus released displaced the Na^+ from exchange

Table 6 : Responses of levels of gypsum, deep tillage, green manuring and FYM on grain and straw yields of pearlmillet and wheat.

Treatment	Pearl millet yield (tonnes ha^{-1})*			Wheat yield (tonnes ha^{-1})**		
	Site-1 (Sandy loam soil)	Site -2 (Clay loam soil)	Pooled	Site -1 (Sandy loam soil)	Site -2 (Clay loam soil)	Pooled
T_1	1.25 (5.59)***	0.87 (4.01)	1.06 (4.80)	1.55 (1.97)	1.11 (1.68)	1.33 (1.83)
T_2	1.72 (7.45)	1.42 (6.38)	1.57 (6.92)	3.07 (3.34)	2.88 (3.14)	2.98 (3.24)
T_3	2.06 (9.26)	1.95 (8.81)	2.01 (9.04)	3.54 (3.90)	3.33 (3.62)	3.44 (3.76)
T_4	1.97 (8.85)	2.07 (9.35)	2.02 (9.10)	3.45 (3.87)	3.09 (3.53)	3.27 (3.70)
T_5	2.35 (10.49)	2.32 (10.75)	2.33 (10.57)	4.03 (4.41)	3.89 (4.54)	3.96 (4.48)
T_6	2.27 (10.10)	2.17 (10.24)	2.22 (10.17)	4.32 (4.85)	3.43 (3.86)	3.88 (4.36)
T_7	2.60 (11.70)	2.47 (11.18)	2.54 (11.44)	4.46 (5.01)	4.11 (4.80)	4.29 (4.91)
CD at 5%	0.16 (0.65)	0.14 (0.45)	- -	0.21 (0.36)	0.34 (0.51)	- -

*pooled yield of two year, **pooled yield of three year, ***figure in paren thesis indicates straw yield.

complex and removal of soluble Na with anions ($CO_3^{=}$ + HCO_3^{-}) through leaching reduced the pH of soil as well as improved the physico-chemical properties of soil. Addition of organic manure and farm waste improved the physico-chemical environment of soil and consequently, greater extraction of water and nutrients by plants from the soil. These results are in conformity with findings of More (1994) and Yaduvanshi and Sharma (2007). Non-significant difference was noticed between deep tillage in summer + GR 25% + green manuring + FYM @ 10 t ha^{-1} (T_6) and GR 50% + green manuring and + FYM @ 10 t ha^{-1} (T_5) on sandy loam soil, whereas on clay loam soil, T_5 treatment significantly increased the yield of pearlmillet and wheat as compared with T_6 treatment. Deep tillage was more effective in sandy loam soil which has low apparent field capacity and higher infiltration rate.

Effect on physico-chemical properties

Application of gypsum @ 25% or 50% GR alone or in combination with GM, FYM and deep tillage in summer significantly reduced the pH and ESP and increased the available N, P, K and micronutrients in sandy loam and clay loam textured soil (Table 7). Maximum reduction noticed in pH was 1.14 and 1.32, ESP 21 and 24, bulk density 0.05 and 0.07 Mg m^{-3}, and increase in organic carbon was 0.4 and 0.3 g kg^{-1} soil, available N 77 and 61 kg ha^{-1}, P 7.1 and 8.8 kg ha^{-1}, K 69 and 51 kg ha^{-1}, Zn 0.33 to 0.24 mg kg^{-1}, Fe 0.9 and 1.0 mg kg^{-1}, Cu 0.05 and 0.06 mg kg^{-1} and Mn 0.6 and 0.5 mg kg^{-1}, in sandy loam and clay loam textured soil, respectively, with the treatment of deep tillage in summer along with GR-50%, green manuring and FYM @ 10 t ha^{-1}. Next highest reduction in pH was 0.92 and 1.01 and in ESP 19 and 22, and increase in organic carbon 0.3 and 0.1 g kg^{-1}, available N 71 and 48 kg ha^{-1}, P 5.4 and 7.0 kg ha^{-1}, K 48 and 45 kg ha^{-1}, Zn 0.31 and 0.20 mg kg^{-1}, Fe 0.8 and 0.9 mg kg^{-1}, Cu 0.03 and 0.04 mg kg^{-1} and Mn 0.5 and 0.4 mg kg^{-1}, under the treatment GR-50% along with green manuring and FYM @ 10 t ha^{-1} (T_5) as compared with control values of pH 8.95 and 9.24, ESP 32 and 38, organic carbon 1.5 and 1.8 g kg^{-1}, bulk density 1.51 and 1.45 Mg m^{-3}, available N 146 and 158 kg ha^{-1}, P 10.7 and 8.5 kg ha^{-1}, K 246 and 231 kg ha^{-1}, Zn 0.48 and 0.55 mg kg^{-1}, Fe 5.2 and 5.0 mg kg^{-1}, Cu 0.28 and 0.24 mg kg^{-1} and Mn 2.3 and 2.2 mg kg^{-1}, for sandy loam and clay loam textured soil, respectively. Non-significant difference was noticed on soil properties between the treatment of T_6 and T_5, respectively. Application of gypsum as per GR and organic manure reduced soil sodicity, improved physico-chemical properties and consequently higher availability of nutrients in soil (Minhas *et. al.*, 1995). Arora *et al.*, (1991) also reported that deep tillage improved the physical properties of soil. The treatment of deep tillage in summer along with GR 25%, green manuring and FYM @ 10 t ha^{-1} was as effective as T_6. Higher reduction in EC, pH and ESP was noticed in sandy loam soil than to clay loam soil. This may be due to the fact that sandy loam soils have low apparent field capacity and higher infiltration rate.

Table 7 : Response of gypsum, deep tillage, green manuring and FYM on physico-chemical properties of sandy loam and clay loam after harvest of wheat (3 year mean).

Treatment	EC_2 (dS m^{-1})	pH_2	ESP	OC (g kg^{-1})	BD (Mg m^{-3})	Available nutrients						
						N	P	K	Zn	Fe	Cu	Mn
						(kg ha^{-1})			(mg kg^{-1})			
T_1	0.53 (0.65)*	8.95 (9.14)	32 (38)	1.5 (1.8)	1.51 (1.45)	146 (158)	10.7 (8.5)	246 (231)	0.48 (0.55)	5.2 (5.0)	0.28 (0.24)	2.3 (2.2)
T_2	0.32 (0.49)	8.29 (8.36)	20 (24)	1.9 (1.7)	1.50 (1.44)	180 (179)	12.7 (12.2)	278 (238)	0.58 (0.65)	5.3 (5.2)	0.28 (0.25)	2.4 (2.3)
T_3	0.33 (0.45)	8.02 (8.26)	16 (21)	1.4 (1.9)	1.48 (1.43)	193 (187)	13.4 (12.5)	286 (254)	0.75 (0.69)	5.5 (5.3)	0.29 (0.26)	2.5 (2.5)
T_4	0.31 (0.44)	8.24 (8.33)	17 (19)	1.7 (1.9)	1.49 (1.41)	194 (189)	13.9 (13.6)	290 (253)	0.78 (0.72)	5.6 (5.5)	0.30 (0.27)	2.5 (2.6)
T_5	0.34 (0.45)	8.03 (8.13)	13 (16)	1.8 (1.9)	1.48 (1.41)	217 (206)	16.1 (15.5)	294 (276)	0.79 (0.75)	6.0 (5.9)	0.31 (0.28)	2.8 (2.6)
T_6	0.27 (0.38)	8.08 (8.21)	14 (16)	1.8 (2.1)	1.46 (1.39)	214 (209)	15.9 (14.7)	290 (271)	0.75 (0.74)	5.9 (5.8)	0.31 (0.28)	2.7 (2.7)
T_7	0.28 (0.39)	7.81 (8.12)	11 (14)	1.9 (2.1)	1.46 (1.38)	223 (219)	17.8 (17.3)	315 (279)	0.81 (0.79)	6.1 (6.0)	0.33 (0.30)	2.9 (2.7)
CD at 5%	0.15 (0.17)	0.27 (0.31)	1.5 (1.7)	1.1 (1.4)	0.03 (0.02)	21 (29)	1.5 (1.7)	26 (31)	0.06 (0.07)	0.14 (0.19)	0.01 (0.02)	0.15 (0.17)

* figure in paren thesis indicates the values of clay loam soil.

Infiltration rate

At the end of Rabi crop both the years, final infiltration rate was measured in all the treatments (Fig. 1). The infiltration rate increased with the application of gypsum @ 25% or 50% with green manuring only or along with FYM @ 10 t ha^{-1} and deep tillage in summer. Maximum cumulative infiltration rate 4.15 cm hr^{-1} in sandy loam and 1.38 cm hr^{-1} in clay loam soil was noticed with the deep tillage in summer + GR 50% + green manuring + FYM @ 10 t ha^{-1} followed 3.85 cm hr^{-1} in sandy loam and 1.15 cm hr^{-1} in clay loam soil by the treatment of deep tillage in summer + GR 25% + green manuring + FYM @ 10 t ha^{-1} and 2.02 cm hr^{-1} in sandy loam soil 0.84 cm hr^{-1} in clay loam soil with GR 50% + green manuring and FYM @ 10 t ha^{-1}, respectively. The application of gypsum to sodic soil has been reported to improve aggregate stability and consequently the infiltration rate (Agassi *et. al.,* 1981; Keren and Shainberg, 1981). Minimum infiltration rate in clay loam soil was due to high exchangeable sodium and electrolyte concentration was primarily associated with an increase in clay tactoid swelling. In these soils, the increase in swelling reduced the diameter of the water-conductance pores and increases the resistance to flow. Similarly, swelling of clay has been found to increase with sodium adsorption ratio (McNeal *et al.,* 1966).

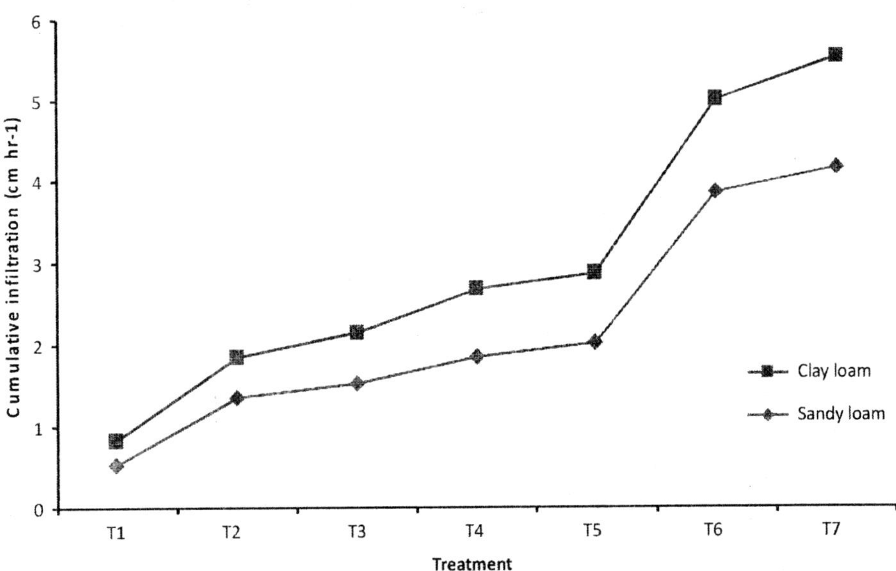

Fig. 1: Response of gypsum, deep tillage, green manuring and FYM on cumulative infiltration (cm hr^{-1}) after harvest of wheat.

Conclusion

In arid and semi arid areas of India, agricultural production is hampered by insufficient availability of good quality irrigation water (Minhas *et al.*, 1995, Choudhary *et al.*, 2002). Sodic water, which constitutes an appreciable proportion of total under ground water resource, adversely affects the plant growth (Yadav and Kumar, 1995). Primary survey indicated that only 31% farmers knew the recommended practices of sodic soil reclamation (gypsum requirement-50%, green manuring and leaching with good quality water), whereas among knowledgeable group only 7% farmers used gypsum along with application of FYM and did bunding of fields for rain water conservation, 11% used gypsum, 22% did bunding of fields for rain water conservation and 33% used FYM. The grain and stover yields of pearl millet and wheat was significantly improved by the addition of gypsum @ 25 and 50% GR with green manuring or along with FYM and deep tillage in summer in sandy loam soil as well as in clay loam soil. Similarly, an increase in content of organic carbon, available N, P, K and micronutrients and a decrease in pH, ESP and bulk density of the soils was observed. Maximum grain and straw yield of pearl millet and wheat was obtained with deep tillage in summer along with gypsum application @ 50% GR + green manuring + FYM @ 10 t ha^{-1} (T_7) in both textured soil followed by GR-50% + GM + FYM @ 10 t ha^{-1} (T_5) and deep tillage in summer along with gypsum application @ 25% GR + green manuring + FYM @ 10 t ha^{-1} (T_6) in clay loam soil and non-significant difference was noticed between T_5 and T_6 in sandy loam soil.

References

Agassi M., Shainberg, I. and Morin I. (1981). Effect of electrolyte concentration and soil sodicity on the infiltration rate and crust formation. Soil Sci. Soc. Am. J. **45** : 845-851.

Arora, V.K., Gajri, P.R. and Prihar, S.S. (1991). Tillage effects on corn in sandy soil in relation to water retentively nutrient and water management and season evaporativity. Soil and Tillage Research. **21** : 1-21.

Choudhary, O.P., Josan, A.S. and Bajwa, M.S. (2002). Role of organic materials in mobilizing intrinsic calcium carbonate to ameliorate sodic irrigations. In: proceedings of the 17th World Congress Soil Science, Symposium No. 34, Abstracts Vol. III,Symposia **22-36** : p. 162.

Jackson, M.L. (1973). Soil chemical analysis. Prentice Hall of India Pvt. Ltd., New Delhi.

Keren, R., Shainberg, I. (1981). The efficiency of industrial mined gypsum in reclamation of a sodic soil: rate of dissolution. *Soil Sci. Soc. Am. J.* **45** : 103-107.

McNeal, B.L., Norvell, W.A., Coleman, N.T. (1966). Effect of solution composition on the swelling of extracted soil clay. *Soil Sci. Soc. Am. Proc.* **30** : 313-317.

Minhas, P.S., Sharma, D.R., Singh, Y.P. (1995). Response of rice and wheat to applied gypsum and farmyard manure on an alkali water irrigated soil. *J. Ind. Soc. soil Sci.* **43** : 452-455.

More, S.D. (1994). Effect of farm waste and organic manures on soil properties, nutrient availability and yield of rice wheat grown on sodic vertisols. *Journal of the Indian Society of Soil Science* **42** (2) : 253-256.

Richards, L.A. (1954) (Ed). Diagnosis and improvement of saline and alkali soil. USDA. Handbook No. 60.

Singh, N.T. (1992). Land degradation and remedial measures with reference to salinity, alkalinity, water logging and acidity in Natural Research Management for sustainable Agriculture and Environment. Deb, D.L. (Ed). Bub. New Delhi, pp 442.

Tyagi N.K. (1998). Management of salt - affected soils Eds. G.B.Singh and B.R.Sharma. fifty Years of Natural Resource Management Research in India. Div. of NRM. ICAR, Krishi Bhavan, New Delhi pp 364.

Walkley, A., Black, I.A. (1934). Estimation of soil organic carbon by the chromic acid titration method. *Soil Science* **37** : 29-38.

Yadav, H.D., Kumar, Vinod, (1995). Management of sodic water in light textured soil. Reclamation and management of water logged soils. (Rao, KVGK, Agrawal, M.C., Singh, O.P. and Osterbaan, R.J. eds) pp. 226-241.

Yaduvansi, N.P.S., Sharma, D.R. (2007). Use of wheat residue and manures to enhance nutrient availability and rice-wheat yield in sodic soil under sodic water irrigation. *Journal of the Indian Society of Soil Science* **55** (3): 330-334.

□□□

Section - 6

Governance and Policy Issues for Community Empowerment

Chapter 33

Resource Conservation for Food and Environmental Security in Central India

P.D. SHARMA AND P.P. BISWAS
Division of Natural Resource Management, Indian Council of Agricultural Research, Krishi Anusandhan Bhavan-II, New Delhi

The central India covering the states of Uttar Pradesh, Madhya Pradesh, Jharkhand and Chhattisgarh is generally characterized by low rainfall, rampant land degradation and poor socio-economic status of people. The region is home to about 55 million tribal people, constituting more than 70 per cent of India's tribal population. As per agricultural statistics (2008), about 40 per cent of population of Jharkhand, Chhattisgarh and Madhya Pradesh and 33 per cent of Uttar Pradesh falls below poverty line. With less than 15 per cent area under irrigation, the rainfed farming dominates the region. The scarcity of water is, the main problem of central India. Apart from the low and erratic rainfall, high evaporative demand and limited water holding capacity of the soils aggravate the problem of moisture availability to crops. The large yield fluctuations due to vagaries of weather are often beyond the risk bearing capacity of the farmers. Most of the rainfall is received during the monsoonal period starting in the month of June and ending in last week of September or sometimes in the first week of October. With undulating topography and low moisture retention capacity of the soil, major portion of the rain water is lost as runoff. The large runoff, in turn, causes soil erosion, flash floods and water logging in low lying areas. The increased soil erosion takes away the top fertile soil, rendering soils unproductive for a long time. Restoring productivity of these lands becomes a costly and difficult proposition.

Soil Resource

The soils of the region are generally categorized as alluvial soils, red soils, red-lateritic soils and black soils. As per soil taxonomy (Velayutham and Bhattacharya, 2000), the soils of Uttar Pradesh belong to Inceptisol (20.9 m ha), Entisol (5.8 m ha), Alfisol (1.6 m ha), Vertisol (0.5 mha) and Mollisol (0.1 m ha). In Madhya Pradesh, the soils fall under Inceptisol (17.9 m ha), Vertisol (9.2 m ha), Alfisol (8.5 m ha), Entisol (8.5 m ha) and Mollisol (0.3 m ha). The soils of Bihar including Jharkhand belong to Inceptisols (6.1 m ha), Alfisols (6 m ha), Entisols (4.7 m ha) and Vertisols (0.1 m ha). The Inceptisols, therefore, constitutes the largest soil order followed by Entisols, Alfisols and Vertisols. The Inceptisols are, generally, coarse in texture (loamy sand to sandy loam) with altered and cambic horizons but no spodic, argillic and oxic horizons. The soils are poor in fertility and water holding capacity. The Entisols are generally immature soils with loamy sand or sandy loam texture. Depth of the soils is not a constraint. These soils have very low clay content and hold water up to 200 mm per meter of soil profile. The soils, generally riverine in nature, are highly erodible. Their nutrient holding capacity is poor. In low rainfall areas monsoon cropping is practiced and in high rainfall areas double cropping is possible. The Alfisols are, by and large, light textured soils which have low moisture holding capacity but high water intake. The rain water falling in such areas gets soaked up and saturates the profile. The soil water percolation is more and therefore, is lost for crop use. Owing to faster intake of water in the profile the surface runoff is limited and soil loss from erosion is relatively low. Soil crusting is a common problem in low rainfall areas. The Vertisols have high clay content and high moisture retention capacity. Owing to its swelling and shrinking characteristics, permeability is low and hence the rate of infiltration of water is minimum. This causes more surface and high soil loss from the top layer owing to surface erosion. It is estimated that 68.5 tonnes/ha/year soil is lost from Vertisols. Due to high clay content, the soils develop cracks during rabi season at flowering stage of crops.

Land Degradation

Land degradation constitutes one of the greatest threats for food and environmental security. Many thriving civilizations were vanished in the past for the simple reason that land care was not accorded the attention it deserved. About 59, 52 and 36 per cent area is degraded in Madhya Pradesh including Chhattisgarh, Uttar Pradesh including Uttaranchal and Bihar including Jharkhand, respectively due to various factors like soil erosion, water logging, salinity/alkalinity and acidity etc. (Table 1). Erosion due to water is a major cause of soil degradation. As per recent estimates (Table 2), over 60 per cent area in Central India is having potential soil erosion rates of > 10 t ha^{-1} year^{-1} (permissible soil loss limit). The degraded soils with very low productivity do not contribute much to the national GDP.

Table 1 : Soil Degradation in Central India (m ha).

State	Water Erosion	Wind Erosion	Physical degradation	Salinity/ Alkalinity	Acidity	Complex problems	Total	Degraded Area(%)
U.P. (including Uttranchal)	11.4	-	2.3	1.4	-	--	15.3	52.0
M.P. (including Chhattisgarh)	17.9	--	0.4	-	6.8	1.1	26.2	59.1
Bihar (including Jharkhand)	3.0	--	2.0	0.2	1.0	--	6.2	36.1

Table 2 : Area affected by potential soil erosion in central Indian states.

State	Area (TGA) (km^2)	Soil erosion (area in percentage of TGA)					
		Moderate (10-15 t/ha/yr)	Moderate severe (15-20 t/ha/yr)	Severe (40-80 t/ha/yr)	Very severe (40-80 t/ha/yr)	Extra severe (>80 t/ha/yr)	Total (>10 t/ ha/yr)
Chhattisgarh	134,805	8.0	6.4	18.2	13.6	19.0	65.3
Jharkhand	79,898	15.5	11.4	20.9	12.2	4.6	64.8
MP	308,641	12.9	9.5	18.9	9.3	8.6	59.2
UP	241,046	27.6	9.9	8.3	13.4	-	59.2

Natural Resource Management Interventions

There is need for macro and micro level land-use planning backed up by appropriate decision support systems consistent with the prevailing land-socio-economic-environmental-market imperatives for sustainable management of our natural resources. The task could be accomplished by employing state of art Remote Sensing, Geographic Information System (GIS) and Information Technology based approaches and tools. The continued degradation of our soil and water resources is telling upon the Total Factor Productivity, decelerating the agricultural growth for quite sometime. The required growth rates per annum for food grains (2per cent) and oilseeds (6per cent) to meet the projected demands of food by 2020 are much higher than the realized growth rates for the last ten years (Chand, 2007). Therefore we must have region-centric all-inclusive agricultural development plans seeking all round development of different areas of the country based on sound ecological principles.

Participatory watershed management: The watershed management would enable sustainable management of soil, water and biotic resources of rainfed areas of Central India, while enhancing land productivity, profitability and livelihood opportunities for rural households. The participatory watershed management ensuring transparent and equitable sharing of the services and benefits amongst the stakeholders is advocated for greater success of the programmes. The concept of integrated and participatory watershed management is emerging as the cornerstone of rural development in rainfed regions, especially in dry and semi-arid regions.

India has made large investments in the area of watershed management over the past three decades and realized significant gains in resource conservation, land productivity, employment generation and environmental upkeep. The watershed management has emerged as potential drought mitigation strategy during the severe drought of 1987 in India. Accordingly, the programme was scaled up at the national level with larger public investments in National Watershed Development Program for Rainfed Areas (NWDPRA), Integrated Wasteland Development Project (IWDP) and many other initiatives supported by World Bank, DFID, DANIDA and SIDA etc. The Rajiv Gandhi Watershed Mission in Madhya Pradesh brought about increase in irrigated area by 59 per cent and in agricultural production by 37 per cent in Kharif and 30 per cent in Rabi season. The ground water improved in over 3000 villages covered under the project. The ICRISAT, Hyderabad has been developing community watersheds in India, to improve land productivity and livelihoods for poor people. The development of 150 community watersheds gave significant yield gains of 35-270 per cent in sorghum, 30-174 per cent in maize, 72-242 per cent in pearl millet, 28-179 per cent in groundnut and 97-204 per cent in pigeonpea. (Dar, 2007).

Studies conducted by CRIDA, Hyderabad on micro watershed basis on Alfisols over three years (2005-08) indicated that an agri-silvi-horti-pastoral system for small and marginal farm holdings having ~1.1 ha area with arable crops (0.4725 ha), agro-forestry (0.3496 ha), vegetables (0.1150 ha), grasses (0.1256 ha) and bushes (0.0890 ha) gives the highest gross income of Rs. 16080/-, net income of Rs. 9793/- and BC ratio of 2.38. The individual enterprises of arable cropping, agro-forestry, vegetables, grasses and bushes contributed 38.2, 10.3, 27.2, 7.1 and 17.2 per cent, respectively to the total net income.

The ICRISAT has carried out meta analysis of 311 watersheds in India in terms of efficiency, equity and sustainability benefits (Joshi *et. al.*, 2005). The mean benefit-cost ratio of watershed program in the country was quite modest at 2.14. The internal rate of return was 22 per cent which compared very well with rural development programs. The watershed programs generated enormous employment opportunities, augmented irrigated area and cropping intensity and conserved soil and water resources. The study concluded that the watershed program was silently rejuvenating and revolutionizing rainfed areas.

The watershed management programme fits very well in the framework of National Rural Employment Guarantee Act (NREGA) of the Government of India. The programme also has great promise in fulfilling the United Nations Millennium Development Goals (MDGs) of eradicating extreme poverty and hunger, promoting gender equality and empowerment of women and ensuring environmental sustainability in the developing world. The greater institutional support to the programme would allow tapping of a multitude of potential benefits of the watershed management programme. There are some limitations that come in the way of successful completion of the watershed management programmes. Some of these are 1) marginal or nil benefits to landless and marginal farmers causing inequities 2) partial meeting of domestic, livestock and ecosystem water needs 3) unsatisfactory restoration and utilization of common property resources 4) little participation of peoples in project implementation phase and not at all beyond for long term management of created resources 5) inadequate monitoring and impact assessment of programmes and 6) emergence of conflicts between different stakeholders with divergent interests. Finding amicable solutions to these problems is of much significance for successful implementation of watershed projects.

To make programme successful, each and every stakeholder in the watershed should be made a beneficiary of the programme. Unless something is accruing directly or indirectly to every stakeholder, the programme is not going to be successful. Even the landless people, who should be important stakeholders need to be given the benefits of the programme. Therefore, all the inhabitants of the watershed need to be involved in the programme right from the planning phase and organized into different committees to look after various conservation programmes. The women constituting main workforce in rural and tribal regions, should be given enough representation in different committees. The kind of conservation programmes and goods to be produced should be in consonance with the needs and aspirations of watershed inhabitants. A number of policy changes have been brought, of late, to have participatory process of watershed management more effective, equitable, transparent and free of conflicts. The monopoly of central and state governments in the implementation of watershed management programmes has been curtailed by recognizing NGOs and registered institutions as equal partners of the programmes. The watershed level institutions have been empowered through decentralization of decision making process. The formation of self-help groups and user- groups is encouraged for wider participation. For financial transparency, the watershed association members are made as co-signatory to all financial transactions. The beneficiaries of the programme are required to make reasonable contributions in cash or kind to ensure their commitment to the project programmes. The landless and socially or economically disadvantaged people are helped through micro-credit and micro-enterprising to ensure equity and social harmony. The salaries of project functionaries are kept minimum so as to make more allocations to project activities. The trainings and exposure visits for local communities and service providers to enhance their skills forms an integral part of all watershed programmes.

Notwithstanding novelties of the innovations, the real benefits of participatory process still remain elusive. The watershed programmes visited by a number of players with divergent interests are faced with inherent conflicts on allocation of resources, activities to be undertaken and sharing of functions and benefits of the programmes. While the Agriculture and Rural Development Ministries would be interested in increasing production, reducing poverty and enhancing employment, the Ministry of Environment & Forests may look only for augmenting forest cover from the watershed management programmes. Similarly, the conflicts may arise between the government and non-government organizations, watershed committees and local elected institutions, upstream and downstream inhabitants of the watersheds holding different perceptions and expectations from these projects. Evolving adequate conflict resolution mechanisms at different levels of governance is, in fact, need of the hour. Most of the conflicts should be resolved through persistent dialogue, flexible approach and improvement in the guidelines. The Government of India has established National Rainfed Areas Authority in 2006 for bringing in much needed convergence of resources with different ministries and departments and helping develop required institutional mechanisms, guidelines and protocols for conflict resolution at various levels of participatory watershed management.

Water harvesting: Even after achieving the total irrigation potential of 140 million ha-m, about 50 per cent of total cultivable area would remain rainfed in India. Rainwater harvesting and management is, therefore, an entry point to successful management of rainfed lands. The rainwater use efficiency for crop production is still low at 30-45 per cent, with attendant loss of 300-800 mm of seasonal rainfall as surface runoff or deep percolation. A wide variety of mechanical and cultural management practices like land leveling, contour bunding, contour trenching, bench terracing, contour farming, intercropping, strip cropping, mixed cropping, mulching, crop geometry and vegetative barriers etc are recommended for in-situ conservation of runoff on sloping lands. The surplus runoff is harvested and stored in water storage structures for recycling as supplementary irrigations during dry spells. The yield increases were over 100-200 per cent even with one supplemental irrigation of 5 cm (Samra and Pratap Narain 1998). The created water resource increased dramatically the productivity of agricultural, forest and livestock resources. The stored water in ponds is also used for rearing of fish. The cheaper ways and means to reduce seepage from the reservoirs have been worked out to enhance water storage efficiency. The small perennial water sources could also be tapped in a participatory mode in certain locations. The technology has been demonstrated in villages of Kalimati, Badasi, Bhopalpani and Sainji, Dehradun, providing irrigation water to a large number of rural families (Sharda *et. al.*, 2006). Water user societies have been constituted for equitable sharing of harvested water.

As groundwater is being increasingly depleted in most parts of the country, its recharge through appropriate watershed interventions is also becoming important. There have been spectacular increases in ground water table (0.2-2 m) in well managed watersheds in different parts of India. The conventional methods of augmenting ground water recharge like percolation ponds and check dams have found a major place in all watershed development programmes. However, the new recharge techniques like recharge pits and shafts and injection wells etc need to be evaluated for best results.

In Jharkhand, rainwater harvesting tank constructed on the mid-land and lined by LDPE (low density polyethylene) film was employed for gravity-fed irrigation in the command area. Multi-tier horticulture system was adopted with litchi as main crop, guava as filler crop and vegetables (including pea, French bean, cucumber and tomato) in the command area. With the supplementary irrigation from the harvested rainwater, rice and cowpea were also taken up. On bunds, bottle gourd, pointed gourd, and maize crops were grown. In all, the total income from vegetables, rice and fish was Rs 18,690 in the very first year.

Management of waterlogged lands: In Madhya Pradesh, nearly 12 mha of rainfed vertisols, are left fallow due to temporary waterlogging during kharif and are cropped only in post rainy season on profile stored moisture. The broad bed and furrow system of the International Crop Research Institute for the Semi Arid Tropics (ICRISAT) for managing rain water in vertisols made it possible to increase crop yields four to five times as compared to normal practice. However, this method could not be adopted widely by the farmers in India because it was costly and labour intensive. The JNKVV centre of the All India Coordinated Project on *Soil Physical Constraints and their Amelioration for Sustainable Crop Production* had developed a cost effective and eco-friendly *Raised-Sunken Bed* technology providing adequate drainage during rainy season to upland crop in raised bed and sufficient ponding of water for paddy in sunken bed. The technology also allowed channelization of the excess runoff water for storage and recycling to provide supplementary irrigation during dry spells and / or rabi season.

The technology not only paved the way for raising a rainy season crop but also helped to reap a bumper harvest with the judicious use of inputs viz HYV, fertilizers etc. The average soybean and paddy yields in raised and sunken beds, respectively, were nearly 3 times more as compared to flat bed (farmers' practice) (Table 3). The economically viable technology has been demonstrated successfully in the region. The cost of land shaping could be recovered within 1-2 years.

Table 3 : Crop yields (kg/ha) under raised & sunken bed system and farmers' practice (average of 16 years).

Raised and Sunken bed system		Flat bed (farmers' practice)	
Soybean (raised bed)	Paddy (sunken bed)	Soybean	Paddy
2222	1013	775	386

Similarly, various configurations of raised and sunken bed system were tested at Barwaha farm, Indore centre for reclamation of alkali vertisols by leaching soils treated with amendments. The cotton (raised bed) and paddy (sunken bed) yields could be sustained with sunken: raised bed ratio of 3:2 with provision of 1-2 life saving irrigations.

Amelioration of acidic soils: In Jharkhand, nearly 0.4 mha of arable land (mainly acid alfisols) having pH < 5.5 distributed in Dumka, Jamtara, East Singhbhum, Ranchi, Gumla, Garhwa districts are having low productivity of < 1 t/ha. The furrow application of lime @ 2-4 q/ha alongwith recommended dose of fertilizers including micro and secondary nutrients (especially S, Zn, B, Mo and Co), FYM, biofertilizers at the time of sowing ameliorates the soils, improving their nutrient use efficiency, productivity and profitability. The effect is more pronounced on pulses and oilseeds compared to cereals and effects savings on chemical fertilizers by 50per cent (Sharma and Sarkar, 2005). Direct application of rock phosphate and phosphogypsum is also beneficial in these acidic soils.

Reclamation of salt-affected lands: Large areas in Uttar Pradesh (1.3 m ha), Bihar (0.4 m ha) and Madhya Pradesh (0.2 m ha) are having low productivity due to presence of salts in soils. Application of gypsum/pressmud as soil amendment alongwith FYM is highly beneficial in alleviating the adverse effect of sodicity on crop productivity. Further, by using sodicity tolerant varieties the rate of application of gypsum could be brought down to 25per cent of gypsum requirement without compromising on crop yields.

Suitable tree species such as *Prosopis juliflora, Acacia nilotica* and *Tamarix articulata* are recommended for plantation on salt affected lands. Promising agroforestry trees for moderately sodic soils or the reclaimed soils are *Populus deltoides, Eucalyptus tereticornis, Acacia nilotica* and *Tectona gradis*. However, Populus-based agroforestry system has proved more remunerative due to faster growth and better market price. Among the fruit trees, Ber, Guava, Pomegranate, Amla, Karonda, Tamarind and Jamoon are promising ones for highly alkali soils with adoption of proper site modification technology including the use of organic and inorganic amendments. Suitable grasses for alkali soils are *Leptochloa fusca, Chloris gayana, Brachiaria mutica and Cynodon dactylon*. Other promising grasses for saline soils are *Aeluropus lagopoides, Chloris barbata, Echinocloa colonum, Dicanthium annulatum, Phragites* spp *and Sida spp.*

Soil fertility management: The central India is confronted with the widespread deficiencies of nutrients, particularly N, P, S, Zn, B and Mo along with low contents of organic carbon. The deficiencies of nutrients are generally ascribed for impaired soil health and stagnating crop productivity in the region. This has further aggravated due to imbalanced use of NPK fertilizers as evidenced by wider consumption ratios (Fertilizer Statistics, 2007-08). The NPK ratio for Uttar Pradesh and Jharkhand are 14.0:3.5:1 and 12.1:6.0:1 against desirable ratio of 4:2:1. The limiting nutrients, not allowing full expression of other nutrients, lower the overall fertilizer use efficiency and crop productivity. As the soils are not being replenished adequately with the nutrients removed by the crops under intensive agriculture, the soils are continually being mined of their nutrient reserve. The increased mining of soil potassium, of late, seems to be a cause of more rampant decline in rice yields in Indo-Gangetic Plain, as revealed by the analysis of data pertaining to rice-wheat cropping system from 24 research stations in IGP . The indiscriminate mining of potassium, however, is going to degrade the K-bearing clays to the point of nil return.

The Integrated nutrient management encompassing conjunctive use of chemical fertilizers including secondary and micronutrients, organic manures, composts/vermicomposts, bio fertilizers and green manures should be the most ideal system of nutrient management.

The system enhances nutrient-use efficiency, maintains soil health, enhances yields and reduces cost of cultivation (Table- 4). We need to augment supplies of organic manures to have integrated nutrient management on a sound footing. The availability of organic manures is not adequate in the region (Table 5).

Table 4: Benefits of site specific nutrient management (SSNM).

Location, State	Grain yield SSNM Kg/ha/yr	Grain yield Farmers' practice (FP) Kg/ha/yr	Change in yield (SSNM-FP), kg/ha/yr	Nutrients applied under SSNM	Extra net returns (SSNM-FP), Rs./ha	BCR of improvement (SSNM-FP) system basis
Sabour Bihar	13,849	8,658	+ 5,191 (60%)	N P K S (4)	33,621	6.8
Ranchi Jharkhand	10,957	6,202	+ 4,755 (77%)	N P K S B Cu Mn Zn (8)	30,195	7.3
Faizabad UP	12,385	7,481	+ 4,904 (66%)	N P K S B Mn Zn (7)	31,414	4.9
Kanpur, UP	14,555	11,605	+ 2,950 (25%)	N P K S (4)	17,612	3.6
Modipuram, UP	16,679	11,334	+ 5,345 (47%)	N P K S Cu Mn Zn (7)	36,917	26.8
Varanasi, UP	12,116	10,996	+ 1,120 (10%)	N P K S B Cu Mn Zn (8)	7,384	2.6

Table 5 : Production of manures (lac tonnes) and biofertilizers (tonnes) in Central India during 2007-08.

State	Rural compost	Urban compost	FYM	Vermi compost	Bio fertilizers
Chhattisgarh	13.2	0.09	20.68	0.67	-
Jharkhand	-	-	-	0.41	202
MP	-	-	-	-	1885
UP	6.26	-	-	0.83	250

The AICRP on Soil Test Crop Response Correlation has brought out ready reckoners for soil test based fertilizer recommendations (NPK) utilizing the locally available resources to achieve the given yield targets of crops. Such ready reckoners depicting target yield equations have been prepared for the country including central Indian states. Further, we need to have a host of customized fertilizers through fortification / coating of micro and secondary nutrients suiting to different soil-crop situations for balanced fertilizer use.

Drought mitigation and contingent crop planning: Central India is subject to high variability in sown areas, yields and output due to aberrations in weather conditions, especially rainfall. Delays in normal monsoons may affect preparatory tillage and timely sowing of crops. The monsoonal delays, have repercussions on the schedule of activities for the entire agricultural year. Even after the timely onset, there may be withdrawal of monsoons causing moisture stress to crops. Some crops are highly susceptible to such mid-season variations in moisture availability, as is the case for rice at the flowering stage.

Depending on the behavior of south west monsoon, the crop production in central dry lands fluctuates widely from year to year. The aberrent weather could be categorized as (i) delayed onset of monsoon, (ii) long gaps or breaks in rainfall, and (iii) early stoppage of rains towards the end of monsoon season. Therefore, to mitigate such weather situations, farmers should have contingent crop planning to ensure some production in place of total crop failure.

Strategy has been evolved for normal onset of rains, breaks in rains, early withdrawal and uneven distribution through selection of situation-specific crops/varieties. If monsoon sets in as late as second week of July, short duration upland rice (variety - NDR-97 and NDR-118) may be included in place of akashi and cauvery. If the rains are delayed beyond this period but start somewhere in last week of July or first week of August, the cultivation of hybrid pearl millet (NHB 3-4, B.J. 104), blackgram (Type 9), greengram (var-Jagriti and Jyoti) may be included. Yet another alternative could be to harvest a fodder of either pearl millet, maize, sorghum or a mixture of cowpea, blackgram and one of the above fodder crops.

In case monsoon rains stop early towards the end of season, normal sowing of short duration upland rice, blackgarm and sesame may be taken up. If the rain stops very early, i.e. by the end of august or first week of September, only fodder crops or grain legumes could be harvested. Depending upon the soil moisture condition, relay sowing of crops like chickpea, lentil, mustard, linseed and barley could be done in rabi season. Efficient cropping systems that can profitably substitute the traditional ones at various locations of central dry land regions of the country have been summarized by AICRPs on dryland agriculture and cropping system research (Table 6).

Table 6 : Efficient cropping systems for central India.

Region	Cropping system
Malwa Plateau of MP	Maize-safflower Sorghum-safflower/chickpea Soybean-safflower
Keymore plateau and Satpura hill zone of MP	Black gram-potato+wheat
Vindhyan plateau zone of MP	Soybean-wheat
Bagelkhand region of MP	Rice-wheat/chickpea Sorghum -chickpea Mungbean-wheat
Subhumid regions of Chhotanagpur Plateau	Rice-linseed Maize-safflower
Alluvial zone of Jharkhand	Rice-potato-onion Rice-berseem-maize+cowpea (fodder)
Plain zone of Chhatisgarh	Rice-tomato
Plains of Eastern UP	Rice-/chickpea
Central plain zone of UP	Maize-potato-wheat Maize-pea-sunflower
Bundelkhand region of UP	Cowpea (fodder)-mustard Sorghum (fodder)-chickpea
Vindhayan plain zone of UP	Rice-potato-greengram

In Indore region, receiving 1000 mm annual rainfall, only single crop can be taken on shallow soils, inter cropping in medium depth soils and double cropping on deep soils.

In dry land of Varanasi region, crops like short duration upland rice, maize, pearl millet, blackgram, greengram, sesame, pigeon pea etc. should be taken up. These crops should be followed by chickpea, lentil, barley, mustard, safflower, linseed etc. on residual moisture during winter season. Inter cropping of vegetables with grain crops was pursued vigorously in some centers such as Varanasi and Phulbani. At both the palces long duration pigeon pea was inter cropped with vegetables such as okra, radish and chilli.

In central plain zone of U.P., maize-potato-wheat was more productive (15.8 t/ha/year) with highest productivity of 43.2 kg grain/ha/day. But, maize-pea-sunflower was found to be more promising giving highest profitability of Rs. 78/ha/day. It was relatively 114per cent higher than the existing rice-wheat system.

In Vindhayan plain zone of U.P., rice-potato-green gram system was more productive (14.3 t/ha/year) with highest productivity (39.2 kg grain/ha/day) and profitability (Rs. 58.2/ha/day). This system was relatively superior to the tune of 69per cent in terms of relative production efficiency and 80per cent in terms of relative economic efficiency when compared to the existing rice-wheat system.

In southern alluvial zone of Bihar, rice-potato-onion was found to be distinctly better with highest system yield of 18.1 t/ha/year in terms of rice equivalent yield, productivity of 49.7 kg/ha/day and relative production efficiency of 135per cent compared to existing rice-wheat system. But, in terms of economics, rice-berseem-maize+cowpea (fodder) system was more remunerative which gave highest net return of 45624/ha/year, profitability of Rs. 125/ha/day and relative economic efficiency of Rs. 224per cent compared to existing rice-wheat system.

In plain zone of Chhattisgarh, rice-tomato system was found to be better than existing rice-wheat system which gave highest yield (15.2t/ha/year) with productivity of 41.6 kg/ha/day and profitability of Rs 168/ha/day.

In Keymore plateau and Satpura hill zone of M.P., the system involving black gram–potato+wheat was better in terms of production (11.8 t/ha/year) and productivity (32.3 kg rice grain equivalent/ha/day).

In Vindhyan plateau zone of M.P., soybean-wheat system was found to be distinctly better in terms of highest rice equivalent yield (7.1 t/ha/year), productivity (19.6 kg grain/ha/day) and profitability (Rs. 62/ha/day).

Integrated farming systems : The vast synergies available with different farm enterprises have remained largely under-exploited due to crop or commodity driven policies. The farming system approach integrating different components of farming, viz crops, horticulture, livestock and fisheries offers greater opportunities of productivity enhancement, employment, income generation, nutritional security and risk bearing abilities for small land holders. It allows the utilisation of primary and secondary products of one component as basic input for another component. Multiple deployment of inputs based on the principles of recycling also minimises loading of environment with pollutants. A number of farming systems suiting to varied socio-economic-environmental conditions could be employed with productivity gains of 3-5 times compared to monocultures.

Investigations carried out in Durg district generated several economically viable mixed farming models integrating the different component like crop, livestock, poultry and duck on 1.5 acre land holding depending on the farmers' resource availability. A model having 2 bullocks + 1 cow + 1 buffalo + 10 goats + 10 poultry + 10 ducks along with crop cultivation generated a net income of Rs 33076 per year against Rs. 7843 per year under arable farming (crop farming alone) with a cost returns of 1: 2.238 and employment generation of 316 days (Ramrao *et. al.*, 2006).

A multi-enterprise model based on an integrated farming system and multiple water use was tested to increase productivity and to provide regular income, employment and livelihood to small land holding farmers of Sharda Sahayak Canal Command, village Kashrawan, Raibareili district, UP. The model of one ha comprised 0.4 ha fish pond, 0.2 ha field crops, 0.2 ha fruit crop, 0.1 ha forage and 0.1 ha vegetable crops. The slopes of embankment and the raised bed were utilized for the Eucalyptus plantation, which served as bio-shield and bio-drainage. The system has B:C ratio of 2.21.

Similarly, an integrated farming system model for marginal and small farmers of Western UP was developed for average farm family of 7 members. The system besides being profitable in terms of net return of Rs.75211/ha/ year, generated 347 additional man days as compared to crops alone (182 man days/year) offering increased employment opportunities to the rural youths. The green fodders were made available whole the year round in sufficient quantities, grown as component crops of different cropping sequences under crop production.

Rehabilitation of wastelands and forage production : The lands not suitable for crop production may be suitable for range/pasture management, tree farming, ley farming, dry land horticulture and agri-silviculture etc. All these systems offering alternatives to crop production are called as alternate land use systems. The alternate land use systems involving multi purpose trees, fruits, crop and grass species have been summarized for different agro ecological zones of India including central India (Solanki, 1999). These systems not only help in generating much needed off-season employment in mono crop dry land areas, but also minimize risk, utilize off season rains going waste as runoff, prevent degradation of soils and restore balance in the ecosystem.

The agrisilviculture system is recommended for soils in land capability Class IV with annual rainfall upto 750 mm. A large number of tree-crop combinations, particularly of nitrogen fixing trees (NFTs) with sorghum, groundnut, castor and pulses were evaluated on alfisols and vertisols. Widely spaced (8m x 8m) *Faidherbia alibida* and *Harwickia binata* combined well with green gram.

The alley cropping, a form of conservation farming system, meets the multiple requirements of food, fodder, fuel etc. besides enhancing soil fertility and preventing soil erosion. In this practice, arable crops are grown in alleys formed by the trees or shrubs, established generally on contours. This system is recommended for areas receiving 500-750 mm rainfall and belonging to land capability classes II to III. Trees or bushes grown in alleys act as live bunds and control runoff. The essential feature of the system is that hedge rows are cut back at planting and kept pruned during cropping to prevent shading and to reduce competition with food crops. Width of the alley and management of the tree crops are the key to success of the system. Permanent alleys provide fodder during drought.

At Jhansi, in *Acacia albida* alleys, significantly higher green (54.3 t/ha) and dry matter (15.8 t/ha) yields were obtained from nandi and signal grasses, respectively. On the other hand, anjan grass registered higher crude protein yield of 1.0 t/ha). *Leuucaena* in such system provided 5-6 t green fodder and 7-9 t fuel wood/ha/annum. During the cropping season, trees are lopped at about 0.5 metre height. These loppings are used as mulch to reduce moisture loss and improve the nutrient status of soil. Arable crops like maize, rice, pearl millet, legumes, oilseeds etc. are planted in the alleys formed by the two rows of trees.

The Agri-horticultural system in medium deep soil area (LCC II to IV) receiving annual rainfall of more than 750 mm, agri-hortiucltural systems consisting of a fruit tree intercropped with annual arable crop is recommended. Ber with clusterbean, cowpea, horsegram or other grain legumes has been widely adopted. Custard apple, Pomegranate and Anola are other fruit crops suitable for this system.

A suitable example of agri-horti-system is growing of cow pea/ green gram/horse gram in inter space of *ber (Zizyphus mauritiaria)* at 6 x 6 m spacing. Phalsa (*Grewia asiatica*) may be planted in between two ber plants in a row with a view to intensify the system. A well managed dry land orchard of *ber would* give 50 kg fruits per tree/ year. There should be 250 plants/ha for optimized production. The grower's income would touch around Rs. 50,000/ha (250 x 50 x 4), assuming that one kg *ber* fetches Rs. 4. One could get an additional income of Rs. 800-1000/- from green gram/cow pea (2.5-3.0 q/ha).

Economic fortification of existing horti-land use system through medicinal plants and aromatic species was tried at Agra (U.P.) by cultivating *Aloe vera* in the interspaces of matured *ber* trees that were planted in 8m x 4m spacings. Planting distance of *Aloe vera* was 0.5 m x 0.5 m and 1 m space was left out after every two rows of *Aloe vera*. The yield of *Aloe vera* and ber was around 47 and 7.5 t ha^{-1} yr^{-1}, respectively. The gross income and net income from the *ber - Aloe* system was estimated to be Rs. 1,23,000/- and Rs. 73,000/-, respectively.

Similarly, intercropping of ground nut has been found to be most profitable under Anola based multi-tier cropping systems in Chotanagpur plateau.

The silvi-pastoral system is recommended for marginal dry lands (LCC V and higher). The system essentially consists of a top feed tree species carrying grasses or legumes (preferable perennial) as understorey crops. Dry land farmers having larger holdings and keeping a land fallow for a longer period for one reason or the other, should go in for this system which could provide both fodder and fuel. The system on degraded wastelands in central India can optimize land productivity at 10 tonnes/ha/year (which includes 45per cent grass, 13per cent timber,18per cent top feed and 24per cent harvested and lopped firewood) from < 1 t /ha (Pathak, 2000). *Stylosanthes hamata* is an improved pasture legume that can be raised on marginal lands and on field boundaries. The system can produce 3-5 t ha^{-1} of dry fodder, second year onwards.

Cultivation of fast growing drought resistant leguminous trees such as *Leucaena leucocephala, Acacia albica, Hardwickia binata, Albizzia spp., S. sesbans. D. sissoo* or *Glyricidia* spp. planted in rows in cultivable wastelands can provide green fodder even in years of severe drought. Such modified and adaptable systems provide quality forage/grazing resource for a longer period of the year normally more than the situation with local/natural grasses of the marginal and sub-marginal lands (Table 7).

Table 7: Forage production under various systems of wasteland utilization.

Land Use System	Dominant species	Forage production (t/ha)	Leaf fodder (t/ha)	Availability Period
Natural grazing land	Sehima-Dicanthium	0.2-1.0	-	July-Sept.
Improved grassland	*Cenchrus ciliaris, C.setugerys+* *Cenchrus ciliaris, Macroptelium antropurpureum (Siratro)*	2.5-6.0	-	July-Nov.
Silvipastoral system	*Cenchrus+Dolichos ausillaris under Albizzia procera*	3.5-8.5	1.5-3.7	Year round

Source : Faruqui and Sunil Kumar (2008).

Conclusion

Central India has a potential to meet the challenges of new millennium in achieving food and environmental security. The major production constraints of the region are land degradation, poor soil fertility, frequent droughts, and poor socio-economic status of the people. The desired interventions for optimizing crop productivity, farm profitability and environmental conservation in the region are soil and water conservation including water harvesting, integrated nutrient management, amelioration of salt-affected and acidic soils, integrated farming /cropping system including agro-forestry and crop diversification, etc. Government support in this endeavor should come through micro credits, subsidies, infrastructure development, crop insurance, remunerative price support and comprehensive market intervention etc. The whole spectrum of activities could be viewed within the perspective of participatory watershed development programmes.

References

Chand, Ramesh. (2007). Demand for food grains. *Economic and Political Weekly* **42**(52): 10-13.

Dar, W.D. (2007). Efficient conservation and utilization of rainwater in drylands for high productivity. In Souvenir Ensuring Water and Environment for Prosperity and Posterity. 10th Inter-Regional Conference on Water and Environment Oct 17-20, pp.1-5.

Faruqui, S. A. and Sunil Kumar (2008) Forage Production Technologies for rainfed areas; An Indian Context. In. Anand Swarup, Suraj Bhan and J.S. Bali (Eds) Conservation Farming: Enhancing productivity and profitability of rainfed areas. 303-321.

Joshi, P.K.; Tewari, L.; Jha, A.K. and Shiyani, R.L.(2000). Meta analysis to assess impact of watershed. Presented at the Workshop on Institution for greater impact of technologies, NCAEPR, New Delhi, India.

NBSS&LUP (2008). Harmonization of degraded lands/wastelands datasets of India - A GIS based approach. NBSS & LUP, Nagpur (personal communication).

Pathak, P.S. (2000). Agroforestry- Issues and Strategies for the New Millenium. *Indian Farming*, **49** (12): 20-26.

Ramrao, W.Y., Tiwari, S. P. and Singh P. (2006) Crop-livestock integrated farming system for the Marginal farmers in rain fed regions of Chhattisgarh in Central India. *Livestock Research for Rural Development*, **18** (7): Article # 102.

Samra, J.S., and Pratap Narain (1998). Soil and water conservation. In: *50 years of natural Resource Management Research* (G.B. Singh and B.R. Sharma, Eds.), NRM Division, ICAR, New Delhi pp. 145-176.

Sharda, V.N., Sikka, A.K. and Juyal, G.P. (2006). *Participatory Integrated Watershed Management – A Field Manual.* CSWCR&TI, Dehradun.

Sharma, P. D. and Sarkar, A. K. (2005) *Managing Acid Soils for Enhancing Productivity.* Natural Resource Management Division (ICAR), Krihsi Anusandhan Bhavan-II, Pusa Campus, New Delhi. *Technical Bulletin* pp.23.

Solanki, K.R. (1999) Agroforestry. *50 Years of NRM research.* G.B. Singh and B.R. Sharma (Eds). ICAR, New Delhi. 447-476.

Tiwari, K.N. Sharma, S.K., Singh, V.K., Dwivedi, B.S. and Shukla, A.K. (2006). Site Specific Nutrient Management for increasing crop productivity in India. Results with rice-wheat and rice-rice systems. PDCSR, Modipuram and PPIC- India programme, Gurgaon. pp. 112.

Velayutham, M. and Bhattacharya, T. (2000). Soil resource management. International Conference on *Managing Natural Resources for Sustainable Agricultural production in the 21st Century* held on 14-18 February 2000, New Delhi.

□□□

Chapter 34

Governance and Policy Issues for Community Empowerment

KRISHNA SRINATH AND H.K.DASH
Directorate of Research on Women in Agriculture (ICAR), Bhubaneswar

I. Governance – the changing paradigm

Worldwide, good governance is recognized as a crucial input for accelerated economic growth and development. This has become more relevant in a country like India which, despite good progress in areas like food production, infrastructure development, social services etc., is still struggling hard to catch up with developed and some developing nations on important development indicators.

In India, governments are the key actors in governance. Others are the NGOs, civil society, research institutes, financial institutions, political parties and other associations. Given its responsibilities and authority in deciding national and sectoral priorities, allocating resources in larger national and people's interest, and the network of institutions its control, government is the major actor in governance. Hence the structure and functioning of government, institution, conduct of officials have great influence on quality of governance.

For a long time governments were the sole player in economic development. It largely controlled and influenced the resource allocation and production related activities depending on nation's development priorities. It was the solely provider of critical basic services to the public. But the state's failure in providing basic services such as health, education, water, sanitation to its people and ineffectiveness of public polices in creating income and employment opportunities and removing poverty triggered a reorientation in the thinking and approach of government to address public issues. The process got impetus under the influence of some enlightened public figures, academician, researchers and international organizations. Not only is the role of the government being redefined, but also other actors such as civil society, research institutions, interest groups, women groups, associations and federations are becoming increasingly important in governance and occupying the space given up by the government. Bottom-up approach to planning, decentralization in decision making and devolution of power to people, representation to wider sections of the society, greater dialogue with civil society, creation of new institutional structures are some of the visible changes taking place to accommodate new values and goal of governance. Towards making governance and its policies people oriented, government of India in budget 2009-10 acknowledges three important challenges.

- To lead the economy back on high GDP growth rate of 9% per annum
- To deepen and broaden the agenda for inclusive development, and to ensure that no individual, community or region is denied the opportunity to participate in and benefit from development process
- To energize government and improve delivery mechanism to provide high quality public services, security and rule of law to all its citizens with transparency and accountability.

The above statements are both a reflection of government's own assessment of country's development scenario and its commitment towards making people's lives better.

Successive governments have introduced a number of policy and institutional reforms to improve public service delivery and spur growth of sectors. Some are;

- RTI act to make greater transparency and accountability in decision making
- Creation of Unique Identification Authority of India (UIDAI) to improve governance with regard to delivery of public services
- Increasing allocation under NREGA to Rs. 39100 crore in 2009-10 and also under other rural development programmes
- National Rural Livelihood Mission in place of SGSY to help weaker sections of society including women.
- National food security mission to give impetus to the growth of rice, wheat and pulses in some lagging districts.
- Rastriya Krishi Vigyan Yojana (RKVY) to achieve 4% growth in agriculture through greater public investment and integrated development of sub-sectors.

While making all efforts to increase resource allocation to key sectors and under important programmes, government of India in this year's budget has underlined the need to focus on efficiency of programmes to achieve the desired development objectives.

II. Government action- past experiences

Policies are the means through which government influences the decision-making behaviour of stakeholders including the community so as to achieve the desired objectives. Policies could either create a favourable environment thereby encouraging action of stakeholders, or an unfavourable environment constraining action of groups, organizations or public at large. In the era of centralized planning, in order to ensure its control, government had to resort to policies, institutions and regulatory frameworks that discouraged public and private action in growth and development of the economy.

Governance and policies together determine the development effectiveness. Even a well meaning and sound polices could prove ineffective if not supported by good governance. As a matter of fact, distortions within the institutions come in the way of implementation of policies and make it difficult for intended actors to take full advantage of good policies. On the other hand, policies are often announced in a hurry or on piecemeal basis without properly and adequately reflecting upon its relevance to people's need and feasibilities of smooth implementation. As a result, outcome of such policies does not turn out to be encouraging and government keep on reviewing and modifying the policies leading to some kind of uncertainty.

Normally in India, we have sector specific policies aimed at growth and development of sectors. As sectors are interdependent or interlinked, it may happen that a policy in one sector may create adverse consequences in other sectors. For example, the recent government initiatives to create income, employment and protect poor against food insecurity have created tension in agriculture sector by squeezing labour supply to the sector, as some studies indicate. Thus there are policy conflicts that may affect growth and development of the economy in medium and long run. In past too we have seen over exploitation of resources, particularly natural and environmental resources due to lopsided government policies and lack of appropriate institutional structures to implement and monitor the policies. In many cases, the private entities have exploited the situation and made huge gains at the cost of environment and welfare of the poor as incase of marine fisheries and forest resources. In other words, private action has led to marginalization of community who heavily depended on common property resources and environmental goods for their livelihood.

III. Need for community empowerment !

In our society which is beset with problems of inequality and poverty, many people have a very poor asset base and lack capability to make use of available assets. While unequal power relations and limited ability constrain them to act on their own interest, market failures causes erosion of their asset base. As a result the asset accumulation process which is crucial for sustained livelihood gets disrupted, thereby pushing the poor into vicious circle of poverty. Thus individual level, poor people, given their resource endowment and capabilities, neither have the ability to overcome the constraints posed by existing institutions nor have the capacity to protect themselves against external shocks. Therefore emphasis is given on community empowerment aimed at developing collective capability of the people to solve the problems.

Lessons from various studies and past experiences have underlined the role of people's participation in development. To enable people participate in development and share the benefits, empowerment of people, particularly the poor and weaker sections, has been an important approach of the governments. Organizing women self-help groups (SHGs) and linking them to development and financial institutions have been the cornerstone of the government's efforts towards community empowerment over successive plan periods

An empowered community can influence the decision making process of the government and contribute to good governance by;

- Systematically articulating community's concern in different fora and drawing support for necessary institutional reforms
- Engaging in continuous dialogue with the government
- Putting forth the genuine demands of the community based on evidence

IV. Process of community empowerment

Community empowerment is a long drawn process. Organization of women self-help-groups and their participation in multifarious activities including management of common property resources taking advantages of government policies presents a brighter picture of community empowerment process in India. However, there are instances of community movements in past that have forced changes in government policies. These movements were organized at the behest of some individuals or groups some of whom could exploit the emotions and weaknesses of people to draw some kind of political and economic mileage out of it. But can we call this as community empowerment? On many occasions such community actions have caused destruction of private and public properties. Community, like government or any other civil society, should respect the rule of law, should be responsible and accountable. Such values need to be spread and imbibed within the community for real empowerment. Credible NGOs, associations and other members of civil society too can play important role in community empowerment process.

Community empowerment pre-supposes the following conditions

- Social cohesion
- Consensus
- Common interest
- Mutual trust

Community empowerment can be a means to certain ends or an end in itself. It is the process of involving, educating, motivating, capacity building of community to take collective action as and when necessary. There is no standard where to begin and where to end. The type of empowerment could vary according to the context depending on nature of public action envisaged, extent of decentralization, pattern of social exclusion and social conflict, and strength and ability of local organization.

The process can start with government initiatives or by any other actors in governance. But given the state of the society characterized by social fragility, breakdown of informal institutions, government initiatives would surely go down with the public at large to create much needed social trust and confidence in government. However government intention must be demonstrated by the transparency in the functioning of the government institutions, pro poor policies and creating people friendly redressal systems. In a particular context what role government has to play and what should be the role of community need to be clearly defined. Importantly, the extent of community empowerment depend on the extent to which government can give up its power of decision making in favour of the community and create conducive environment for community action appropriate policy and institutional framework.

Developing credible community leadership is essential for community empowerment. It can start with identifying individuals from different social and economic background, having some education, good attitude and a sense of concern for community. These individuals can be groomed as community leaders through regular orientation and capacity building programmes. These leaders can understand community's concerns and support the community for collective action. This is critical to avoid confrontation between the community and the government and bridge the gap between government and the community.

V. Areas of community empowerment

Basically there are three broad areas of collective action that warrant community empowerment. These are:

- Increasing efficiency and effectiveness of programmes
- Protecting against adverse impact of social, economic and natural shocks
- Preserving the environment, conservation and management of the resources

Even though the governments of the day are quite aware of these daunting tasks and their role and responsibilities thereof, past experiences and achievements in the said areas clearly suggest that without people's involvement nothing much could be achieved in a stipulated time frame.

Increasing efficiency and effectiveness of programmes

With increasing outlay, improving the outcomes of the programmes has become an important concern of the government and the civil society. Some key issues are better targeting of the programmes in terms of area wise and community wise coverage, ensuring people's participation in such programmes, creating programme monitoring and social auditing mechanisms.

As discussed above, a number of programmes have been launched for agricultural and rural development. Under NREGA, 100 days of employment is provided to BPL people with a wage rate of 100 per day. For the year 2009-10, Rs39,100 crore has been allocated for the. Under RKVY nearly Rs 25,000 crore would be spent in this plan period. Similarly, National Food security mission has been launched to boost production of important crops like rice, wheat and pulses in the marginal areas by adoption of improved technology.

Various studies and audit report on NREGA suggest that there is wide spread leakage and mismanagement in respect of quality of work done, choice of beneficiary, irregularity in disbursement of wages to the labourers etc. All these indicate, unless remedial measures are taken such programmes may fizzle out as far as their impact is concerned. Community empowerment in this context means involvement of people in deciding the type and place and timing of work, keeping the community and individuals updated about information such as funds available, monitoring of disbursement of wages, beneficiaries etc. at panchayat level.

In agriculture and allied sectors too, a number of schemes and policies are being implemented. In order to improve their development effectiveness, community participation in their implementation is essential. Take the case of programmes; RKVY or NFSM. To create wider impact the benefits should percolate down to all sections of farmers. This necessitates that the technologies should be adopted by large number of farmers in an area. Some of steps are:

- Informing the community and creating public awareness about the programme
- Mobilizing the community for participation
- Capacity building of the community to get access to relevant information, knowledge and technology.
- Encouraging community to take collective action by providing incentives
- Engaging the community in deciding the area in which public investment should be made
- Developing the plan of action based on community's need and preferences

While there is adequate provisions under the project to address issues like informing the community and meeting the information needs of the community, other points need good deal of attention. Considering that access to quality seed by farmers is an important issue in achieving the target, government can take a decision whether to take major responsibility in supplying seeds or empower community in quality seed production. For community empowerment, a detail programme has to be chalked out indicating the role of government and that of the community. Some issues are appropriate seed policy that encourages farmer and women groups in seed production w.r.t. certain identified crops and institutional mechanism for providing extension and training, services for seed certification on demand and at reasonable cost.

Government programmes generally contain two aspects; short term measures for supporting production and income generation, and creation of individual and community assets and institutions. While the first one is important to protect the poor against poverty and hunger, the second aspect aims at strengthening people's capacity to generate income. Therefore need based programmes and policies should be worked out to strengthen the asset base of poor and expand opportunity of asset accumulation and consolidation with people's participation. Obviously one important aspect in this context is maintenance of the assets and their use which depends on community's capability.

Reducing the adverse impact of shocks

If past experiences are any indication, governments, in many occasions, have failed to contemplate and implement quick action o reduce the impact of shocks- social, economic and natural. Such shocks result in loss of human lives, crop and livestock, public and private properties, and after all, loss of income of the people. In these situations, not only individuals, but also community as a whole may suffer from livelihood insecurity because of loss of asset base aggravated by lack of capabilities to take timely collective action. In the event of shocks the governments generally have some or all of the following responsibilities depending on situation.

1. Reaching the affected area
2. Identifying the problem and its cause
3. Identifying the intervention
4. Planning the intervention
5. Mobilizing resources for intervention in the area
6. Organizing/launching the intervention in the area
7. Involving people in implementation of the intervention

For various reasons, government of the day may not be able to discharge all these responsibilities effectively. Starting from the first responsibility, immediately reaching the area is a precondition to undertake other activities for tackling the situation. But if due to reasons such as remoteness of the area, lack of communication facilities, government apparatus fails to reach the affected area in time, then considerable time elapse before actions are really taken in the field. To overcome such problems, community empowerment offers the mechanism to reduce the response time in planning and implementing collective action.

But the question is how to go for community empowerment? Since different situations warrant different kinds of action and need different kinds of preparedness, strategy for community empowerment could vary according to the situation. Some recurring problems such as incidence of pests and diseases in crop and livestock, outbreak of human diseases, flash floods may be taken for developing community empowerment models as given in table.

Shock	Possible impact	Solution	Government role	Areas of Community empowerment	Policy issues
Incidence of pests and diseases in rice	Destruction of crops, reduction in good production, food insecurity	To tackle pest or disease menace as quick as possible and prevent its spread to other areas	-Mechanism to collect feed back immediately as pest incidence occurs.	Educating and empowering farmers to identify the pest or disease.	Policy regarding pricing and distribution of pesticides.
				-How to share the information with community	Policy framework to provide incentives for collective action by mobilizing resources from within the community, to include poor farmers who can not afford such measures, protecting the community or the group in case there is loss to individuals during community action, provision of compensation, group insurance
			-Developing a plan of action (may be IPM) to control the menace in the area and to prevent its spread to other areas.	-Access to such information about the plan of action by the community	
			-Making available the critical inputs such as pesticides, bio-agents and other equipments in the area	-capacity to purchase or access such inputs and knowledge and skill in using the inputs following appropriate procedures	
				-All farmers should come under the intervention for better outcome	

Thus strategies can be developed to involve community in different situations to create synergy between government action and community action. Appropriate policy framework clearly spelling out what is to be done and not to be done, and creating a reward system that make it win-win situation for both government and community would certainly empower the community for collective action.

Preserving the environment, conservation and management of bio-resources

This is another important area that demands community empowerment and participation. Over years, while attempting to achieve high economic growth and development, we have done a great damage to the environment, Degradation of natural resources such as forests, water and soil have reduced our capacity to produce ore food. Moreover, mindless activities have created an unhealthy environment for people to live in. In this situation there are two pronged action that are being worked out. First, restoring the environment to the extent possible and second, adopting resource conservation measures to stop further degradation of environment. In both cases, community has a larger stake.

During past years some of good cases of community empowerment in different areas such as joint forest management, watershed development have been tried. Recently, participatory irrigation management is being tried.

Joint forest management with community was successful to the extent that it not only reduced forest related offences and conflict between official and forest communities but also could protect the forest and enhance the productivity of lands. But much could have been achieved through need based policies for forest management. There is criticism that many of the forest communities have become marginalized in absence of policies related to property rights and sharing of benefits. Even though there is good potential of market of NTFPs, non-existent of sound marketing and pricing policies in overall framework forest conservation and management failed to bring desired benefits to tribal people. As a result dependence on forest economy has considerably reduced and people from periphery of forests are migrating out to other areas.

Similarly participatory watershed development with involvement of women have yielded significant results in terms of productivity of land and enhancing livelihood base of households. But sustainability aspects have not been adequately addressed. Hence appropriate policy framework is essential to harness optimum benefits and create institutional mechanism for holistic management of watersheds with participation of people.

Recently, participatory irrigation management is seen as an innovation for judiciously using available water resources. Water users associations are being formed and given necessary orientation and training to increase productivity of water use through scientific water management. Even though reports of group conflict and break down of mechanism of functioning of WUAs are coming in, strengthening efforts of community empowerment in water management would go a long way in achieving the objectives. In this context, appropriate policy framework for encouraging judicious use of water through water budgeting and pricing and appropriate selection crop combinations should be given attention.

Few days back government of India has taken a decision to transfer green bonus to the state for their efforts to increase forest cover. This decision should encourage states to come up with meaningful policies to promote aforestation measures by communities. At the same time issue of transferring property rights to people, and conflict between people and environmental requirement should also be addressed.

In agriculture, conservation and scientific management of resources have become crucial to maintain productivity of lands. As large areas under different agro-ecosystems are affected by water scarcity, deteriorating soil and water quality, it has become vital to undertake large scale conservation measures with people's participation. While government and its research institutions have the role of developing and disseminating appropriate technologies in these areas, farming community should have capacity to access and adopt such technologies in field. This requires long term measures such as soil and water reclamation, drought proofing measures, changing cropping pattern which could only come through sustained government campaign involving people and rational policy approaches.

Conclusion

No doubt, the past years have witnessed significant changes in government's approach to address issues of public importance. New institutions are being created to bridge the gap between government and the people. Civil society is being strengthened to work for people. Policies are being recast with wider public consultation and participation to make them more effective for creating a favourable environment for public and private action. All these are intended to empower people, particularly the poor and weaker sections of the society. However, a blue print of action for empowering the community to take collective action in specific areas has become crucial today to complement government action in big way.

Further Readings

Assets, Livelihood and social Policy Editors -Caroline Moser and Anis A. Dani

Empowerment and Poverty Reduction -A Source book: Edited by Deepa Narayan

Rural Livelihoods and Poverty Reduction Policies :Edited by Frank Ellis and H. Ade Freeman

□□□

Chapter 35

Capacity Building of Farmwomen for Management of Natural Resources for Sustainable Livelihood

SUMAN AGARWAL[1] AND HARAPRIYA NAYAK[2]
Directorate of Research on women in Agriculture, Bhubaneswar
[1] Principal Scientist, [2] Research Associate

Over the years, there is a gradual realization of the key role of women in agricultural development and their vital contribution in the field of agriculture, food security, horticulture, processing, nutrition, sericulture, fisheries, management of natural resources and other allied sectors. Evidence is growing that gender-sensitive development strategies contribute significantly to income creation as well as to equity objectives. It has been increasingly realized that women have vast entrepreneurial talents, which could be harnessed to create employment opportunities. However, rural women in India face several constrained that come in the way of their economic development. In India, over 90% of the rural women workers are unskilled, there are certain qualities required for making of an entrepreneur. These qualities may be induced through education and training. Sustainable livelihood approach integrates environmental, social and economic issues into a holistic framework and begins not only with community needs assessment but with community strengths and assets assessment. This is especially true in identifying not only the types of assets which people use, but also how existing livelihoods can be strengthened with new and appropriate technologies. Keeping this in view, under the NATP (National Agriculture Technology Project) project on empowerment of women in Agriculture, a study was undertaken in Orissa for empowering rural women through natural resource management by providing skill development trainings for sustainable livelihood.

Methodology

The study has been designed to improve livelihood options for women through their capacity development in use of natural resources. The focuses of the study was toward poor rural women who have not benefited from any other programme and need support for their livelihood. Forty Women Self Help Groups (SHGs) were formulated consisting of six hundred farm women belonging to small and marginal farm families for economic empowerment. A Participatory Rural Appraisal (PRA) was conducted to assess the availability of natural resources in their native area/village for sustainable livelihood options. Based on the available resources, Skill development trainings in different areas were organized to develop the capacities of farmwomen.

Water is life and has a profound influence on human health and wellbeing, so management of this resource is important. Hence, trainings were imparted to the women SHGs for fresh water pisciculture. So that they can use family owned or community ponds for income generation. Besides, Skill development trainings were organized for floriculture, vegetable growing, and nursery raising for appropriate use of homestead land and water especially using the waste water of the household. Women were also trained for bee-keeping, vermi- composting, agro- processing, value addition in fruits and vegetables, poultry bird rearing, and product making from coir-yarn. They have utilized the agro and animal based waste for livelihood options in different ways as detailed below in Table 1.

After building the capacities of farmwomen, SHGs were facilitated to set-up the enterprises depending upon the facilities and resources they had. The enterprises were given a strong footing as emphasis was placed on technical and technological feasibilities, cultural acceptability and economic viability.

Table 1 : Livelihood options from Agro and animal based waste.

S. No	Agro and animal based waste	Livelihood options
1.	Paddy Straw	Mushroom cultivation and vermi-compost using used paddy straw.
2.	Coconut waste	Coir, rope, doormats, broom, toys and decorative household articles
3.	Farm and animal waste	Vermi compost
4.	Vermi compost	Kitchen gardening, nursery raising, floriculture etc.
5.	Used/ waste water from the household	Kitchen gardening, nursery raising and vegetable growing.

Results and Discussion

Each SHG was encouraged to set-up 2 to 3 enterprises so that they can have round the year income. Details regarding the number of groups involved in different enterprises have been given in Table 2.

Table 2 : Involvement of Self Help Groups (SHGs) in different enterprises.

S. No	Name of the enterprise	Skill development trainings	Number of SHGs involved in micro enterprises	
			No	%
1.	Floriculture	Marigold cultivation	10	25
2.	Vegetable Growing	Cauliflower, brinjal, tomatoes, green leafy vegetables	16	40
3.	Nursery raising	Vegetables and flowers	1	2.5
4.	Mushroom cultivation	Paddy straw and Oyster mushroom	10	25
5.	Pisciculture	Fresh water pisciculture	17	42.5
6.	Poultry rearing	Backyard poultry	4	10
7.	Agro-Processing	Paddy parboiling, processing of cereals , pulses and nuts (Ground nut), badi and papad making	11	27.5

8.	Value addition in Fruits & vegetable	Preparation of jam, jelly, squash, sauces, ketchup, pickles	1	2.5
9.	Bee-keeping	Honey	4	10
10.	Vermi-composting	Vermi- compost	3	7.5
11.	Coir work	Rope, doormats, broom, toys and decorative household articles	2	5

An analysis of the involvement of these SHGs in different enterprises during September, 2002 to October, 2004 indicated that 27.5% SHGs are involved in Agro- processing, 25% in floriculture, 25% in mushroom cultivation, 10% in each Bee-keeping and poultry bird rearing, 7.5% SHGs are involved in vermin-composting, 5% are involved in coir work and 2.5% are involved in nursery raising and value addition in fruits & vegetables (in each case).

From the above enterprises the self help groups earned money and also got some profit. The average net profit earned per month per group ranged from Rs. 161-796. The highest profit per month was obtained from bee-keeping enterprise. The total profit earned by each group (in six months to one and a half years duration depending on the time of setting-up of enterprises) ranges in between Rs. 2559 to Rs. 9562 (Table 3).

Table 3 : Detailed profit earned by the SHGs from the above enterprises.

S. No.	Name of enterprise	Average net profit earned per month per group (Rs.)	Total duration of enterprise as on October 04 end	Average total profit earned so far by per group (Rs.)
1.	Floriculture	600	6 months	3578
2.	Vegetable Growing	213	1 year	2559
3.	Nursery raising	459	6 months	2755
4.	Mushroom cultivation	639	6 months	3838
5.	Pisciculture	463	1 year	5559
6.	Poultry rearing	348	1 year	4183
7.	Agro-Processing	457	8 months	3660
8.	Value addition in Fruits & vegetable	161	1.5 year	2900
9.	Bee-keeping	796	1 year	9562
10.	Vermi-composting	600	6 months	3600
11	Coir work	500	1 year	6000

Conclusion

Women constitute almost half of the population of the country and are an important natural resource. If farmwomen's skills are developed appropriately they are able to express their potential to the benefit of the entire household and community. They contribute significantly in the management of natural resources like, land, water, soil, plants and animals which, provide crucial and interrelated livelihood options for their economic upliftment.

□□□

Chapter 36

Spatial Data Base for NREGS and Conservation of Natural Resources

S.K. SAHA, P. LAGHATE, R.K. SHARMA AND S.N. DAS
Soil and Land Use survey of India, IARI Buildings, New Delhi-110012

Judicious use of natural resources plays an important role not only helps to sustain food security of a country but ensures conservation of vital natural resources. Our population is projected to be about 1.4 billion by 2020. For adequate and sustained food production vis-a-vis food and environmental security, efficient use of land resources is of paramount importance.

Watershed development planning with scientific data base has been found to a proven tool for efficient utilization of natural resources of the country in the context of ever burgeoning population. Database, especially the spatial database is pre-requisite for planning, implementation, monitoring, evaluation and accountability of any developmental programme confining to land resources. Many schemes for development of agriculture, horticulture, water resources, forest and watershed are in operation in the country.

Huge investment of government funds are made annually through various schemes by different departments and ministries towards conservation of natural resources that are essential for sustaining food production, ensuring livelihood security and employment to the rural people of the country. It is perhaps the best way to utilize the funds optimally considering scientific data as base.

Modern tool in the form of Remote Sensing and GIS pave the way to acquire the real time data of natural resources and managing the same in more scientific manner for manipulation an analysis for customized deliberations. Geo-spatial techniques have been applied to general digital spatial data base on microwatersheds, soil and land use for Dawas District, Madhya Pradesh for planning of conservation of natural resources that may serve as a tool for the development of the district.

The paper deals with a case study of generation of data base for watershed development planning for Dewas district of Madhya Pradesh using LISS III IRS data and SOI topographical sheet on 1:50 K. ARC-INFO software has been utilized to convert the conventional data into digital format. Various important themes on soil and land resources have also been developed out of digital spatial data base. Thus, data sets could be effectively utilized by the district planners to enhance livelihood security by providing strong safety net for the vulnerable groups through National rojgan employment gara(NREGS) for sustainable agricultural development in the district.

Geographical Perspective of the Study Area

Dewas District in M.P. is situated on the Malwa plateau in the west-central part of Madhya Pradesh and lies between 22° 17" 37' to 23° 19" 15' North Latitude and 75° 53" 56' to 77° 07" 32' East Longitude. It falls in Ujjain Revenue Division. The district is bounded by Ujjain in the north, Indore in the west, West-Nimar in the south-west, East Nimar in the south, Hoshangabad in the South East, Sehore in the east and Shajapur district in the North-East. The tropic of cancer passes through the district near Nemawar village, south of Khategaon town. The location map of the district is given in Fig. 1. The district has been subdivided into six blocks and comes under NREGS in second phase. The demographic information and the NREGA statistics of the district is presented in Table 1.

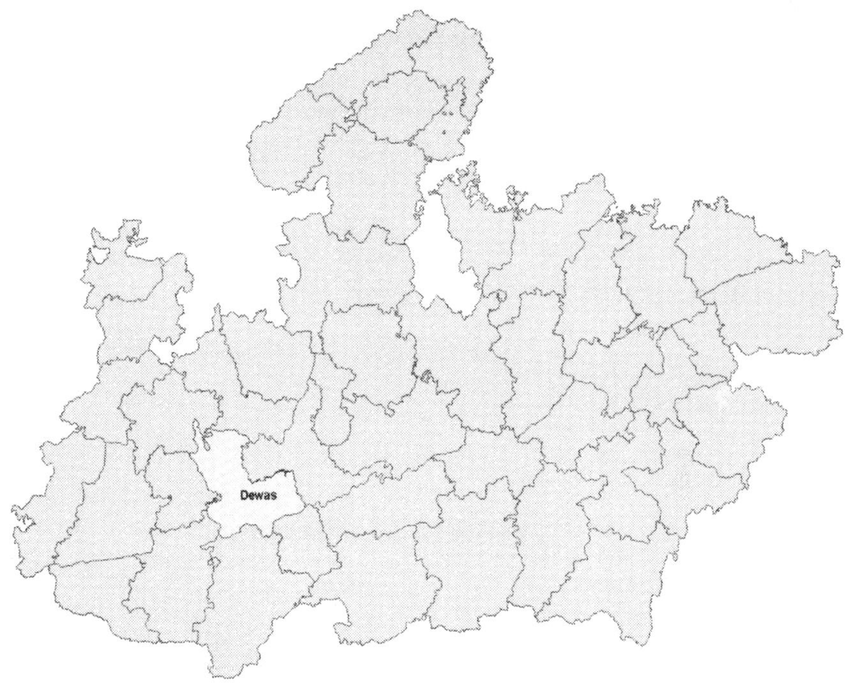

Fig. 1: Location map of Dewas district of Madhya Pradesh.

Materials

The digital database of Deawas district of Madhya Pradesh has been generated using the following base materials.

Data Sets Used

- IRS- 1C/1D LISS III geocoded FCC
- SOI Topographical sheet on 1:50 K
- Watershed Atlas of India

Hardware

- CONVEX Wide Image Scanner
- HPxw4300 workstation
- HP Designjet Plotter

Software

- ArcGIS 9.2- GIS Software
- Microsoft Access 2003- RDBMS software

Table 1 : Census Data of Dewas District of Madhya Pradesh.

Population	1981	1991	2001
Population	795,309	1,033,807	1,308,223
Sex ratio (females per 1000 males)	929	924	930
Scheduled Caste (SC) %	18.2	18.2	18.26
Scheduled Tribe (ST) %	13.8	15.0	16.45
Males Literacy Rate %	49.4	61.1	75.74
Females Literacy Rate %	15.7	25.6	45.03
Human Development Index (HDI)	-	-	0.569 (1998)
Rank in Madhya Pradesh : HDI	-	-	7 (1998)
Gender Related Development Index (GDI)	-	-	0.600 (1998)
Rank in Madhya Pradesh GDI	-	-	18 (1998)
NREGA Staistics as on July 2009			
Employment provided to the household households	1.3465 lakh		
Persondays (in Lakh)			
Total	46.06		
SCs	16.79 (36.45%)		
STs	13.92 (30.22%)		
Women	17.74 (38.51%)		
Others	15.35 (33.33%)		
Total fund Rs.	118.68 Crore		
Expenditure	79.54 Crore		
Total works	18905		
Works completed	1169		
Works in progress	17736		

Methodology

The digital data base on important soil and land characteristics for watershed development planning of the district has been generated through Rapid Reconnaissance Survey on 1:50 K scale. The steps involved in preparation of district microwatershed map for watershed development planning are outlined below.

- Preparation of drainage map, delineation and codification of micro-watersheds.

- Development of Erosion Intensity Mapping Units (EIMU) from visual interpretation of satellite data followed by ground survey and preparation of EIMU map.
- Development of digital layers of micro watershed, EIMU and Administrative boundary and their integration using GIS platform.
- Computation of Sediment Yield Index (SYI) using RDBMS, priority categorization and gradation of microwatersheds.

Development of Digital Layers

All the drainage maps, codified hydrologic unit maps up to microwatershed level and EIMU maps (Toposheet wise) are traced on 0.75 micron tracing film and scanned using CONTEX Wide Image Scanner. The digital database for all these layers is generated using ArcGIS 9.2 GIS software using the following processes.

- Geo-referencing
- Digitization of Arcs
- Editing
- Labeling
- Edge Matching
- Appending
- Topology Building
- Rectification of Polygon Errors
- Attachment of Attribute Data

The area of each microwatersheds and EIMU has been deduced out of digital database. All these digital layers of the district have been integrated on GIS platform with reference to the district administrative layer.

Computation of Sediment Yield Index (SYI)

Computation of SYI has been achieved by running a Microsoft Office Access based RDBMS program in the entire catchment in order to generate subwatershed and microwatersheds wise priority on 1:50K scale. The database is then customized on district basis for the ease of the policy makers.

Sediment Yield Index (SYI) Model : The prioritization leads to adopt a selective approach that helps to phase out the planning in a systematic manner. The identification and demarcation of watersheds in the catchment that are prone to yielding higher sediment yield is the primary task envisaged by the Natural Resource Management Division, Department of Agriculture and Cooperation, Ministry of Agriculture, Govt. of India towards catchment area treatment.

Commensurate with the needs of database, Soil and Land Use Survey of India (SLUSI), Department of Agriculture and Cooperation, Ministry of Agriculture developed Sediment Yield Index (SYI) model for prioritization of watersheds in the catchments of River Valley Project (Anon, 1990). It's a predictive model based on the soil and land characteristics of reconnaissance level that are acquired through Rapid Reconnaissance Survey.

Grading of Microwatershed for Priority Fixation : Identification and demarcation of various priority categories of microwatersheds have been achieved using SYI values. Higher value of SYI suggests higher priority and vice-versa. Priority microwatersheds are thus selected from very high and high priority categories of microwatersheds which need immediate attention for suitable soil and water conservation measures. Microwatersheds of the district have been grouped in five categories namely, Very High, High, Medium, Low and Very Low on the basis of SYI.

Results and Discussions

Watershed Development Programme : Watershed Development provides the way for regeneration, conservation and the judicious use of all the natural resources along with the human resources within a watershed. It involves enhancement and stabilization of production, maintaining ecological balance, creating employment opportunities for rural poor in the rainfed areas. Thus, watershed development has increasingly chosen as the viable mechanism for rural development to provide sustainable means of livelihoods to the rural poor in these areas.

The data base generated on the important catchment characteristics of the district provides strong foundation for the sustainable agricultural production through watershed development programmes. The data base on priority (very high and high) microwatersheds may be utilized by the policy maker to design various work scheduled based on personsdays under NREGS.

The database on priority microwatersheds is a vital base for strategic watershed development planning. Besides, it provides spatial distribution of various soil and land attributes that will help to know terrain condition in terms of slope, soil depth, soil erosion, land use and land cover condition. These attributes would serve as a special aid for planners dealing with soil and water conservation, aforestation, agriculture, horticulture and eco-development.

All the 881 microwatersheds of the district have been grouped in five categories namely, very high, high, medium, low and very low on the basis of SYI values which have been summarized in Table 2.

Table 2 : Priority Categorization of Microwatersheds.

S. No.	Priority Categories	No. of Microwatersheds	Area (ha)
1.	Very High	141	72,835
2.	High	103	73,315
3.	Medium	150	2,27,368
4.	Low	241	1,33,112
5.	Very Low	246	1,89,892
	Total	**881**	**6,96,522**

The data reveals that about 20.98per cent of the total area of the district is falling under very high and high priority categories (Priority Area) and needs immediate attention for soil and water conservation measures. The spatial distribution of different priority categories of microwatersheds is shown in Fig. 2. Priority categorization using microwatershed as base allows better abstraction of vulnerable area.

Considering Rs. 5000.00 per ha as cost of treatment under watershed development programme of which Rs. 3000.00 is assumed for wages requirement and rate of unskilled wages being Rs. 100/- per day, 30 man days could be generated for each ha of vulnerable treatment area. Thus, the present study reveals that 34,41,870 persons days could be generated under NREGS from 1,14,729 ha of treatment area out of 1,46,150 ha of very high and high priority areas in the district.

Soil Erosion Control through NREGS: Soil and water conservation practices should receive top priority in watershed development planning as they form the foundation of the sustainable agriculture production. Attempts need to be made for storage of rain and run-off

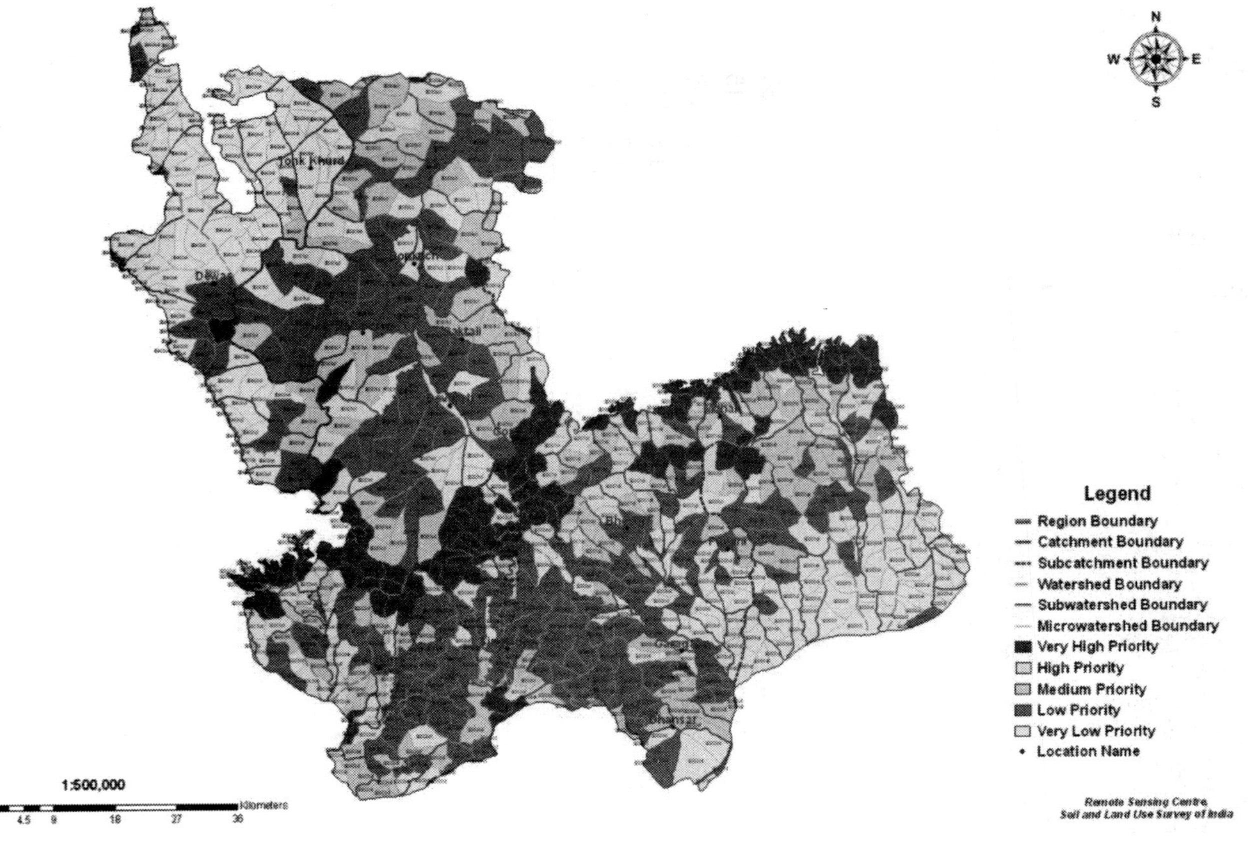

Fig. 2 : Spatial Distribution of Priority Categories of Microwatersheds in Dewas, Madhya Pradesh.

water by raising structures such as bunding, check dams, water storage ponds / tanks for effective uses. The first step to check soil erosion is the in-situ method, especially contour farming. It is estimated that 1,07,80,440 mandays may be generated for treatment of 3,59,348 ha of moderate to severe and severely eroded areas of the district through various engineering and vegetative measures assuming Rs. 3000.00/ ha as cost of labour charge and Rs.100.00 as daily wage per labour. The spatial data on soil erosion will be helpful for strategic planning of conservation of soil erosion in the district which are given in Table 3 and shown in Fig. 3.

Table 3 : Spatial Extent of Soil Erosion Classes.

S. No.	Erosion Classes	Area (ha)
1.	None to Slight	46,880
2.	Slight to Moderate	1,62,491
3.	Moderate	1,10,619
4.	Moderate to Severe	3,37,834
5.	Severe	21,514
6.	Miscellaneous	17,184
	Total	**6,96,522**

Employment generation through Eco Development: The eco-hazard zonation map of a district may attract the environmentalist (Fig. 4). The spatial distribution of eco-hazard zonation class has been derived by appreciating important soil land attributes that has been summarized in Table 4. The objective behind such thematic map is to identify the hazardous zone in relative term so that upper reaches could be rehabilitated with forest vegetation and other vegetative measures. It may serve as an additional input for planners.

Table 4 : Spatial Extent Eco-hazard Zonation Classes.

S.No.	Eco-hazard Classes	Area (ha)
1.	Very Slight	1,58,707
2.	Slight	2,29,466
3.	Moderate	1,82,186
4.	Severe	1,08,979
5.	Miscellaneous	17,184
	Total	**6,96,522**

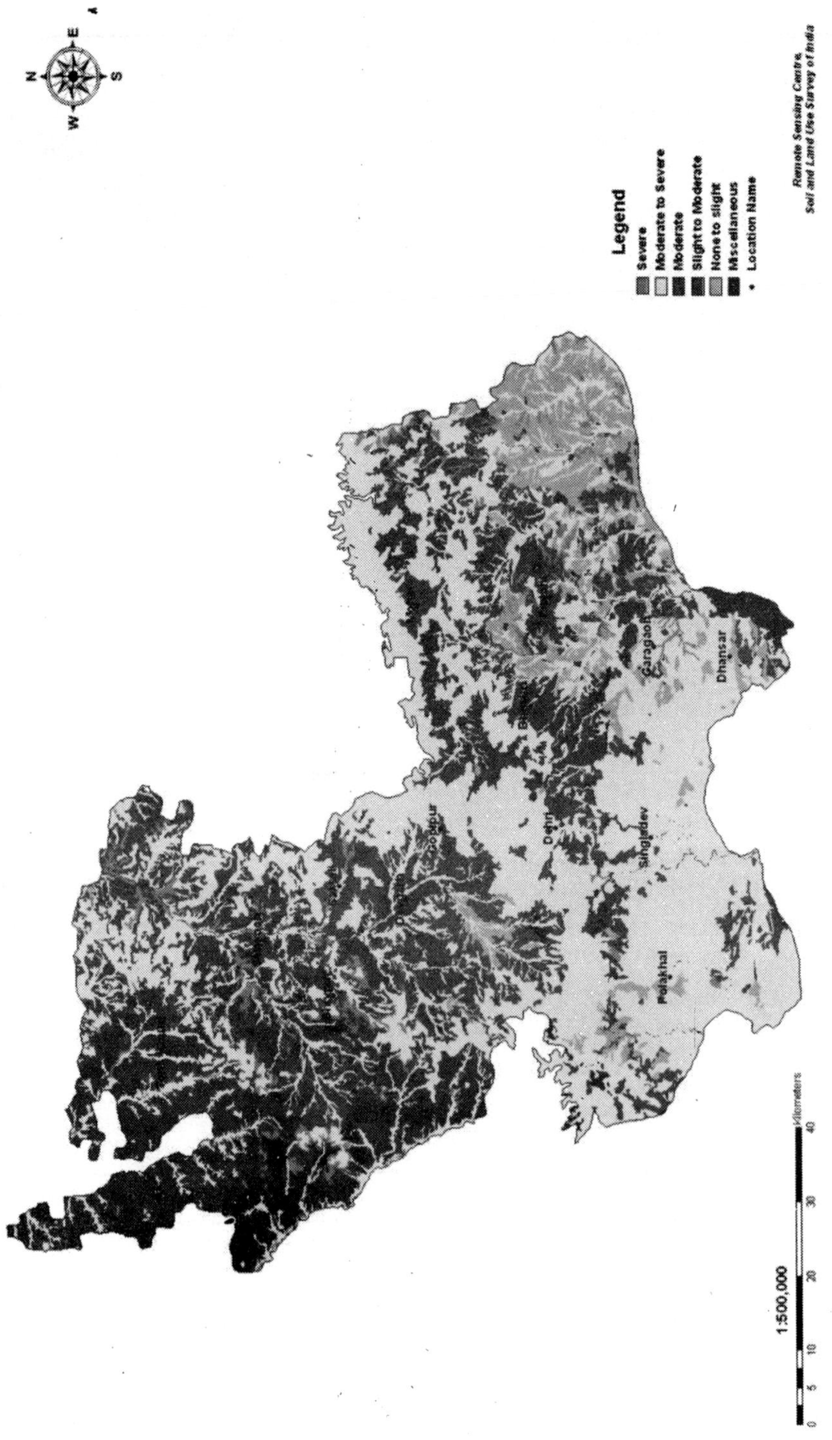

Fig. 3 : Spatial Distribution of Erosion Classes of Microwatersheds in Dewas, Madhya Pradesh.

Fig. 4 : Spatial Extent Eco-hazard Zonation Classes in Dewas District, Madhya Pradesh.

It is worth mentioning that there is an increasing awareness about the interaction between rural, urban and industrial development and ecological balance of the area. Eco development refers to the process of regeneration, protection and conservation of natural resources through active participation of local people to maintain and restore ecosystem. It involves checking of land degradation, deforestation and loss of biodiversity, augmentation of the availability of fuelwood, fodder and grasses from the regenerated areas, improvement of quality of life and self-sustenance aspect of people living in and around forest areas. The present study reflects that about 3269370 mandays may be generated for treatment of 1,08,979 ha of severe eco-hazard zonation class of the district assuming the aforesaid criteria.

Suggestive Action Plan for Soil and Water Conservation Measures: The suggestive action plan for soil and water conservation measures has been formulated primarily based on LCC and presented in Table 5. The spatial distribution of various kinds of measures has been shown in Fig. 5. The measures suggested are tentative as it is based on reconnaissance level of information. However, it may be a very useful for the planner while preparing the working plan.

Table 5 : Suggestive Action Plan for Soil and Water Conservation Measures.

S. No.	Suggested Conservation Measures	Area (ha)
1.	**SCM-01:** Strip cropping, Crop rotation, Stubble mulching	1,58,221
2.	**SCM-02:** Land leveling and field bunding, Contour bunding, Stubble mulching	1,11,821
3.	**SCM-03:** Land leveling, Stone line bunding / Contour farming, Gully Plugging Stubble mulching, Agro- forestry	59,662
4.	**SCM-04:** Bench terracing, Contour farming, Gully Plugging, Agro-forestry	5,738
5.	**SCM-06:** Contour trenching, Contour vegetative hedges, Reforestation	18,456
6.	**SCM-07:** Contour trenching, Contour vegetative hedges, Intermittent orchard terracing, Reforestation	1,05,120
7.	**SCM-08:** Contour trenching, Contour vegetative hedges, Intermittent terracing, Hillside ditches, Lock and spill drain, Reforestation	61,973

8.	**SCM-09:** Gully plugging with grassed waterways, Contour farming, Check dam/storage pond	12,289
9.	**SCM-10:** Gully plugging with grassed waterways, Check dam, Controlled/ rotational grazing	211
10.	**SCM-12:** Land leveling and contour bunding, Strip cropping, Gully plugging along with Check dam / Storage pond	88,334
11.	**SCM-14:** Gully plugging with grassed waterways, Contour /Stone line bunding, Contour farming, Agro forestry	13,934
12.	**SCM-15:** Gully plugging, Pasture development	24,749
13.	**SCM-16:** Gully plugging, Contour trenching, aforestation	18,830
14.	Miscellaneous	17,184
	Total	**6,96,522**

Microwatershed Atlas for Planning, Implementation, Monitoring, Evaluation and Accountability of Various Development Programmes: Monitoring and evaluation are vital components of watershed development programme. Advanced technology of Remote Sensing and Geographic Information System offer a sound scientific platform in this endeavor. Microwatershed atlas generated for the entire district (Fig. 6) provide valuable input in identification, evaluation and monitoring of the soil and water conservation measures taken up for watershed development. It plays an important role in reducing duplication of work operated through various schemes with similar objectives under different ministries. Thus, microwatershed atlas of the district may serve as Watershed Development Information System. It is worth noting that only 20,089 ha of area have been found as overlapping area among the three various treatments suggested under watershed development, erosion control of area suffering from moderate to severe and severe water erosion area and eco-development of the district (Fig. 7).

Conclusion

The district-wise digital data base on microwatershed, soil and land characteristics provides a sound base for various aspects of watershed development planning. The aim of microwatershed atlas is

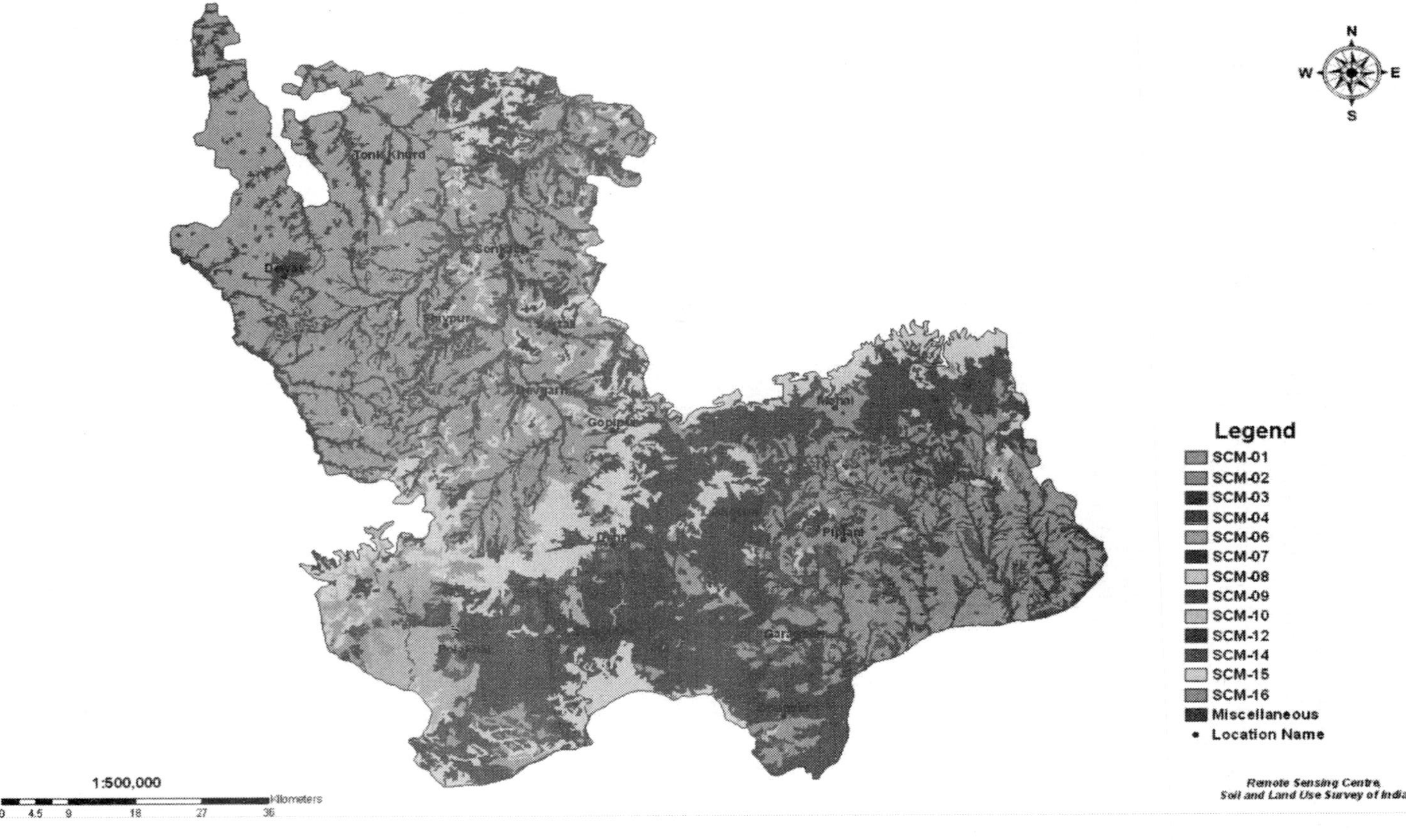

Fig. 5 : Spatial Distribution of different SCM Classes in Dewas District, Madhya Pradesh.

Fig. 6 : Digital Microwatershed Atlas of Dewas District of Madhya Pradesh.

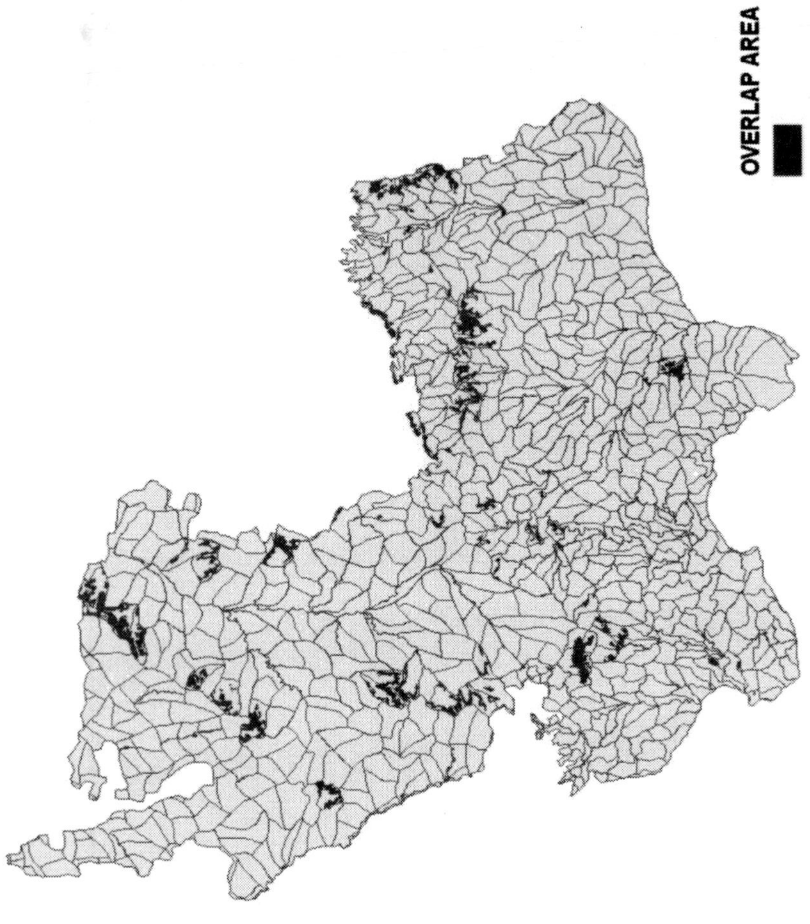

Fig. 7 : Overlap Area of Different Types of Treatment of Dewas District of Madhya Pradesh.

to provide a scientific window to the policy and decision maker to view issues related to the natural resources with socio-economic fabric to sustain agriculture production, ensuring livelihood security through employment under NREGS, eco-development and accountability of funds being used for various programmes in the district through many government schemes.

❑❑❑

Chapter 37

Lac Integrated Agriculture : Means for Efficient Resource Utilization

R.K. SINGH, R. RAMANI AND BANGALI BABOO
Indian Institute of Natural Resins and Gums, Ranchi - 834 010, India

The Indian lac insect *Kerria lacca (Kerr.)* (*Homoptera :Tachardiidae*) is cultured for commercial production of lac mainly on host plants *viz., palas* (*Butea monosperma*), *ber* (*Ziziphus mauritiana*) and *kusum* (*Schleichera oleosa*) in India, scattered in the forest and personal holdings of the lac growers. Besides, there are some other plant species (tree as well as bushy), which can be used as lac hosts such as *Ficus* spp., *Acacia* spp. and *Flemingia* spp. In India Chhattisgarh, Jharkhand, West Bengal, Madhya Pradesh, Orissa and Maharashtra are the main lac producing states. Chhattisgarh and Jharkhand are the two states which contribute the bulk of lac production among these states. The average national production of lac during last three years was 20,348 tonnes in which contribution of Chhattisgarh was 7,744 tonnes (38.1 per cent), followed by Jharkhand which accounted for 5,958 tonnes (29.3 per cent).

The advantage of lac production lies in the fact that it can be taken up on degraded, marginal and poor fertile soils and its cultivation does not require much management practices. Moreover barren, cultivable wastelands, pastures and ravine lands can safely be put under lac cultivation. The lac cultivation is generally practised by the marginal and small farmers as means of meeting cash expenses towards family needs and their household requirements.

In the present scenario, where sustainability, conservation and profitability have become buzzwords, system approach has become obvious choice. Integration of agricultural components coupled with judicious natural resource management is imperative in such approaches. Lac integrated agriculture is one such approach which includes lac cultivation with crop farming such as growing cereals, pulses, oilseeds, vegetables, tuber crops, spices etc. While, integrating the lac culture in agricultural enterprises, food security, technical feasibility and socio economic condition of the farmers have to be taken into account. To address food security and profitability at the same time, farmers may adopt lac cultivation with various enterprises alongside agriculture. Since lac cultivation is mostly confined to poor degraded lands of the lac producing states; one has to see whether integration of lac is beneficial to minimize the extent of land degradation alongwith increasing the income of lac growers. For this, potentiality of lac hosts from its adaptability, its role in mitigating the natural resource deterioration and its other uses are required to be looked into, before its integration with other enterprises. An attempt has been made in this paper to visualize the role of integration of lac with agriculture for better economic return and conservation agriculture.

Potentiality of lac hosts in integration with agricultural crops/other enterprises

Alley Cropping

Alley cropping is an agro-forestry system integrating trees and shrubs with annual food crops. In this system arable crops are grown in the spaces between rows of planted woody shrubs or trees, which are pruned during the cropping season to provide *in situ* green manure and to prevent shading of crops. The beneficial effects of the system in reducing erosion, surface runoff and soil moisture loss depend on the proper choice of the protective species. Perennial woody lac host plants like *ber, palas* and *kusum* can be fitted very well under the system. An alley farming trial in Nigeria was compared with the ability of fallow and mulches of *Flemingia macrophylla, Cassia siamea* and *Gliciridia sepium* to control weeds. In a 120 days test of the decomposition rate of foliage from the first cutbacks from these hedges, it was reported that decomposition rate for *Cassia* 46 per cent, *Flemingia* 58 per cent and *Gliciridia* 96 per cent (Yamoah *et al.*, 1986).

Crop diversification

In the face of shrinking natural resources and ever increasing demand for larger food and agricultural production arising due to high population and income growth, crop diversification is the main course of future growth of agriculture. Lac hosts may prove to be very good source of crop diversification. Since most of the lac hosts are leguminous in nature and are well adapted to the marginal lands by virtue of their biological nitrogen fixation, these may fit well in different agroforestry systems such as agri-silviculture, silvi-pastoral, agri-silvi-pastoral and agri-hort-silvipastoral for optimizing lac productivity as well as improving soil condition. Diversification towards these high value and labour intensive crops can provide adequate income and employment to the farmers.

Experiments carried out at Indian Institute of Natural Resins and Gums (IINRG), Ranchi have indicated that under silvi-pastoral approaches, dinanath grass (*Pennisetum pedicellatum*) in combination with either tapioca or turmeric or both can be grown as a compatible intercrop in between *palas* plantation for increasing productivity and profitability without affecting the lac yield on *palas*. Another experiment showed that *urd* (blackgram) may be taken as intercrop successfully during initial phases of mixed plantation of *ber* and *kusum* (Anonymous, 2007). Growing turmeric and ginger as intercrops not only ensures high return from the *kusum* orchard but also covers the risk of lac crop failure. Similarly, winter *kusmi* lac production on *Flemingia semialata* with integration of vegetable crops like okra, garlic and bitter gourd round the year has been developed with enhanced returns at IINRG, Ranchi. The results revealed that sticklac yield to the tune of 17.5-19.6 q/ha can be obtained from the winter season lac crop with integration of vegetable inter crops. The yield obtained was just double to that of without vegetable integration. Irrigation with fertilizer application to the vegetable crops might have affected the broodlac yield (Anonymous, 2007). In an another experiment being carried out at IINRG, Ranchi, horticultural crops like aonla, papaya and guava have been raised with the integration of lac hosts *viz., Zizyphus maurtiana* and *F. semialata*. Summer *kusmi* lac crop *(jethwi)* has been successfully raised on one-year plantation of *F. semialata* with good yield (broodlac obtained/ broodlac used) ratio of 1:10, with integration of vegetables like bitter gourd and cucumber under irrigated condition.

Natural resource management

Out of total geographical area of 329 million ha of the country, about 187 m ha or 57 per cent is affected by various forms of land degradation processes. As per the available records, about 148 m ha is affected by water erosion, 13.5 m ha by wind erosion, 13.8 m ha by chemical degradation and 11.6 m ha by physical deterioration (Sharda, 2006).

The seriousness of resource degradation in India can be gauged by the fact that about 5333 million tonnes of soil, associated nutrients, seeds and plant propagules are eroded every year. Of this, nearly 1582 million tonnes (30per cent) are transported to the sea, about 480 million tonnes (9per cent) are deposited in storage reservoirs causing their eutrofication and pollution (Dhruva Narayana *et. al.*, 1983) and rest are redistributed within land scape.

Hence our focus should be to conserve and manage natural resources efficiently for sustainable agriculture. Diversity of crop species and the diversity of varieties within the species have traditionally strengthened the resilience of agriculture (Pimentel *et. al.*, 1992).

Deep penetration by tree roots or other vegetation makes the soil more permeable to rainwater so that runoff is slower and more uniform than on cleared land. Because of protective ground cover, lac hosts are important in reducing storm flows while improving recharge of underground aquifers as well as reducing the load of nutrients and pollutants entering water bodies. Lac host plants/trees with suitable agricultural crops will be instrumental in soil and water conservation measures. The spreading, densely-matted root system of the lac host plants stabilizes eroding land. Its rapid early growth, even on infertile sites and tolerance of both highly acidic and alkaline soils make it popular for stabilizing and revegetating mine spoils. One of the bushy hosts *Flemingia macrophylla* has substantial amount of nutrients (N-2.35 to 2.83per cent, P-0.19 to 0.25per cent, K-0.98 to 1.40per cent) (Yamoah *et al.*, 1986a). Pruned biomass of these trees may prove to be very good source of composting.

Role of lac host plants in agro-forestry system may also be viewed as a sink for carbon dioxide through the process of photosynthesis and hence can be used to mitigate global warming to a given extent, if proper mechanisms are put in place.

Socio-economic security

Lac cultivation provides both social and economic benefits to the livelihood of rural communities. At the subsistence level, lac cultivation normally address livelihood strategies like secure provision of food, health care needs, concerns to reduce risk factors etc. The demand for these types of services from lac from the forest and personal holdings is normally modest and rarely constitutes a threat to biodiversity of the region.

Lac hosts usually provide essential fodder, fuel and other related domestic requirements to rural populations. They are particularly important in relieving hunger periods in the agricultural cycle, can provide employment during slack periods and act as a buffer against risk and house-hold emergencies.

Use in fodder

Most of the lac host trees and shrubs can be used as forage for cattle, sheep, goats and wildlife (animals, birds and insects). One of the recently found promising lac hosts, *Prosopis juliflora* produces palatable and nutritive pods and leaves that are eaten by camels, cattle, sheep and goats in Egypt and Sudan (Sudan Update, 1997). *Flemingia macrophylla* at 10,000 plants in one hectare produced a yearly average of 12.4 tons of leaf dry matter over quantity cutting intervals (Yamoah *et al.*, 1986b). *Flemingia* appears to have some value as a dry season bin.

Other uses

Intensification in agriculture, which is critical to overcome food security, requires measures to protect watershed and arrest land degradation (FAO, 2003). Controlling desertification and land degradation using windbreaks and shelter-belts with lac host plants can easily be achieved.

One of the bio-waste product of lac *i.e.,* lac mud is an industry effluent obtained while washing of lac. It can be used as organic manure since it contains high quantity of N, P, K and organic carbon (2.8 per cent N, 0.4per cent P and 1.3per cent K and 38.8per cent OC) (Sarkar *et. al.*, 2006).

Conclusion

It can thus be seen that lac hosts have very good potential for incorporation in agricultural systems with multifold advantages. Integration of lac with agriculture and other enterprises will be more helpful in soil improvement and checking soil erosion. If the vast areas of degraded, waste and sloping lands of the country are put under lac cultivation system, they will not only become productive but also provide extra remuneration to the resource poor farmers. Lac hosts are particularly important to rural communities for culturing lac to provide subsidiary income, fodder for livestock and related domestic requirements. If the food crops of the farmers fail due to drought or other vagaries of monsoon, then income through lac provides relief to them, as lac hosts are hardy and can, thus, sustain the lac crop through such difficult conditions.

References

Anon. (2007). *In-situ* moisture conservation for lac hosts. *ILRI Annual Report*: p 9.

Anon. (2007). Integrated lac production system. *ILRI Annual Report*: p 9.

Dhruva Narayana, V.V. and Babu Ram, (1983). Estimation of soil erosion in India. *J. Irrg. Drainage Engg.*, **109** (4): 419–434.

FAO, (2003). Forestry outlook study for Africa: regional report for opportunities and challenges towards 2020. *FAO Forestry Paper* 141. FAO, Rome.

Pimentel, D., U. Stachow, D.A. Takacs, H.W. Brubaker, A.R. Dumas, J.J. Meaney, J.A.S. O'Neil, D.E. Onsi and D.B. Corzilius. (1992). Observing biological diversity in agricultural/forestry. *BioScience* **42** (5): 354-362.

Sarkar, P.C. and Kumar, K.K. (2006). Lac industry waste. Chapter 38. In :*Environment and Agriculture.* K.L. Chadha and M.S. Swaminathan (Eds.). Malhotra Publishing House, New Delhi, India.

Sharda, V.N. (2006). Land degradation. P 51-80. In :*Environment and Agriculture.* K.L. Chadha and M.S. Swaminathan (Eds.). Malhotra Publishing House, New Delhi, India.

Sudan Update. (1997). Desert tree is a victim of its own success. *Sudan Update.* **8**(22): 11/97.

Yamoah, C.F., A.A. Agboola and K Maiongoy, (1986a). Decomposition, nitrogen release and weed control by prunings of selected alley cropping shrubs. *Agrof. Syst.* **4**:239-246.

Yamoah, C.F., A.A. Agboola and K. Malongoy. (1986b). Nutrient contribution and maize performance in alley cropping systems. *Agrof. Syst.* **4**:247-254.

□□□

Section - 7
Regional Experiences and Future Thrust

Chapter 38

Hydrologic Continuum in a Watershed that Promotes Conjunctive Use of Rain Water for Attaining Food, Livelihood and Environmental Security

D.C. DAS

C-18, East End Apartments, New Ashok Nagar, Delhi - 110096 (U.T.)
National Fellow of IASWC, SCSI, ISAE and Fellow of Insititution of Engineers, Indian Association of Hydrologists and School of Fundamental Research ; Life Member and Ex-Vice-Phesident for India of World Association of Soil and Water Conservation.

Perceptions of path ways how the rain water becomes available to a consumer/ user or a receptacle static or moving have varied widely under different agro-climatic or agro-ecological situations. This is true equally among traditional communities as well as scientific communities.

Communities have however, evolved a wide range of systems to ensure that water becomes available to him through out the year even though precipitation comes in spells/seasons and/or supply is grossly inadequate. Traditional modes include withdrawing from natural sources like rivers, lakes, *kunds* etc and a host of man made systems such as diversion channels like *kuhl,* seepage tanks(*naula /bawli*), step wells, *tanka* etc to reservoirs and barrages, tanks, open wells, shallow to deep tube wells, low lifts from channel flows, etc. In their effort to meet increasing demands, these devises at places have been scaled up while at other sites some of these have been put to use in combinations. In recent decades all watershed development projects have used a wide variety of systems to facilitate more infiltration followed by prolonged percolation and seepage, interception–impounding-redistribution, interception-lifting-delivery, lifting from seasonal/ perennial channel-storage-distribution, harnessing natural seepage flows-storage-lifting/delivery. They are exploiting thus augmented water etc for irrigation, drinking water supply, meeting domestic and livestock needs etc. Most of these systems make conjunctive use of available and augmented surface and ground water on-site or near site in the watershed itself.

For example, in Kumaoon hills, drinking water was coming from *naula*/infiltration well while *gadheras* (small order streams) *were* for live stock, irrigation and other domestic needs. In HP mid hills, baulis(Naulas) are for drinking water, diversion channels *kuhls* stapping small streams are used for irrigation while ponds are for livestock. Whereas, in Sahibi catchment wells are the main source for drinking water and ponds are for live stock. Wells are also the main source for irrigation. Traditional *Johads* have come back as one of the main sources of water, but wells continue to be the direct source as well.

With improved land husbandry and increased population ,demand for water from these sources are also rising while changing land uses have been reducing water volumes at these sources. It is necessary to find ways and means not only to sustain these sources at the present level of water availability but also to enhance the water availability further.

To improve planning and designs and thus improve cost effectiveness, understanding and quantifying the hydrologic cycle with its sub-processes in operation in the land mass of the watershed would be of great help. This is also needed to impart a higher sustainability to these systems. This paper is an attempt to bring together many such processes supported by results obtained by a wide range of research and scientific studies sponsored primarily by international agencies.

Materials and Method

The hydrologic processes briefly recaptured above are neither unilateral nor linear in process analysis and then in application of the results obtained. Perhaps it would have been very difficult to formulate a comprehensive research plan due to multiplicity of subject matter wise principles involved besides, large geographical net working, long period and large investment required.

The author has been pursuing the theme since 1966 through a research project on "Disposition of rainfall under different land uses-Small Plot study" at CSWCRTI, Research Centre, Ootacamund. The results of this study were published in IHP's proceedings of International Symposiurm held in New Zealand (Das *et. al.*, 1970). After a long gap UNDP's study on Techno-economic feasibility of traditional seepage tank (*Naula*) and its modified version, namely Infiltration wells with hand pump in Kumaon Hills (Grassroots, 2000) allowed the expansion of the perception gained through the above study. Then came the results of evaluation studies of WB supported IWDP (Plains) in Gujarat (AFC,1998) Study of MOA on assessing changes in hydrologic status of watersheds of Sahibi catchment, Rajsthan carried out by IEG(Chopra and Subba Rao,1998) and IWDP(Hills) also supported by WB in Himachal Pradesh, Punjab and Uttarakhand (CES, 2005) which validated the same. The study of UNDP (CES, 2005a) on drought proofing strategy and perspective action plan for Chattishgarh helped to quantify some of these through scientific case studies.

Finally, the voluminous data generated for rivers and streams of Kumaon Region for valuation of water by Institute of Economic Growth and GBPant Institute of Himalayan Environment and Development helped validate many for watersheds of various sizes (Kadekodi *et al*, 1997). Re-computed and/or re-arranged data thus obtained have been used to present the case with specially drawn flow charts/diagrams and in tables.

Watershed -The master fresh water reservoir

There cannot be any difference in opinion that none including rivers and reservoirs can have the supply unless precipitation water is routed through the land mass, rather land surface, soil profile and substrata below, which together serve as the master fresh water reservoir. In the context of a river and stream it is a geo-hydrological area unit such as basin, catchment, sub-catchment, watershed, sub-watershed, micro watershed etc. (Das and Pandey, 1989, Kuffman, 2004 and Das, 2008).

Now, globally two third of fresh water (precipitation) is routed through the land mass out of which about 30 or 40 per cent is used for green mass production or goes as ET and the remaining bulk moves to rivers, lakes, reservoirs and ground water(valley deposits and aquifer found en-route). For the water resource development that would ensure ecological rather environmental security to largest possible geographical area while providing and sustaining social and economic security to the larger sections of our population, the sub-processes of the hydrologic cycle in operation on land are vital. These are important to put the concept of hydrologic continuum in totality for weighing options available for augmenting fresh water supply for the target house holds when the per capita availability is declining fast. These facts seem to get little attention or get shadowed by huge volumes in rivers, reservoirs etc. which are easily visible, even though these represent only 3per cent of country's fresh water and also these depend on the water in the land mass for their volumes as well as sustainability. This could be visualized from the summary of results of a study (Box-1) conducted at ISRIC (Kauffman, 2004).

Box 1 : Green and Blue Water for Food and Ecosystems

What's the problem? Sixty five per cent of the world's fresh water resource is green water. Water resources, their use and misuse, and its scarcity is the dominant theme across most of Africa, the Middle East, China, the Indian sub-continent and Australia. And yet, the largest component of fresh water resources, green water, held in the soil, has been neglected in policy, action and research. Managers and policy makers lack basic information on the resource, its availability, quality, the management options open and their likely consequences.

Definition of green water is rainfall that is stored in the soil and available plants. Globally, it makes up some 65 per cent of fresh water resources and it is the basic of rain fed farming and all terrestrial eco-systems.

Rain falling on the land may be intercepted by the ground cover, runoff the ground surface, or infiltrate into the soil. Intercepted water is evaporated directly to the atmosphere. Infiltrating water may be held in the soil – green water, or drain to ground water and stream base flow. Runoff, stream base flow and ground water may be dubbed blue water. An important distinction between green water and blue water is that green water may be used only in situ: blue water may be tapped, transported and used elsewhere – for irrigation, urban and industrial use, and as environmental flow in streams.

Significanntly nearly all the world's attention, research and development efforts have gone into blue water. In the case of irrigation, this is dealing with the management of only 3 per cent of global fresh water resources. Green water deserves attention from policy makers, planners land users and water engineers and for these reasons.

1. Rainfed agriculture contributes most of the world's farm production; 95 per cent of sub-Saharan Africa where in makes use of only 15-30 per cent of rainfall, the rest is lost, mostly as destructive runoff;
2. The partitioning of rain water is a dynamic process(governed by rainfall intensity, terrain, land cover, micro topography and soil conditions);
3. Soils process several times more water than they retain; while soil erosion by runoff and bank erosion by peak flow contribute nearly all the sediment load of streams, leading to the siltation of reservoirs and water courses. This means that management of green water is also management blue water;
4. Finally, agricultural demand for water is in competition or, even, conflict with the needs of industry , urban populations and the environment.

By Kauffman, International Soil Reference and Information Centre (ISRIC) 2004

Hydrologic status and hydrologic response

A stream or river will be finally fed by the water present in the land mass of a watershed. Thus quantum of water present in a watershed would be a determinant of the channel flow in quantity as well as how it reaches to the stream besides, distribution of total flow over 12 months of the year. Hydrologic Status (HS) refers to this quantity of water present in the watershed at any point of time. Water is stored in land mass and it also traverses through it till a destination is reached. Watershed retention can be as soil moisture, ground water deposits and can be water on movement specially, in substrata below soil profile (Das, 2005). The third form is not taken into account and the balance of infiltrated water minus the part retained in soil profile is taken as ground water recharge.

Water storages within land mass

Water is stored in the soil profile in different storages where two types play directly in marinating the hydrologic continuum or hydrologic cycle. These are retention storage (RS) or the difference between wilting point and field capacity and detention storage (DS) or the difference between field capacity and maximum water holding capacity (Fig. 1). These two storages as well as other types are intermixed and not compartmentalised through out the soil profile or the depth that permits hydrological processes to operate and are supported by physical, chemical and biological forces in operation in this depth. The volumes vary with land form and land uses so also are the operating hydrologic depth (Das *et al*, 1970 and Das, 2002). Sponge effect and soil expansion at the surface, such as in forests with good leaf litters on floor, creates additional or temporary DS that help reduce overland flow and prolongs the process of infiltration and subsequent seepage and percolation. For increasing volume of fresh water in land mass more RS as well as more DS, including higher soil expansion per cent, are needed which is possible through better land husbandry. The conceptual illustration of soil moisture storages and the forces in play is given in Fig. 2.

The storage below the soil profile or in the substrata is called variously such as soil penetration volume (Kadekodi *et al.*, 1997) or interflow storage (Mutreja, 1986). As biological activities are nil or very minimum in this zone the volume is largely subject to gravity for its disposition. Though in broader sense all water that crossed the ground surface is ground water, the penetration volume or interflow storage is distinct from the portion of absorbed rain water that finally reaches ground water storages, such as the valley deposits or aquifers build up en-route above an impermeable layer.

Whereas, hydrologic response (HR) is the path way through which the water in land mass moves out. These are out flows into rivers, lakes and reservoirs; evapo-transpiration and leakage or transfer from the watershed. Both these could be to a reasonable extent modulated to meet our purpose. The storages of water that moves out to channel systems are illustrated in Fig. 3.

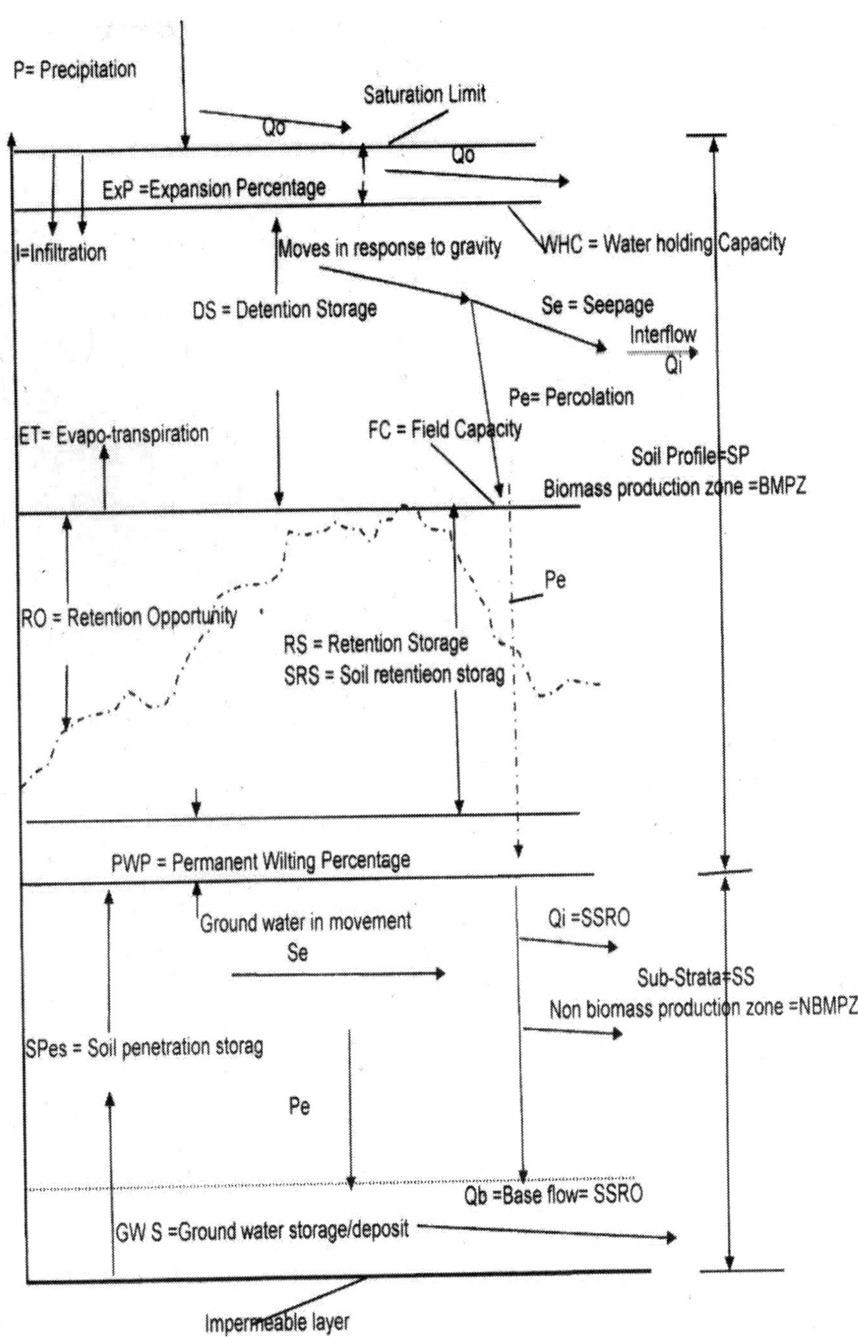

Fig. 1 : Water and Moisture Sorages in Land mass and disposition of infiltrated volume from soil profile and sub-strata below.

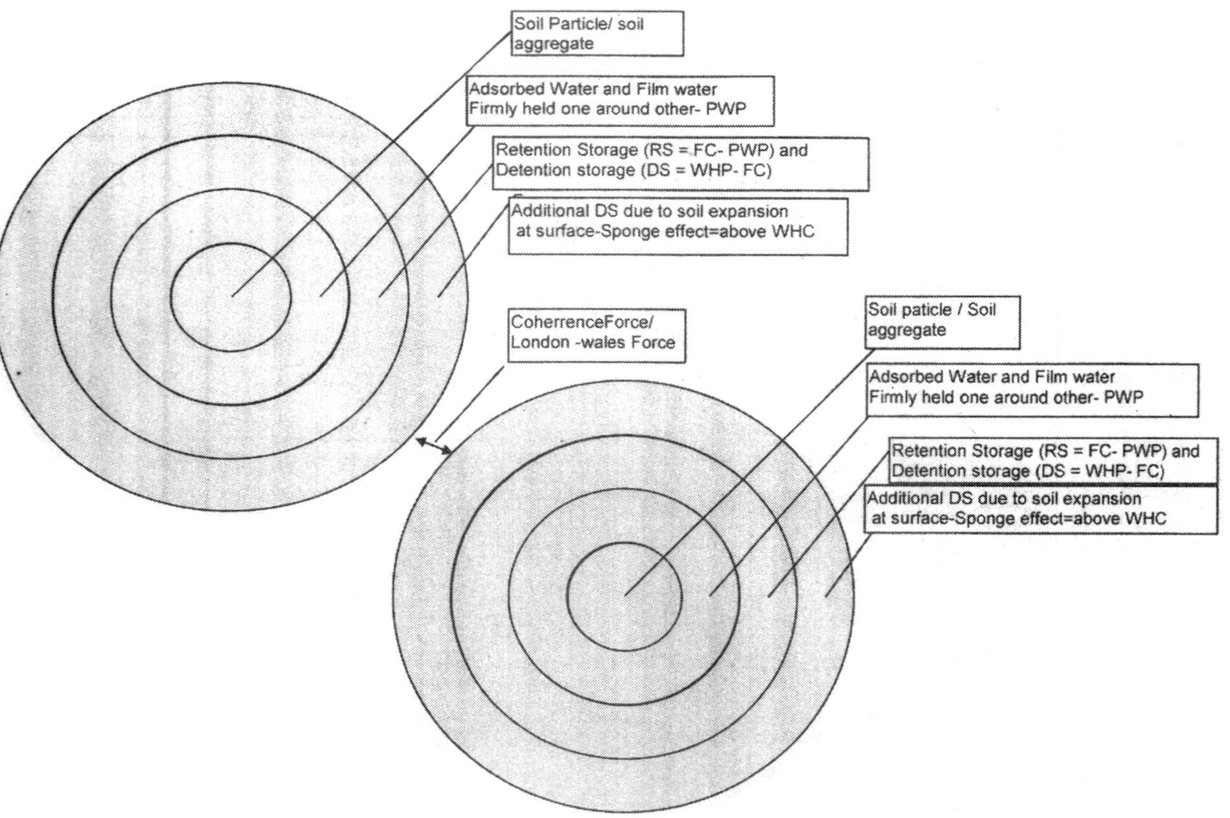

Fig. 2 : Illustration how moisture/ water held in soil profile and how water moves within soil profile.

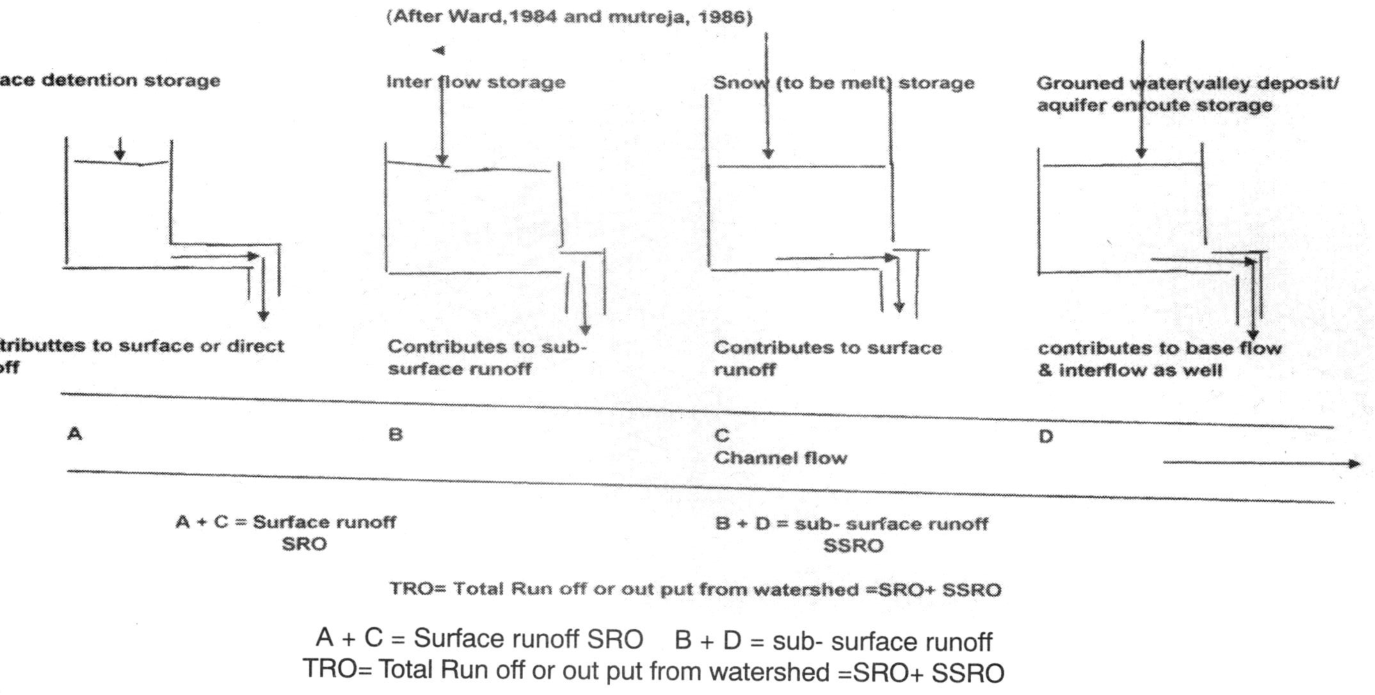

A + C = Surface runoff SRO   B + D = sub- surface runoff
TRO= Total Run off or out put from watershed =SRO+ SSRO

Fig. 3 : Storages on land surface and below that finally cause outflows or contribute to Blue water.

Water movement within land mass

Land mass whether soil profile or sub-strata simultaneously is serving as the store house as well as passage for the water which have got into it across the land surface. Even though this volume is ground water in broader sense its storage and movement at two levels are different. Within the soil profile the movement is regulated by soil suction, gravity, soil porosity and pore geometry besides the characters of infiltration processes. Like the storages movements, in response to soil suction/evapo-transpiration or gravity, are also concurrent though vary in speed and volume. The movement in response to gravity perhaps is tortuous, not continuous and slow to very slow comparable to that of sloths'. Even then due to geology and gravity water moves vertically when it could be called percolation as well as laterally due to a gradient and when it could be termed as seepage. Destination of seepage could be ground water storage and channels/surface storages. Generally, seepage is used for both processes and it could be from rainfall as well as from channel flows and surface storages such as ponds, tanks etc. as a secondary input into land mass.

Whereas, in the substrata it is geology and gravity are in play and the line of percolation and seepage could be faster than the movement through soil profile lying above. The gradient for seepage flows is due to dip, fault etc which also generate transitory/translatory accumulation at sites that facilitates out flows as **push-through** as in case of springs or as **interflows** into channel sections (Hewlett and Hibbert, 1967; Ward, 1984 ; Mutreja, 1986 and Das, 2002). Its final exit from land mass is mostly controlled by geology and the slopes on the hill sides or along the foot hills. The movement terminating variously including interflow takes place even when the soil /land mass are not saturated. Following new entries of rain water sections of hill slopes could get saturated and in between due to recession these would turn into unsaturated sections. The flow dynamics are therefore, not uniform and are very complex. The possible flow patterns are illustrated in Fig. 4 (Ward, 1984).

The process analysis of the location of springs and of small order streams and *naula's* (traditional seepage tanks) has established the distinct identities of water in land mass both in terms of storages and movement as well as final out flow either into channels or to ground water deposits/aquifers. Further seepage, as a hydrologic process and soil profile and sub-strata as the transitory storage, are the real means

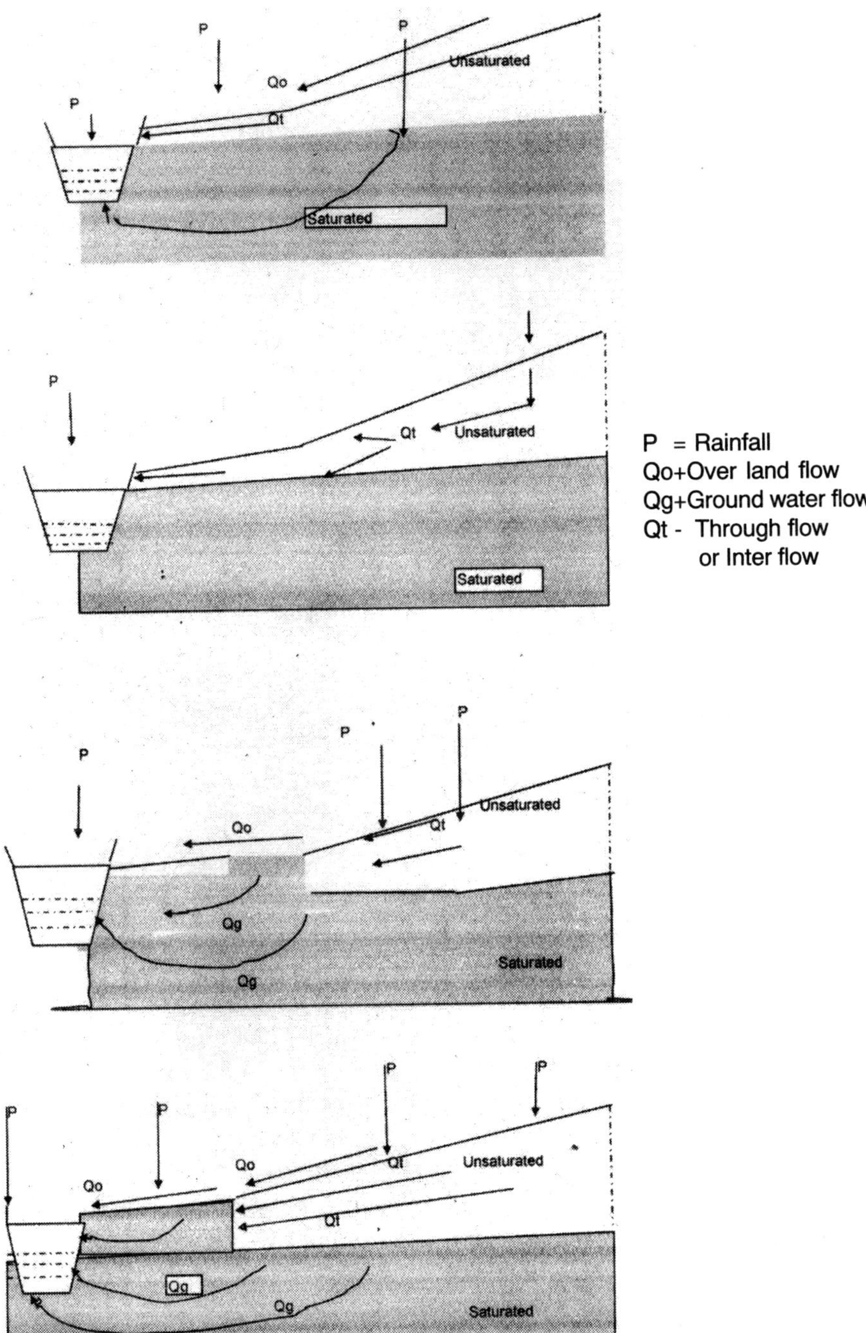

Fig. 4 : Possible flow paths of incident rainfall from land surface and land mass to channe in humid head water areasl(after Ward,1984).

and immediate source respectively for dependable and equitable sub-surface runoff in the channel during the whole year besides the base flow (Hewlett and Hibbert, 1967 and Das, 2002). These could be used to augment water availability to plan for a land use combination that could meet the needs for sustainable development as well as of environmental security.

The percolation lines on the other hand reaching impermeable layer build up ground water as valley deposits or as clear aquifer elsewhere enroute as well. The percolation and seepage flows follow a diminishing pattern and could have variable recession times depending upon the supply of absorbed water, length of flow and en-route transitory accumulation facilities (Hewlett and Hibbert, 1967 and Mutreja, 1986). These could also suffer severe reduction in excessively rainfall deficit years as found by site specific scientific water budgeting in Chattishgarh (CES, 2005a).

Watershed area and channel flow dynamics

The problem is in developing a pattern of flow for a channel and sources of water variously present within land mass that contribute to the flow. If this can be done then options to improve the link between land mass with the sustainability of various water sources here, specially channel flows, can be worked out.

Some hydrologists such as Mutreja (1986) have considered that the channel system may be getting water from four different storages operating on the contributing catchment/watershed namely, 1) Surface detention storage, 2) Channel (runoff channels or small order streams) storages, 3) interflow and 4) base flow or flow from ground water (valley deposits or aquifer else where). The first two due to similar hydrologic response time and recession pattern are taken as surface runoff and thus reducing the number of components into three where snow melt has not been considered.

The channel flow for Nepalese river basins on the other hand has been separated in four components as under (Chyurlia, J.P.,1984 quoted from Bruijnzeel and Bremmer,1989):

- Direct Flow (Q_d) : The portion of rain water that enters directly into stream without being stored. It is also referred to as immediate or surface runoff.

- Snow melt runoff (Q_s) : Snow melt water mostly as surface runoff
- Base Flow (Q_b) : Inflow of ground water from valley deposits & other aquifers
- Interflow (Q_i) : The portion of the precipitation , that is temporarily held in storage in soil mantle before being released in stream.

The results from a study of a water balance of East Ramganga catchment of Kumaon hills (Kadekodi *et. al.*, 1997 and Das, 2002; 2008) showed that for better hydrologic status and favourable equitable channel flow through out the year, endeavour should be to increase the volume of fresh water in land mass, increase subsurface flows/ runoff against surface or immediate runoff. Where these are not occurring either due to natural attributes or because of distortions caused by anthropogenic interference, interventions have to be devised to achieve these goals.

The computation of water balance for a 60per cent probability or 40per cent deficit rainfall year the water outflow as seepage/inter flow varies substantially from clay to sandy soils when district inter flow ranges from 10per cent in Mahasamund district to 16per cent in Sarguja district. This amount would be an supply for collection into debris/ponds or at a stop or check dam which could be effectively used irrigation (Das *et al.*, 2009).

Carry over or carry forward water in land mass

The concept in the form of antecedent moisture or precipitation has been in use in describing catchment's wetness vis-à-vis runoff. But value of antecedent moisture condition (AMC) or antecedent precipitation index (API) is decided as informed guess-estimation and no methodology has been evolved based on hydrologic processes such as infiltration, evaopo-transpiration etc. (Mockus, 1969).

Similarly, in computation of net irrigation requirement, contribution of rains received in preceding rainy days such as winter rains or snowmelt for summer cropping is deducted. This contribution is called input from soil water reservoir (Doorenbos and Pruitt, 1975).

Herbst *et. al.*, (1966) mooted this concept as a hydrologic process that suggest that notwithstanding the low or high magnitude of such preceding rainfall amount and aberration in distribution, there is a carry over or carry forward amount. They advanced a method to compute such amount and used the same for drought studies. In India this was used for drought study and then weekly water balance study of Tatarpur and Pithalpur watersheds of Sahibi catchment in Rajsthan which received treatments (Das quoted from Chopra and Subba Rao, 1998). From rainfall of 500.20mm received in 15 weeks carried over amount was 217.8mm for Tatarpur and from 500.14 mm received in 19 weeks carried over amount was 215.8 mm for Pithalpur warshed. The observed runoff from Tatarpur was 169.8mm and retention 47.2 mm against a runoff of 43.2 mm and retention of 172.6 mm for Pithalpur. The enhanced retention for second watershed was due to more soil conservation treatments including farm ponds.

The Dynamic water balance data obtained for physical accounting of water for the period from Apri,1996 t0 March, 1997 for East Ramganga catchment (23,600 ha catchment area) where following components were used *viz.* rainfall (P), snowmelt input (SM), surface or direct/immediate runoff (SRO), base flow or Subsurface runoff (SSRO), evapo-transpiration (ET). (Kireet Kumar and G.S. Satyal (quoted from Kadekodi *et al,* 1997) are shown in **Box-2**. The river is both rain and snow fed one. Here SSRO has not been separated further between base flow and interflow. Whereas, Mutreja (1986) has shown three methods for hydrograph separation between base flow and direct runoff (Surface runoff + interflow) from total runoff hydrograph.

Box-2

a) With rainfall alone, Total input	=	rainfall	5461.78 mm
Total withdrawal/output	=	SRO +SSRO +ET	
	=	1249.28 +4840.27 +448.52	6538.38
Excess output over input			+1076.6
b) With rainfall +snow melt total input	=	rainfall 5461.78 + Snowmelt 5610.29	11072.07
Total output	=	SRO +SNRO + SSRO +ET	
	=	1249.28 +5610.29 + 5540.43 + 448.52	12848.52
Excess output over input			+1076.5

c)	Soil moisture accretion	Total during the year	2100.93
	Soil moisture depletion	Total during the year	3177.53
	Excess depletion/withdrawal during the year		+1076.6

The excess output has come from carried over rain water absorbed and penetrated in land mass (soil profile + sub strata) which could be from more than one preceding year. The amount went to meet 448.52 mm of ET and 628 mm of total run off in the form of SSRO (Das, 2007).

The basic data generated for working out water balance of the Haigad watershed, having an area of 950 ha spread over between the elevations of 2338 m and 1160 m and 48per cent under forest and 36per cent classified as barren land, are reproduced in Table 1. Disposition of total input or rainfall as evapo-transpiration from land surface and cropped area and the portion went as outflows in the form of base flow, inter flow and surface flows is shown the flow chart at Fig. 5.

Examination of the Table 1 and the Fig. 5 revealed that about 91.73per cent 0f rainfall infiltrated into land mass and was in play in the soil profile where as about 1224 mm or 68.4per cent of the infiltrated volume flowed down to the sub-strata. Out of this amount 262.72 mm went out as interflow or as sub-surface runoff from water on move in the sub-strata primarily and 125.60 mm went out as base flow. Besides, 161.43 mm of rain water which did not infiltrate went out as surface runoff. In absence of a methodology for separating interflow and SRO from direct run off of 424.35 mm was done from the quantum of rainfall, change in soil moisture recorded for the months (Table 1).

An amount of 1115.30 mm or 57.17per cent of rainfall went out of the watershed while 835.55 mm or 42.83per cent of rainfall was retained by the land mass. This amount entirely was taken as ground water storage in working water balance. Considering the fact that total volume of base flow was only 125.60 mm as against 262.72 mm as sub-surface runoff or interflow, bulk could be considered to be in the landmass and yet to reach a destination either valley deposit or aquifers en-route or as delayed interflow. An investigation on source areas (long hill slopes) of a few springs was carried out by the Earth Centre of University of Rookee to detect the presence of water table or a build up of an aquifer could found only moist and parched patches (UOR quoted from Das, 2002). This also agrees to suggested interflow

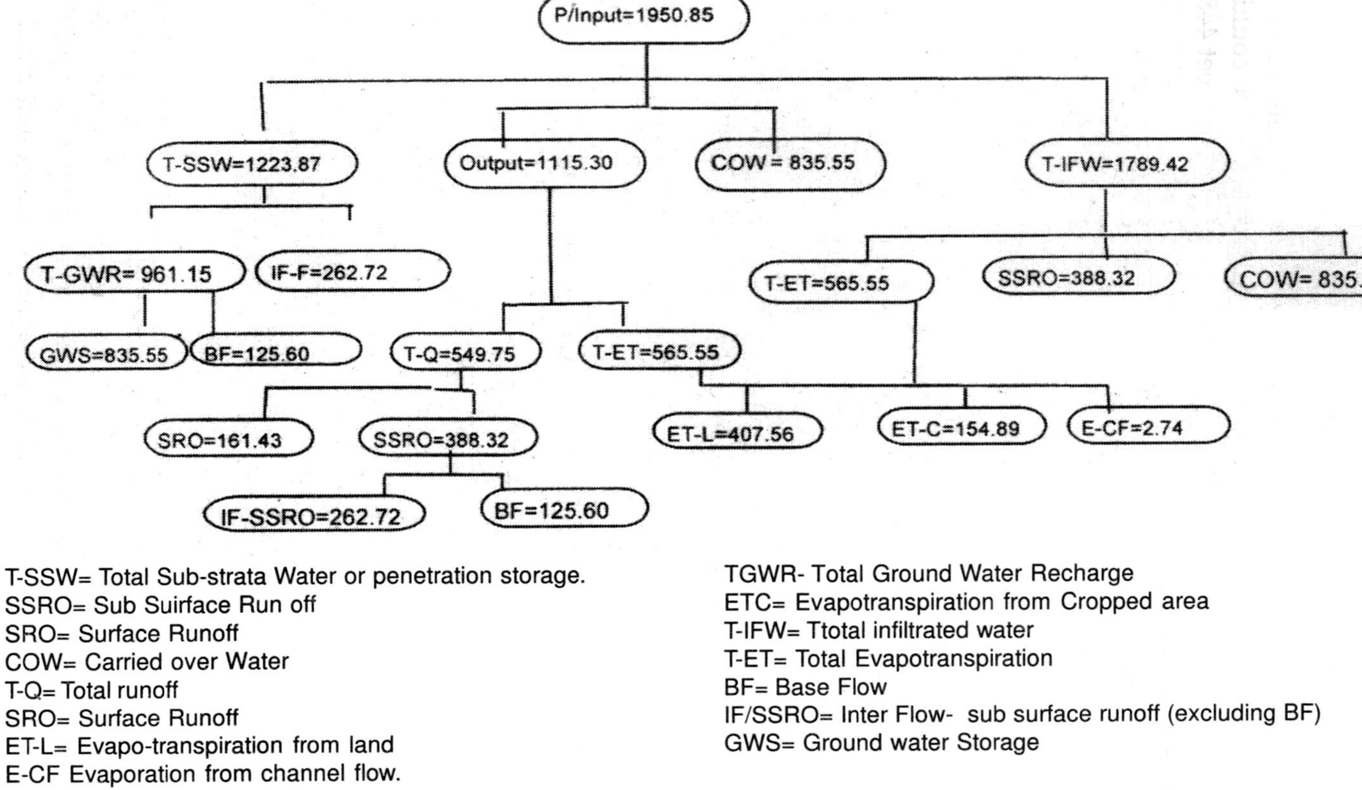

T-SSW= Total Sub-strata Water or penetration storage.
SSRO= Sub Suirface Run off
SRO= Surface Runoff
COW= Carried over Water
T-Q= Total runoff
SRO= Surface Runoff
ET-L= Evapo-transpiration from land
E-CF Evaporation from channel flow.

TGWR- Total Ground Water Recharge
ETC= Evapotranspiration from Cropped area
T-IFW= Ttotal infiltrated water
T-ET= Total Evapotranspiration
BF= Base Flow
IF/SSRO= Inter Flow- sub surface runoff (excluding BF)
GWS= Ground water Storage

Fig. 5 Main and Sub-loops in Hydrologic Cycle or Hydrologic Contnuum in watershed hydrology.

modes under unsaturated condition as shown in Fig. 4 (Ward, 1984). In any case this amount would be available as initial stock or carry forward stock for the water balance of the succeeding year.

Table 1 : Water balance data of Haigasd watershed Kumaoon Hills.

Month	Rainfall) (P)	Total runoff Total	Base flow	Inter flow*	Surface Runoff*	ET	ds SM	Remarks
April	7.87	4.64	4.64	nil		66.93	-55.86	* surface runoff (SRO) & interflow were shown together. SRO was taken as 50 % of the total and only for months which received rainfall & had recorded high Sm accretion
May	76.62	5.34	5.34	nil		55.50	18.67	
June	308.84	27.79	5.35	11.32	11.32	60.60	222.17	
July	416.78	69.06	5.50	31.78	31.78	65.66	285.38	
August	961.75	99.63	5.64	46.99	46.99	94.01	771.96	
September	152.64	85.8	6.82	39.49	39.49	72.44	-1.29	
October	26.35	81.51	9.30	40.17	32.05	42.69	-93.85	
November	nil	73.61	10.91	62.70		17.53	-99.05	
December	nil	38.06	14.10	23.96		20.91	-58.97	
January	nil	22.84	16.92	5.92		21.21	-44.05	For October being the mosoon's receeding month, 45 % of the total was taken
February	nil	19.1	19.09	0.01		25.80	-44.9	
March	nil	22.37	21.99	0.38		42.29	-64.66	
Total	1950.85	549.75	125.60	262.72	161.63	565.55	835.55	

Again, surface runoff occurred in 5 months when rainfall was received while sub-surface runoff other than base flow was for 10 months and base flow for all 12 months. Thus for equitable flow pattern and much larger volume SSRO (Base flow plus interflow) was found more dependable and have greater potential of harnessing and recycling as was already reported for East Ramganga catchment.

In a separate publication it was illustrated how the goal of increasing infiltration, promote voluminous and prolonged seepage with consequent increased sub surface runoff to make water availability more equitable for attaining greater security of water, food and environment over extensive area and covering large number of families (Das, 2007).

Conclusion

The important conclusions are drawn from the deliberations are as under :

- Seepage and consequent interflow are distinct processes and phenomena in the hydrologic continuum or hydrologic cycle of a watershed and could used to assess of water availability and options to use the same.

- Water storages in soil profile and that in the sub-strata as well as movement of water at two levels are different but both if large have beneficial influence upon the disposition of incident rainfall
- Similarly carry over rainfall or infiltrated water over time is also a hydrologic phenomenon and very useful to meet part of channel flow and plant available moisture in rainfall deficit years. This should be a component in working out annual water balances.
- There is a need to research programme with a net work of watersheds in different agro-climatic regions to develop modes to quantify the dispositions specially separating direct runoff between interflow and surface runoff and help in optimum utilisation of the same for food and environmental security.

References

AFC: (1998). Final Impact Evaluation of IWDP (Plains), Gujarat (WB supported), Agricultural Finance corporation, Ahmedabad Centre.

Bruijnzeel, A and Bremmer C N. (1989). High lands-Low lands interaction in Ganges-Brahmaputra River Basins. Occasional Paper-10, ICIMOD Kathmandu, Nepal.

Chopra, K and Subba Rao, D.V. (1998). Assessment of Hydrologic Status Changes due to Watershed Management in Sahibi Catchment, Rajsthan (the Chapter by D. C Das) Institute of Economic Growth, Delhi.

CES (2005). Final Evaluation Reports on IWDP (Hills) World Bank supported a) Himachal Pradsh, b) Punjab, and c) Uttatranchal. Consulting Engineering Services (India) P. Ltd, New Delhi.

CES (2005a). Drought Proofing Strategy and Perspective Action Plan for Chhattishgarh UNDP sponsored Study Report. Consulting Engineering Services (India) P. Ltd, New Delhi.

Das, D.C., Raghunath, B and Thomas, P.K. (1970). Rainfall disposition studies on small plots at Ootacamund (India) Results of Representative Experimental Basins (IHD), IASH-UNESCO. Pub. 96:296-308,.

Das, D.C. (2002). Hydrology of soil profile and sub-strata to improve hydrologic performance of watersheds. *J. Soil & Water Conservation*, **1**(2&3): 182-196.

Das, D.C. (2005). Assessing Hydrologic status of a watershed from on land retention and ground water recharge. *J. Sol & Water Conservation*, **4** (1&2): 25-34.

Das, D. C. (2007). Emerging Trends in Watershed management to meet Global Challenges. Key Note Address. Souvenir. National Seminar IASWC, Dehradun, Chandigarh.

Das, D. C. (2008). System approach to watershed modeling for sustainable development, including forestry. International Forestry Review. **10** (2): 281-291.

Das, D.C., Sarkar, T.K. and Mukhopadhaya, D.P. (2009). Drought in Chattishgarh. *J. Soil & waer Conservation*, SCSI, New Delhi, **8** (1): 25-32.

Doorenbos, J. and Pruitt, W.O. (1975). Guidelines for predicting crop water requirements. Food and Agriculturer Organization (FAO) of The United nations. *Irrigation and drainage* Paper-24.

Grassroots-UNDP (2000). Techno-economic feasibility of Infiltration well (Modified *Naula*) in Kumaoon Hills.

Herbst, P.H., Bredenkamp, D.B. and Barker, H. M.G. (1966). A technique for evaluation drought from rainfall data(excess deficit rainfall). *J.Hydrology* Amsterdam, **IV** (3): 246-258.

Hewlett, J.D.and Hibbert, A.R. (1967). Factors affecting the response of small watersheds to precipitation in humid areas. .Forest Hydrology (Eds:W. E. Sopper & H. W. Lull Pergamon, Oxford, 275-290.

Kadekodi, G.K., Murthy, K.S.R., Kreet Kumar,Adhikari, S.K., Satyal, G.S. and Subba Rao, D.V. (1997). Valuation of Water-The case of Kumaoon region. Instt. Of Economic Growth, and G. B. Pant Institute of Himalyan Environment and Development, Delhi.

Kauffman, Sjet. (2004). Green and Blue water for food and Ecosystem, ISRIC, Wageningen University Research Centre, the Netherlands.

Mutreja, K.N. (1986). Applied Hydrology.TATA MaGraw-Hill Publishing Co. Limited, Company, New Delhi.

Mockus, Victor (1969). National Engineering Hand Book-Section-4 Hydrology. Soil Conservation Service, United States Department of Agricuklture (USDA), Washington D. C.USA.

Ward, R.C. (1984). On the response to precipitation of head water streams in humid areas. *Journal of Hydrology*, **74**:171-189.

□□□

Chapter 39

Experiences of Watershed Development Programme in Central Indian : Success Stories and Probable Ways for Further Up Scaling

V.S.TOMAR
Rajmata Vijayraje Scindia Agricultural University, Gwalior (M.P.)

For enhancing the food productivity and alleviation of hunger, a regionally differentiated strategy based on agro-climatic regional planning which taking into account agronomic, environmental and socio-economic conditions will help to adopted to realize the full potential of growth in every region. Integrated watershed management is considered as an effective unit for overall development of agriculture for obvious regions because it is the most rational approach in preventing deterioration of ecosystem, restoration of degraded lands, efficient water harvesting and improving the overall productivity in rainfall areas. The recent years have witnessed a marked increased in the level of interest among R&D organizations, administrators and schemes launched by the Ministries of Agriculture, Rural Development and Environment & Forestry, committed local leadership and voluntary agencies are coming forward to help implementation of watershed development programmers by motivating and involving people's participation.

The Indo-UK project involved Operational Research on dryland farming on black soils (Vertisols) with multidisciplinary integrated approach. Under this project, the dryland technology, evolved by through research, was demonstrated on farmers' fields to improve the productivity per unit land and livestock during 1975-84. The objectives of the project were To test, evaluate and demonstrate usefulness of dryland technology evolved through research so as to make the most efficient use of soil and water resources for Improving overall socio-economic condition of the farming community including crop and livestock management of the selected villages. The project area (2712 ha) covered three villages Jamburdi Hapsi, Rinjlai and Nainod covering eastern part of two sub-catchments of Gambhir River.

Success of Indo –UK projet

Before the inception of the project, mono-cropping pattern predominantly in *Rabi* season was in practice by most of the farmers of the study area. Farmers used to keep most of their cultivated area (60-90%) fallow during *Kharif*. To replace the fallow by *Kharif* crops, preparation of crop plans for individual farmer on the basis of catchment and land use was done. Due to improvement in land and water resources, they started growing more crops in the *Kharif* as well as *Rabi* season and also started adopting a regular crop rotation.

Model Watershed Development work at Hingonia Pipliyatapha watershed

The success of Indo-UK project, prompted ICAR to start Operational Research Project on dryland agriculture at 8 locations in the country including Indore during 1986. It was given additional responsibility of developing model watershed by the Government of Madhya Pradesh in the year 1987. The Operational Research Project, College of Agriculture, Indore, a constituent campus of the Rajmata Vijayaraje Scindia Krishi Vishwa Vidhyalaya, Gwalior has implemented the program of watershed research and development.

The project area is located in the village Hingonia and Pipliyatapha. The site is located at about 26 km in the North-West of Indore city and the watershed covers 424-hectare area. Geographically, it is located at 76.4°E longitude and 22.7°N latitude with a mean altitude of 540 m above mean sea level. These villages are situated in lower reaches of river Gambhir, a tributary of river Chambal.

Characteristic of the Watershed Area

The main characteristics of watershed are presented in Table 1.

Table 1 : Characteristics of watersheds

Features	Hingonia	Pipliyatapha
Area	258.00	165.90
Maximum length (m)	1875.00	1560.00
Maximum width (m)	1800.00	1050.00
Coefficient of compactness	–	3.96
Slope percent	1.00	0.90
Slope factor	0.82	0.71
Maximum relief (m)	15.00	13.50
Time of concentration (min)	36.00	36.60

Land Use Pattern

The project area covers 258 hectares of village Hingonia and 165.9 hectares of village Pipliyatapha out of 424 hectare or about 89.87 percent is cultivable. As per the land use capability classes, project has 4 class of land 37.5% of class II, 49.7% of class IV and 7.6% of class V lands. The detailed land use is presented in Table 2.

Table 2 : Land use of project area (ha)

Particulars	Hingonia	Pipliyatapha	Total	Percentage of total area
Total geographical area	258.05	165.90	424.00	–
Total cultivable land	227.80	153.30	381.10	89.80
Land capability classes				
Class II	64.60	78.30	143.00	37.50
Class III	134.00	15.60	189.70	49.70
Class IV	-	19.30	19.30	5.00
Class V	29.00	-	29.00	7.60
Cultivable waste	41.80	22.50	64.30	15.17
Orchard	2.00	3.70	5.70	1.36
Cultivable land	183.90	127.00	310.90	73.34
Pasture and grass land forest	19.80	8.70	28.50	6.75
Village road, Nallah etc	10.40	3.90	14.30	3.38

From the above table, it is observed that 64.3-hectare area is under cultivable waste, which can be utilized for orchards, pastes, grassland or energy plantation etc. Class V land needs to be brought under permanent pastures and grassland etc., whereas, class IV land area is found suitable for construction of soil conservation structures.

Major Soil Types

The major soil group found in the area is typical Vertisol. The soil is synthesized from basaltic parent material. This comes under the category of clay loam soils having gray brown colour. The soils found in the area have been divided under two sub groups as:

(A) *Kamliya Kheri Series*: This series include shallow dark grayish brown gravelly clay soils having depth of about 30 cm. The soils are overlain with weathered basalt. These soils are predominantly found in the village Hingonia section of the project.

(B) *Sarol Series*: Medium and deep both soils having a depth of 30 cm to 90 cm are included in this series. These soils are deep gray in colour. The clay percent decreases with depth. The Pipliyatapha section of the project comprises of this type of soil.

The soils are moderately fertile and rich in calcium carbonate nodules. They are good at moisture retention. These soils become impermeable to water in saturated hard and difficult to operate agricultural implements. On drying deep and wide cracks are formed. This results not only in moisture loss but also mechanical injury to the roots of standing crops.

These soils are vulnerable to runoff during the event of intense rains. The pH of soils ranges from 7.2 to 8.5 with no salt hazards. These soils are low in Nitrogen, low to medium in Phosphorus and high in Potassium content. The organic carbon content is very low i.e. to the extent of 0.3 to 0.4 percent.

Topography

Both the villages have nearly leveled topo-sequence. The poor management has disturbed the natural terrain of the area. Due to this one to three percent slope has developed. Though the crops do not suffer from water logging, but the problem of sheet erosion exists in some of the fields. The rills have also developed where sheet erosion has not been controlled. Gullies have developed into natural drains (Nallah).

Land Holding

The size of holdings varies considerably in the project area. The composition of holdings reveal that around 48 percent of land holdings are of less than 4 hectares, 40 percent of holdings lie in the group of 4 to 8 hectares and remaining 12 percent are in the group of more than 8 hectares. The complete profile of different groups comprises the majority of small farmers. Such a profile provides an appropriate canvas to carry out the programme for resource poor dryland farmers. Considerable attention has been paid in the formulation of the programmes and their implementation in which major emphasis has been given to the farmers with smallholdings. Details of land holdings are depicted in Table 3.

Table 3 : Distribution of land holdings

Holding groups	Families Number	Percent	Holding Area (ha)	Percent
Small (0-4 ha)	34	47.8	86.2	23.2
Medium (4-8 ha)	28	39.4	163.0	45.3
Large (7-8 ha)	9	12.6	110.8	30.8
Total	**71**	**100**	**360**	**100**

Keeping characteristics of the watershed area and analysis of troubles and possible studies in view, a detailed programme with regard to soil and water conservation and agronomical treatment of the area was prepared and implemented with the farmers' participation to improve the productivity of the area. Thus, an integrated approach for increased agricultural productivity in these dryland areas of Vertisols was adopted.

Treatment of the Area

Soil and Water Conservation Measures

A watershed management plan for soil conservation works was made in 1989 after a detailed topographical, soil and socio-economical surveys of the area during the year 1987-1988. This plan involved construction details of bunding, various engineering measures and strategy to develop other resources.

Field bunding was done in an area of about 250-hectare. Considering the soil type, topographical and climatic conditions 1.0

sq.m. cross-section of 0.75-m high bunds had been constructed. Minor levelling has also been carried out at few locations. In all, 33 waste weirs, 8 loose boulder structures and 3 gabion structures were constructed at suitable sites. Waterways of about 1800 m length were provided in the area for safe disposal of excess water.

As a result of different soil conservation measures adopted (Table 4), runoff has been completely controlled and diverted safely which was the main cause of sheet erosion. The waterways have been provided to carry away the diverted runoff safely away from the cultivated fields. Thus, gully head advancement phenomenon has been completely controlled by avoiding the concentration of flowing water from a particular location. The silt particles are being trapped in the field along with bunds, waste weirs, loose boulders and gabion structures and diversion drains.

In this watershed, the highest runoff was recorded from the untreated portion and amounted to about 30-35% of the rainfall causing runoff. This untreated area recorded 50 percent more runoff than the treated area (about 5 ha) of almost identical dimension. The reduction in runoff from the treated area may be attributed to increased depositional characteristics of the individual soil and water conservation measure. The graded bunds, diversion drains, terracing etc. reduced the runoff velocity with concurrent and consequent increase in time of concentration. At the same time, grassed waterways drained out the excess water at non-erosive velocity.

Table 4 : Soil and water conservation measures adopted in the watershed.

S. No.	Particulars	Nos./m length/ ha	
1.	Graded bunding	1500	m
2.	Leveling	8	Ha
3.	Gabion structures	3	No.
4.	Loose boulder structures	8	Nos.
5.	Waste weirs	33	Nos.
6.	Water ways	1800	m
7.	Diversion drains	1375	m
8.	Terracing	5	Ha.
9.	Spurs	2	No.
10	Water harvesting tanks	2	No.

Soil loss was commensurate to runoff in both treated and untreated area of the watershed. The highest soil loss of 1500 kg/ha from the untreated and the lowest of 350 kg/ha from the treated part of the watershed.

The gabion structures constructed in the gullied portions are working satisfactorily. In most of the gullies, the silt is being deposited every year. Provision of gabion structures not only stopped the gully head advancement phenomena but also stabilized the gullies.

At the same time, loose boulder structures are providing obstruction in flow of runoff water and are trapping the silt on the upstream side. These are working as grade stabilizing structures. The rainwater harvested through various resource development programme in the watershed is being stored in two water harvesting tanks of 10000 cu m and 17000 cu m storage capacity respectively.

These tanks were developed in the gullied portions for their effective utilization. The gabions used, as inlet and outlet in these tanks are very cost effective and have a very high demonstrational value as well. These tanks are getting filled every year because of excess runoff from the large chunk of the watershed. The catchment area of these tanks is about 50 and 55 ha respectively. However, the former is getting less water from cultivated fields with low runoff potential, whereas, the later is receiving larger runoff because its catchment area is stony fallow having higher runoff potential.

During pre-project period, the life span of shallow wells used to be only 2-3 months after the cessation of monsoon rains. However, in the post-project area, wells do not dry up till as late as March in the subsequent year. There was increased water availability in the watershed due to impounding and recharging of water. As such, irrigated area has been increased from 28 hectare during 1987-88 to 168 hectare during 1994-95.

All the wells of the area exhibit the natural phenomenon of water level variations. Water level in wells starts receding after the monsoon and reaches to lowest level before the monsoon. During the monsoon, water level rises due to well recharge. The difference between maximum and minimum depth of water level in wells recorded before the monsoon and after monsoon, respectively. This difference reflects the total rise in ground water as a result of recharge. In cases when water table goes down even below the bottom of wells, rise in ground water table was calculated considering maximum water table depth at bottom of well. Close look of data reveals that even in the year of scanty rainfall, water remains available in most of the wells throughout the year (Table 5).

Table 5 : Improvement in water potential.

Particulars	Base Year 1987-88	Year 1995-96
Surface water storage (cu m)	Nil	27000
Total number of open dug wells	19	29
Shallow/ Deep wells (water up to October-November)	3	27
Deep wells (water up to February)	6	6
Deep wells (live year round)	3	3
Total tube wells	3	70
Live up to January	3	18
Live up to April	–	8
Live year round	1	4
Totally dry	–	40
Sprinkler sets	–	2
Total irrigated area (hectare)	28	168
Irrigated area (%)	6	40

As a result of soil and water conservation programs, moisture availability in soil profile and water availability in well increased. A significant improvement in water potential was recorded in this area. During pre-project period, the life span of shallow wells used to be only 2-3 months after the cessation of monsoon rains. However, in the post-project era, wells do not dry up till as late as March in the subsequent year.

There was increased water availability in the watershed due to impounding and recharging of water. As such, the irrigated area has increased from 28 hectare during 1987-88 to 168 hectare during 1994-95 (Table 6). This increase in irrigation potential in wells due to water recharge resulted in increase in area under *rabi* (Table 7).

Table 6 : Annual rainfall, rise in ground water and percentage change due to recharge.

Year	Rainfall (mm) R_1	Average fluctuation in water level in well (m) W_1	Percentage change over previous year (%) Rainfall R_2	Percentage change over previous year (%) Water level W_2	Percentage change over average value (%) Rainfall R_3	Percentage change over average value (%) Water level W_3
1989	624.0	2.30	--	--	-36.15	-66.12
1990	1046.0	4.92	+6.70	+113.05	+7.0	-27.83
1991	779.8	3.70	-25.41	-24.46	-20.2	-45.51
1992	583.7	3.45	-25.10	-6.75	-40.2	-49.18
1993	1062.5	7.82	+82.02	+126.62	+8.7	+15.17
1994	1445.4	13.57	+36.04	+73.52	+47.8	+99.85
1995	946.7	6.29	-34.49	-53.64	-3.1	-7.36
1996	1331.6	12.31	+40.66	+95.70	+36.2	+81.29
Average	977.4	6.79	--	--	--	--

Table 7 : Cropping pattern in the watershed.

Crops	Area (hectare) Before project (1987-88)	After project (1996-97)
Kharif		
Sorghum	48.0 (18.40)	3.5 (1.10)
Maize	4.5 (1.70)	—
Pigeon pea	4.5 (1.70)	7.0 (2.20)
Soybean	132.0 (50.60)	243.0 (78.00)
Mixed crop	9.0 (3.40)	—
Fodder	24.5 (9.40)	40.0 (12.80
Inter cropping	3.0 (1.10)	9.0 (2.80)
Vegetables	7.0 (2.60)	6.0 (1.90)
Miscellaneous	28.0 (10.70)	3.0 (0.90)
Total *kharif*	260.5 (83.70)	311.5 (97.90)
Rabi		
Wheat	56.0 (33.7)	233.5 (77.8)
Gram	77.0 (46.3)	33.0 (11.0)
Fodder	1.5 (0.6)	9.5 (3.1)
Linseed	11.0 (6.6)	4.0 (1.3)
Vegetables	8.0 (4.8)	3.0 (1.0)
Total *rabi*	**166.0 (53.3)**	**300.0 (94.3)**

Figures in parenthesis are percent to total area.

Before project, only 166 hectare was under *rabi* crops while in 1996-97, it increased to 300 hectares. An increase in area under wheat by 44 per cent clearly reflects the impact of programme with increase in water availability in wells. Introduction of new varieties, improved management practices coupled with more available water in wells resulted in increase in land productivity of *kharif* and *rabi* crops (Table 8).

Table 8 : Yield of local and improved varieties of *kharif* and *rabi* crops (kg/ha).

Crop	Local variety	Grain yield (kg/ha)			Improved Varieties	Grain yield (kg/ha)		
		87-88	91-92	93-94		87-88	91-92	93-94
Sorghum	Vidisha 60-1	680	600	1430	CSH-5, SPV - 475	1210	3250	3780
Maize	Satha	390	900	1250	Ganga 5	410	2560	2900
Soybean	Kalituar	320	850		JS 71 -05, PK 472	480	1450	1940
Wheat rainfed	Pissi	280			Sujata	560		700
Wheat partially irrigated	Lok -1				Sujata, Lok-1	960	940	2200
Wheat irrigated	WH 147				WH 147, Raj 1555	2200	3300	4260
Gram rainfed	Ujjain 21	320	590		Ujjain-21	380	800	1340
Gram irrigated	Ujjain 24	470		1130	Annagiri	560	1560	1840
Linseed	-				ILS 73-85		460	920

Thus, various soil conservation measures in the watershed enhance ground water recharge and ultimately increased crop intensity and crop productivity. Although, annual ground water recharge in a particular year depends upon other factors but a close estimate for the year can be drawn by the average annual recharge in water table in the past years.

Agronomic Measures for Soil and Water Conservation

The earlier cropping systems were substituted by soybean + arhar, soybean-wheat, soybean-linseed, soybean-chickpea and soybean-safflower. A complete package of conservation agronomic practices with improved/high yielding varieties, fertility and plant protection management techniques for different crops has been demonstrated in the watershed.

Reduction in runoff and soil loss and improvement in production have been achieved through proper land use, selection of suitable crops and varieties, crop rotation and cultural practices. To avoid any soil erosion from the fields, various agronomic practices like key line sowing and contour cultivation and mixed cropping mere adopted. Various soil conserving crops and their varieties were introduced in the watershed area (Table 9).

Table 9 : Improved varieties of different crops introduced

Crop(s)	Variety
Sorghum	CSH-5, SPV-475
Maize	Ganga-5
Soybean	Pb-1, JS 71-05
Wheat rainfed	N 4, Sujata
Wheat (partially irrigated)	Sujata
Wheat irrigated	WH 147, Lok-1
Gram	U-21, Annagerri
Linseed	L.L.S. 73-25

Soybean + arhar, soybean-wheat, soybean-linseed, soybean-chickpea and soybean - safflower substituted the earlier cropping systems. The utility of FYM, compost and other fertilizers was discussed with farmers and they were motivated to use fertilizers. Seven years nutrient use reveals that the consumption of N and P increased by about four and seven times from 1987-88 to 1994-95.

The area under double cropping and increase in intensity of cropping depicts the impact of soil and water management practices adopted in the area. On the other hand, increase in the production of crops with the introduction of improved varieties (Table 1.6) was a result of close association of the scientists and farmers. This not only increased the production of crops but also brought more area under double and multiple cropping.

Forestry and Horticulture

For economic utilization, degraded lands put under forestry and horticulture. The area covered was about 5.5 ha under these systems.

Alternate Land Use System Tested

On the slopy and uncultivated lands of farmers, plantation of fruit plants like guaua, custard apple, aonla and phalsa was done. The plantation of Eucalyptus around the Hingonia water harvesting tank and that of *Jatropha curcas (Ratanjyot)*, *Glyricidia*, *Subabul* and *Eucalyptus* around the Pipliyatapha water-harvesting tank was done.

Animal Husbandry

The construction of water harvesting tanks at Hingonia and Pipliyatapha villages has facilitated the drinking water availability to the animals. Earlier, it was made was available from wells, tubewells or hand pumps. This facility has resulted in increase in the number of milch cattle in the project area. Small, marginal farmers and landless labours that could not afford to keep milch cattle have gone for goat keeping. This subsidiary profession has provided farmers with the additional and alternate source of regular income by selling milk and animals.

At the end of plan period, 590 liters of milk was produced in both the villages. As a result of increase in the milk production, the wholesalers have started milk collection centre and milk producers are not required to run individually to city for selling their milk. This has enabled them to spare more time for their agricultural and other social activities. The details of cattle population is presented in Table 10.

Table 10 : Cattle population and milk production in the watershed area.

S. No.	Breed of animal	Number of animals		Total	Milk per day per animal (litre)
		Hingonia	Pipliyatapha		
1.	Buffaloes	135	112	247	6.80
2.	Cows	84	16	100	3.4
3.	Cows (Crossbred)	17	18	35	12.6
4.	Goat	87	16	103	–

Impact Assessment

The data on progress indicators (Table 11) clearly indicate the overall development of the area in terms of natural resource conservation and crop productivity due to implementation of watershed approach.

An area of 424 ha was treated with various soil and water conservation measures beside introduction of improved varieties of various crops and agronomical systems. The increased recharge due to runoff control is being exploited for irrigation. Analyses of water table data collected from the open wells located in the watershed reflect average annual rise of 10 m in water table due to recharge. The enhanced availability of water led to increase cropping intensity from 133 to 165 per cent. Increase in availability of irrigation water in wells has resulted in 44 per cent increase in *rabi* cropped area and almost doubled the productivity of *rabi* and *kharif* crops. The total area under *kharif* and *rabi* has been increased by almost 75 per cent. The integrated area had gone up from 28 ha to 67 ha during 1991-92. The soil conservation measures viz., graded bunds, diversion drains, waste weirs, loose boulder structures and gabion structures not only checked soil loss from the cultivated fields but also stabilized the gullies.

Thus it is very clear from the impact study that various soil conservation measures in the watershed enhance ground water recharge and ultimately increase crop intensity and crop productivity. Further, the integrated approach applied has a definite impact on the overall socio-economic conditions of the farmers of the watershed area.

Table 11 : Progress indicators of Hingonia-Pipliyatapha watershed.

Particulars	Base 1987-88	Year 1995-96
Net cropped area (ha)	311.0	318.5
Kharif cropped area (ha)	253.0	305.0
Rabi cropped area (ha)	166.5	303.0
Double cropped area (ha)	108.5	290.5
Total cropped area (ha)	419.5	603.0
Net irrigated area (ha)	27.0	168.0
Cropping intensity (ha)	133.0	191.0
Percentage to net cultivated area (*kharif*)	81.3 %	98.0 %
Percentage to net cultivated area (*rabi*)	53.5 %	97.4 %

Fertilizer consumption per hectare cropped area

Particulars	Base 1987-88	Year 1995-96
Nitrogen	15.760	48.900
Phosphate	18.900	52.600
Potash	0.001	5.200

Area under crops

Particulars	Base 1987-88	Year 1995-96
Sorghum (ha)	48.3	3.5
Maize (ha)	4.5	1.0
Soybean (ha)	132.0	232.0
Pigeonpea (ha)	4.5	9.5
Kharif vegetable (ha)	7.0	9.0
Kharif fodder (ha)	24.5	38.0
Wheat irrigated (ha)	8.0	137.0
Wheat partially irrigated (ha)	30.0	78.0
Wheat rainfed or sequential (ha)	18.0	10.0
Chickpea rainfed or sequential (ha)	56.0	21.0
Chickpea irrigated (ha)	21.0	32.0
Linseed (ha)	11.0	5.0
Rabi fodders (ha)	1.5	5.0
Rabi vegetable (ha)	12.5	12.0
Floriculture (ha)	1.0	3.0

Average yield (kg/ha) of crops

Particulars	Base 1987-88	Year 1995-96
Sorghum	1220	3400
Maize	390	2200
Soybean	320	1750
Wheat rainfed	280	730
Chickpea rainfed	320	1120
Wheat partially irrigated	960	2400
Wheat irrigated	2200	4180
Chickpea irrigated	–	1850
Linseed	–	860

Water resources development

Particulars	Base 1987-88	Year 1995-96
Total number of open dug wells	19	29
Shallow/Deep wells (water available up to October-November)	3	27
Deep wells (water available up to February)	6	6
Deep wells (live year round)	3	3

Total tube wells	3	70
Live up to January	3	18
Live up to April	–	8
Live year round	1	4
Totally dry	–	40
Sprinkler sets	–	2

Soil and water conservation

Particulars	Base 1987-88	Year 1995-96
Water harvesting tank		
a) Storage capacity 10000 cu. m.	–	1
b) Storage capacity 17000 cu. m.	–	1
Gabion	–	5
Graded bund	–	10 (155 m length)
Grassed waterway	–	5 (1800 m length)
Loose boulder structures	–	9
Spurs	–	2
Diversion drain	–	4 (1375 m length)
Agro-forestry	–	3.5 ha

Agricultural implements

Particulars	Base 1987-88	Year 1995-96
Improved plough, blade-harrow, Mahakal Dufan etc.	12	56
Sprayer and duster	9	27
Power thresher	2	19
Tractor	–	6
Bullock cart	14	32

Milch cattle

Particulars	Base 1987-88	Year 1995-96
Buffaloes	88	240
Cows	163	103
Crossbred cows	2	26
She-goat	35	108
Milk production - Mean lit/day of buffalo	4.8	6.2
Milk production - Mean lit/day of cow	2.5	3.3
Milk production - Mean lit/day of crossed cow	8.0	10.8
Gobar gas plant	–	8.0

Further, these experiences were shared with the farmers of other villages viz. Baroli (1998- 2004), Jaitpura (2004- 2006) and Panod (2007 onward) located in Malwa region under ORP. Many significant achievements were made on developing agronomical technologies for efficient management of dryland crops. The results of these integrated approaches and activities are found very promising and have been accepted and appreciated by one and all as it has led to improving the socioeconomic condition of the farmers of the area.

Probable Ways for Further up Scaling

Though adoption of soil and water conservation technology can principally be recommended for all the potential sites, but insisting on adopting same design on each site i.e. generalization of design, size and even budget should not be made. The details regarding design of soil and water conservation measures be made separately for each site.

It is also found that each watershed is unique in nature but also not comparable with each other. Therefore detailed watershed planning is to be determined separately location wise because of high variability of natural resources.

Extensive training and capacity building in soil and water conservation techniques, farming practices based on resources available may induce farmers to adopt on their own but through technical guidance from resource personnel.

Government must execute soil and water conservation works in common property lands/government lands wherever it is required considering the feasibility, suitability and practicability.

Approach

From our experiences, it is suggested that the following important components for watershed management should also be considered which may become the backbone for the success of the programme:

Farmer participatory approach

- Use of new science tools for management and monitoring of watersheds.
- Link on-station and on-farm watersheds.

- Approach to improve livelihoods of people through integration of farming practices and conservation of soil and water.
- A consortium of institutions for technical backstopping of the on-farm watersheds.
- A micro-watershed within the watershed where farmers conduct strategic research with technical guidance from the scientists.
- Low-cost soil and water conservation measures and structures.
- Bigger and appropriate soil and water conservation measures including water-harvesting tanks in government as well as farmers lands without restrictions on the cost.
- Amalgamation of traditional knowledge and new knowledge for efficient management of natural resources.
- Emphasis on individual farmer-based conservation measures for increasing productivity of individual farms along with community-based soil and water conservation measures.
- Continuous monitoring and evaluation by the stakeholders.

❏❏❏

Chapter 40

Efficient Moisture Conservation and Fertilizer Practices for Profitable Productivity of Pearl Millet Under Semi-Arid Inceptisols

S.P. SINGH[1], G.R. MARUTHI SANKAR[2], P.K. MISHRA[3] AND P.K. SINGH[4]

[1,4] All India Coordinated Research Project for Dryland Agriculture (AICRPDA), R.B.S.College, Bichpuri, Agra, Uttar Pradesh. [2,3] AICRPDA, Central Research Institute for Dryland Agriculture, ICAR, Santoshnagar, Hyderabad-500059

Two field experiments were conducted on pearl millet during *kharif* 2007 and 2008 on experimental site of All India Coordinated Project on Dryland Agriculture (AICRPDA) at RBS College, Bichpuri with objectives to assess the superiority of five soil moisture conservation practices viz., (i) flat bed + broadcast sowing; (ii) flat bed + broadcast sowing + gurr; (iii) line sowing; (iv) line sowing + gurr; (v) ridge & furrow sowing; and five fertilization practices viz., (i) control; (ii) 100per cent recommended dose of fertilizer (RDF) at sowing; (iii) fertilizer N of RDF in 2 splits (½ + ½); (iv) fertilizer N of RDF in 3 splits (½ + ¼ + ¼); (v) fertilizer N of RDF in 3 splits (1/3 + 1/3 + 1/3) in a split plot design with four replications keeping moisture conservation practices in main plots and fertilizer practices in sub plots. From the rainfall point of view, the two seasons were significantly different (219.3 mm in 2007 and 646.8 mm in 2008) during crop growth. A mean pearl millet grain yield in the range of 1546 to 2288 kg/ha during 2007 and

2611 to 2966 kg/ha during 2008 were recorded under flat bed + broadcast sowing and ridge & furrow sowing treatments, respectively. Similarly, a mean grain yield in the range of 1251 to 2236 kg/ha during 2007 and 1937 to 3552 kg/ha during 2008 were attained under control and application of N of RDF in 3 splits (1/3 + 1/3 + 1/3) treatments, respectively. The moisture conservation treatments had a coefficient of variation of 15.0 and 12.9per cent, while the fertilizer treatments had a coefficient of variation of 13.8 and 10.4per cent during 2007 and 2008 respectively. The moisture conservation and fertilizer treatments were significantly different in both 2007 and 2008 seasons at $p < 0.05$ level. The treatments gave a significantly higher yield and monetary returns in 2008 compared to 2007 in the study (Fig.1). The ridge & furrow system gave a maximum mean grain yield increase of 25.1 per cent compared to flat bed + broadcasting method. Based on the least significant difference criteria, ridge & furrow system gave a significantly higher grain yield compared to all treatments in both the years. The ridge and furrow system gave a significantly higher net returns of Rs. 13761/ha in 2007 and Rs. 17487/ha in 2008 with a benefit - cost ratio ratio of 2.73 and 2.36, respectively. Compared to this, the flat bed + broadcast method of sowing gave the lowest net returns of Rs.7573/ha in 2007 and Rs.15924/ha in 2008 with a benefit - cost ratio of 2.03 and 2.27, respectively (Table 1). Among fertilizer treatments, application of recommended N in 3 equal splits gave maximum net returns of Rs.11932/ha with benefit - cost ratio of 2.33 in 2007 and Rs.22171/ha with benefit - cost ratio of 2.74 in 2008, while control gave a significantly lower net returns of Rs.4966 and Rs.10129/ha with benefit - cost ratios of 1.68 and 1.97 in 2007 and 2008, respectively.

Thus, ridge and furrow system was the most efficient method followed by line sowing + gurr, line sowing, flat bed + broadcast sowing + gurr and flat bed + broadcast sowing method among different moisture conservation methods examined. Among different fertilizer treatments, application of recommended N in 3 equal splits was the most efficient treatment; followed by recommended N application in 3 splits viz., 50per cent as basal and 25 per cent each as top dressing at two stages; application of N in 2 splits *viz.*, 50per cent as basal and 50 per cent as top dressing; 100per cent recommended fertilizer at sowing; and control (Fig. 2). The yield increase due to splitting of fertilizer over basal application was about 24.8 per cent under 3 equal splits (1/3 + 1/3 + 1/3); followed by 19.4 per cent under 3 splits of

Table 1 : Analysis of variance of effects of moisture conservation and fertilizer treatments on yield and monetary returns of pearl millet under semi – arid inceptisols at Agra.

Treatments	2007					2008				
	Yield	CC	GR	NR	BC	Yield	CC	GR	NR	BC
Moisture conservation treatments										
FB+BCS	1546	7325	14898	7573	2.03	2611	12553	28477	15924	2.27
FB+BCS+Gurr	1720	7725	16547	8822	2.14	2734	13033	29165	16132	2.23
LS	1862	7325	17576	10251	2.40	2792	12553	29450	16897	2.34
LS+Gurr	2032	7725	19246	11521	2.49	2866	13033	30036	17003	2.30
R&F	2288	7925	21686	13761	2.73	2966	12853	30340	17487	2.36
Sem	63					80				
CD (5%)	196					247				
CV (%)	15					12.9				
Fertilizer treatments										
Control	1251	7325	12291	4966	1.68	1937	10417	20546	10129	1.97
100% RDF at sowing	1792	8650	16893	8243	1.95	2522	12553	26700	14147	2.13
RDF N (1/2 + 1/2)	2076	8850	19614	10764	2.22	2927	12653	31068	18415	2.46
RDF N (1/2+1/4+1/4)	2140	9000	20221	11221	2.25	3230	12753	33922	21169	2.66
RDF N (1/3+1/3+1/3)	2236	9000	20992	11932	2.33	3552	12753	34924	22171	2.74
Sem	58					65				
CD (5%)	165					183				
CV (%)	13.8					10.4				
CRF	219.3					646.8				
ARF	289.9					756.8				

CC : Cost of cultivation (Rs/ha)
GR : Gross monetary returns (Rs/ha)
NR : Net monetary returns (Rs/ha)
BC : Benefit – cost ratio
FB : Flat bed
BCS : Broadcast sowing
LS : Line sowing
R & F : Ridge and furrow
RDF : Recommended dose of fertilizer
Sem : Standard error of mean
CD : Critical difference
CV : Coefficient of variation
CRF : crop seasonal rainfall (mm)
ARF : Annual rainfall (mm)

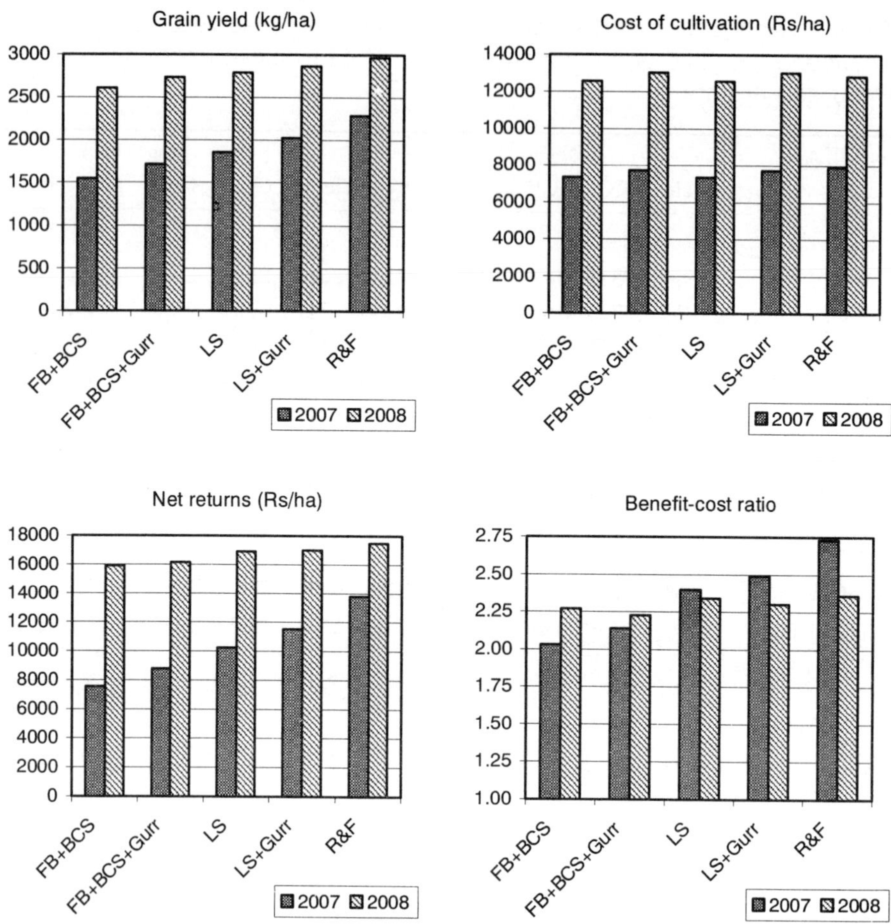

Fig. 1 : Effect of moisture conservation on yield and monetary returns of pearlmillet at Agra.

50 per cent as basal and 25per cent each as top dressing at two stages (½ + ¼ + ¼); and 15.8per cent under 2 splits of 50per cent each as basal and top dressing (½ + ½) in the below normal rainfall of 219.3 mm from 11 rainy days during 2007. Under above normal rainfall of 646.8 mm from 51 rainy days during 2008, the yield increase with splitting of fertilizer over basal was about 32.9, 28.1 and 16.1per cent under the 3 splits of (1/3 + 1/3 + 1/3), (½ + ¼ + ¼) and (½ + ½), respectively (Table 1). The response ratio due to N fertilizer application (kg grain/ kg fertilizer) was maximum of 13.1 under 3 equal splits (1/ 3 + 1/3 + 1/3), followed by 11.8 under 3 unequal splits (½ + ¼ + ¼), 11.0 under 2 equal splits (½ + ½) and 7.2 kg/kg under 100per cent

RDF as basal application at sowing under below normal rainfall situation in 2007. The response ratios were 18.9, followed by 17.2, 13.2 and 7.8 kg/kg under 3 equal splits (1/3 + 1/3 + 1/3), 3 unequal splits (½ + ¼ + ¼), 2 equal splits (½ + ½) and 100per cent RDF as basal application, respectively under above normal rainfall situation in 2008.

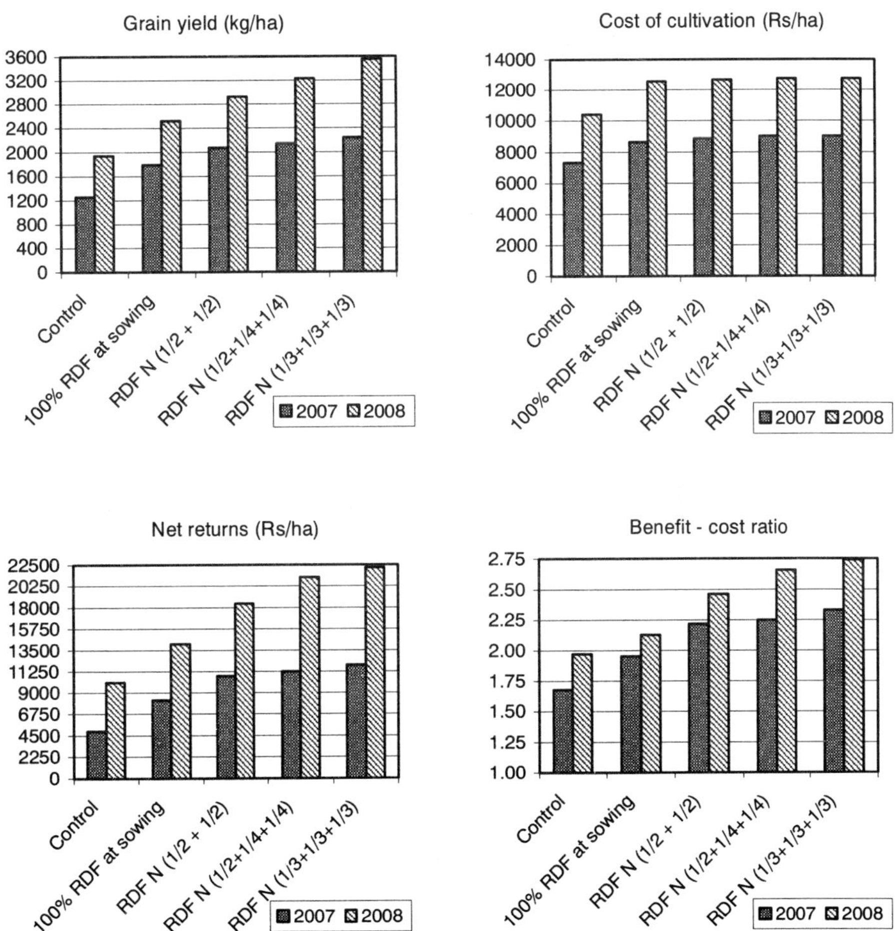

Fig. 2 : Effect of fertilizer on yield and monetary returns of pearl millet at Agra.

□□□

Chapter 41

Drought Management through Water Harvesting Systems and Contingency Crop Planning

J. SOMASUNDARAM, K. RAMESH, BRIJ LAL LAKARIA R.S.CHAUDHARY AND A. SUBBA RAO
Indian Institute of Soil Science, Berasia road, Nabibagh
Bhopal - 462 038 (M.P)

The importance of conserving soil and water resources has been recognized ever since the dawn of settled agriculture about 10 to 13 millennia ago, when agriculturists become less mobile than hunters and gatherers. Soil management was important to the agrarian culture of the Harappan Civilization of south Asia ~3200 BC (Kenoyer, 2003; Lal, 2008). Subsequent to the fall of the Harappan Civilization, the Vedic literature in Sanskrit dating back to 1000-2500 BC, contains several references to methods of tilling soil, collecting water, predicting rainfall, and urging farmers to observe good land husbandry (Wasson 2006). The importance of soil and water conservation practices or measures have been emphasized in many literature. In order to sustain the soil productivity and to meet growing food demand, the best management practices (BMP) with reference to soil and water conservation practices in the region is the need of the hour. However, adoption of best management practices depends on many factors like farmers' attitude, awareness, farm characteristics, etc.

India has about 47 million hectares of dry lands out of 108 million hectares of total rain fed area. 'Dry land' known as lifeline of agriculture, feeds half of our country's population and also supporting 60% livestock population and contributes more than 45% of the total food grain production of the country (Sharda and Ojasvi, 2005) due to inadequate and highly erratic rainfall while 37% of the irrigated area accounts for 55% of the total food production. As the scope of expanding irrigation potential is limited, the effective soil and water conservation techniques coupled with management practices are essential to increase the present production level 241 MT to 300 MT by 2020. Unprecedented global warming with a drastic change in rainfall pattern with erratic and high intensity trends; prolonged dry spells are characteristic features of the current monsoon trend. To have at least one good harvest either in *kharif* or *rabi*, farmers' are facing difficulties under changing climatic conditions and soil moisture stress during critical crop growth stages in the region. The gap between water supply and demand necessitates harnessing available water resources with efficient water conservation and management techniques.

The per capita availability of water in India is reducing exponentially owing to increasing population pressure and has declined from 5300m^3 in 1955 to 2200 m^3 in early 1990's against the world's average of 7400 m^3 and Asian average of 3240m^3. It is projected to be 1463 m^3 and 1235 m^3 by the years 2025 and 2050, respectively (Navalawala, 2000; Sharda and Ojasvi, 2005). India receives about 400 x10^6 ha-m of rainfall annually, most of which is received in 100 hours over a span of 25 nonconsecutive rain days. Thus, 45% (180 x 10^6 ha-m) is lost as runoff or blue water. Some of the water stored in soil as green water is lost by soil evaporation, and the productive green water used as direct transpiration is rather small. Impendence to deep root penetration due high bulk density, low porosity and hard-setting are among important factors responsible for low fraction productive green water (Lal, 2008).

Madhya Pradesh (MP) state is known for its numerous river basins and it spreads over about 30.7 m ha, out of which nearly 50 per cent area is culturable. In spite of numerous rivers and river basin in the state only 20% of the agricultural area is under some form of irrigation.

Farming in the state is mainly depends upon the south-east monsoon rain. The mean annual rainfall shows wide variation, it varied from 700 mm in Jhabua, Ratlam, Dhar and Khadwa in the west to 1300 mm in Shahdol, Balaghat and Mandla in the east necessitating varied cropping patterns across the state. Madhya Pradesh state has a long history of traditional water harvesting systems (WHS). The *Pani Roko Abhiyan* (Save Water Rainwater Conservation Programme) with an aim to harvest and conserve rainwater during 1999-2000, on-farm water conservation activities in a *Do It Yourself* (DIY) mode during 2001 and Jalabhisek programme during 2003-04 are the glaring examples of the state initiatives on creating awareness among the people and conserving water for mankind.

Principles of Rainwater Harvesting

The principle of rainwater harvesting is to concentrate, collect and conserve rainwater where it falls and storage, recycle properly and use efficiently according to needs and geological condition. In addition to that it also supplies drinking water for people, livestock and wild life. WHS not only provide supplemental irrigation for growing food crops but also increase the productivity of arable lands by enhancing crop yields and by reducing the risk of crop failure especially in the semi-arid region. The term water harvesting was probably used first by Geddes (1963) who defined it as "the collection and storage of any form of water, either runoff or creek flow for irrigation use" (Samra *et al.,* 2001). The importance of water resource development was recognized by many ancient civilizations which emphasized on various mechanisms of water harvesting and its utilization. The traditional WHSs remain sustainable until the beginning of large scale abuses on them leading to damage beyond repair. The consequences are frequent water scarcities, drought, poverty, migration, famines, etc.

Water harvesting though an age-old practice (Table 1) is emerging as a new paradigm in water resource development and management due to the recent efforts of both government and non-governmental organizations to promote water harvesting and groundwater recharge in urban and sub-urban areas. The outstanding efforts and success stories of Sukhomajri, Bunga, Relgaon Siddhi, Tarun Bharat Sangh and other similar projects are telling in loud and clear terms the importance of water harvesting and watershed management. These works have demonstrated that the benefits of resource conservation and management have increased biomass production and resulted in flood and drought control, groundwater augmentation, employment generation and improvement in socioeconomic conditions of the local people. Water harvesting can no longer be labeled a technology of the past and fit only for the poor, but it is as relevant for urban areas as it is for rural areas. It is a universal technology, which is relevant to all situations with location specific applications.

Table 1 : Some glaring examples of ancient water harvesting systems

Period	Activity
3rd millennium B.C.	Dams built of stone rubble were found in Baluchistan and Kutch
3000 – 1500 B.C.	Indus - Sarasvati Civilization had several reservoirs to collect rainwater runoff. Each house had an individual well
321 – 291 B.C.	Archeological evidence for dams, lakes and irrigation systems in the time of Chandragupta Maurya's rule
3rd Century B.C.	Kautilya's Arthasastra mentions irrigation using water harvesting systems
1st Century B.C.	Sringaverapura near Allahabad had a sophisticated water harvesting system using the floodwaters of the Ganges
2nd Century A.D.	Grand Anicut or Kallanai built by Karikala Chola across the river Cauvery to divert water for irrigation is still functional
11th Century A.D.	King Bhoja of Bhopal built the largest artificial lake (65,000 acres) in India fed by streams and springs
12th Century A.D.	Rajatarangini by Kalhana describes a well- maintained irrigation system in Kashmir

I. *In-situ* Rainwater Harvesting

In-situ water conservation is a more potential, feasible and practical proposition under most situations. *In-situ* rainwater harvesting can be achieved by increasing infiltration rate by agronomic measures (summer ploughing, contour cultivation, mulching, strip, mixed and inter cropping, ridging and furrowing) and engineering measures (contour bunding, graded bunding, land leveling, conservation bench terracing (CBT), contour trenching, conservation ditches and broad bed and furrow system and sunken and raised beds) for very gentle to gentle sloping (3-6% slope) fields and also for undulated fields. Besides harvesting moisture from between and within the crop rows, run off losses are considerably reduced. The store water is used for providing life saving irrigation to the crops grown in the surrounding areas. Agronomic measures are preferred as they involve least disturbance of the land and are farmer's friendly. However, in situations where agronomic measures are inefficient due to topographic or climatic constraints, mechanical measures are recommended to improve soil moisture regime and conserve runoff water. In addition to this, check dam, tanks and other reservoirs are also constructed to store runoff water in the lower catchment area.

Black soils in central India are characterized by low infiltration rate; they often generate fairly large volume of runoff during monsoon season. The low intake of rate of black soils make them ideal for storage of runoff and construction of ponds. The major parts of black soil receive medium to high rainfall (750-1250mm). For designing farm ponds in semi-arid black soil regions, it has been estimated that for 1 ha catchment, the capacity of farm pond may be taken as 2.5-3.3 ha-m based on runoff as 50% probability. Apart from dugout ponds, other WHSs in this region include check dams, gully plugs/ *nala bunding* and inter-terrace conservation measures (Sharda and Ojasvi, 2005).

II. *Ex-situ* Rainwater Harvesting

The *ex-situ* rainwater harvesting technologies are highly location specific and practices evolved in a given region have limited applicability in the other regions. This can be achieved by earthen embankment, farm pond, baories, submergence bunds, drainage line treatments (spillways /check dams/ Anicut / nala bunding), etc.

Traditional water harvesting structures

The traditional WHSs in the state play an important role in conserving rainwater with the Concept of *'Catch Water Where It Falls.* It is the process of collection and storage of the precipitation runoff for a variety of purposes in the semi-arid region where water shortage are common because of scanty and erratic distribution of rainfall. In addition to that it also supplies water to the life on the earth. WHSs not only provide supplemental irrigation for growing food crops but also increase the productivity of arable lands by enhancing crop yields and by reducing the risk of crop failure especially in the dry region.

But this ancient and dying wisdom needs to be revived to meet the modern freshwater needs adequately, equitably and sustainably and modernized with inputs from science and technology. Many of the traditional WHSs have either fallen to disuse due to variety of physical, social and economic factors and also due to the decline of Institutions that have nurtured them. However, some of the traditional WHSs like *Haveli, Kund, earthen checkdam* and *Johads* and prove promising among the systems in the present context.

Haveli System

A unique cultivation system based on water harvesting and runoff farming is still prevalent in the eastern parts of Madhya Pradesh. It was developed in fertile plains to northwest of Jabalpur city (Agarwal and Sunita, 2001). As early as 1902, in Hoshangabad district, several blocks of land were embanked by the government so as to demonstrate this system. In this system rainwater is held in embanked fields (Fig 1 to 3), enclosed on four sides until sowing time. Soil from field is taken to construct embankment or *bundhan* up to 1 m height in the field for conserving rainwater. Water is let out about one months after monsoon is over and as soon as land is dry the fields are sown, After this there is no requirement of water for the crops and its gets matured on the stored soil moisture (green water) in the profile.

Fig. 1 : A farm woman bringing her fields under *Haveli* system at Balaghat district

The system is suitable for two broader soil groups; *kabar* (heavy black clay soil) and *mund* (lighter soil, less fertile than kabar). This soil is unfit for *kharif* crops like paddy or cotton but it is excellent for wheat crop. It controls *kans* (*Saccharum munja*), a major weed in the region.

Fig 2 : Completed *Haveli* system in Balaghat district

Fig. 3 : View of *Haveli* system after the crop season at Balaghat district

In Jabalpur improved system of *Haveli,* i.e partial collection of *Haveli* water within or with in the vicinity of *Haveli* fields and recycling the harvested water for supplemental irrigation to *rabi* chickpea resulted in 21% increased yields over farmers' practice of growing chickpea (1.59 tonnes/ha) on conserved moisture. For *in-situ* moisture conservation on sloping lands, improved method of line of sowing across the slope and furrowing at 1.8 m interval recorded higher soybean yield (1.37 tonnes /ha) as compared to farmers' method of broadcast sowing without furrowing (9.3kg /ha).

Earthen checkdam

Balaghat region experiences erratic and low amount of rainfall during the last 2-3 years. Earthen checkdam is one measure to harvest rainwater in the water scarce regions (Fig 4). The post-evaluation study of water conservation project revealed that farmers used to face severe water scarcity during *kharif* season. It was very difficult to harvest even single crop in the region before construction of checkdam.

Fig. 4 : A View of Earthen Checkdam

However, after these interventions, farmers are getting atleast one assured crop per year apart from drinking water for people, livestock and other domestic purposes.

Survey of a few selected villages in Balaghat district revealed acute water scarcity during crop growing period. Farmers were unable to harvest 2-3 q/ac of rice. Construction of earthen dams resulted in 2-3 additional irrigations from the stored water, which helped to improve production of crops. During *kharif,* farmers now prefer to grow rice, red gram and maize. Yield of rice crop have increased from 3-4 q/ac to 8-10 q/ac due to water availability. During *rabi* crops like gram, wheat, *linseed* and *lakodi* are being taken up by farmers due to ensured water availability.

From the evaluation, it was inferred that water-harvesting structures serve as a life-line in the water scarce region. Therefore, harvesting rainwater during monsoon season through farm pond/ tank or checkdam will be very much useful for villagers during post-rainy season.

In three villages, a total of 3000 Nos. of livestock are getting drinking water facility from water stored in the earthen dam (Fig 5). These villages have more number of unemployed laborers. However, *Dimar* tribal people now trying fish farming in the stored water, which is another income generating activity for the local people.

Fig. 5 : High Livestock population in the villages

Pat System

Bhil tribal of Bhitada village, Jhabua district of MP have devised a system taking advantage of peculiarities of the terrain to divert water from swift-flowing hill streams into irrigation called '*pats*'. It is practical and ecological sound method of water harvesting/management. The stream is bunded to provide a static head of 30-60cm, sufficient to divert water in to irrigation channel. The gradient of the channel is less than the gradient of the streambed. Over a distance the channel dips less than the streambed and climbs to a height which varies from 3-25m. (Agarwal and Sunita, 2001).

The diversion bunds across the stream are made by piling up stones and then lining them with teak leaves and mud to make them leak-proof. The pat channel has to negotiate small *nullahs* that join the stream off and on and also sheer cliffs before reaching the fields. The villagers irrigate their fields by turns. The channel requires constant maintenance and it is the duty of the family irrigating the fields on a particular day to take care of the pat on that particular day. It takes about two weeks to get the pat flowing and the winter crop is sown in early November.

Tanks

Sanchi Hill in MP is well known for ancient rainfed WHS: There are three ancient tanks on Buddhist site of Sanchi, dating back to 3rd B.C. One tank is situated high on the hill while the other situated on the downhill (Fig 6). The slopes of the hill once formed serve as natural

Fig. 6 : Bird's eye view of tank located at the top of Sanchi hills

catchment area. Archaeological survey of Inda (ASI) has renovated the tanks of Sanchi to restore the its ancient WHS. Renovated tank provides 2.27 million litres of water.

Water harvesting/conservation practices

In Central India, Vertisols have a large potential of increasing agricultural production, provided appropriate technologies for conservation and management of natural resources, particularly soil and rainwater management practices were adopted and implemented.

Conservation tillage practices for soybean-wheat system

Long-term tillage management studies with different nitrogen levels on soybean-wheat cropping system showed that the yield of soybean and wheat under conservation tillage (i.e., no tillage with crop residues retained on the surface and direct drilling of seed, and reduced tillage i.e., residue retained + 1 sweep tillage) has been as effective as conventional tillage (i.e., residue removed + 1 summer tillage by sweep cultivator + 2 tillage by sweep cultivator) treatment (IISS-Two decades of Soil Research by Subba Rao *et al.*, 2009).

Broad Bed and Furrow (BBF) - An improved land surface treatment

A comparative study on broad-bed and furrow (BBF) (1 m wide broad bed and 0.5 m furrow on 0.4-0.6 % slope) and flat–on grade (FOG) land surface treatments showed that the crop yield (soybean, maize and pigeon pea) is more on BBF than FOG. The runoff and sediment losses from the BBF are lower than FOG. The BBF treatment retains more moisture than FOG during later part of the cropping season after withdrawal of monsoon. The BBF is identified as an improved land treatment for the deep Vertisols of Central India (IISS Two decades of Soil Research by Subba Rao *et al.*, 2009). Similarly, Mandal *et al* (2005) reported that a quantum of 250–325 mm rainfall at the square initiation to peak flowering stages was found to influence yield in cotton. Under climatic aberration, the high fluctuation of cotton productivity in deep Vertisol (Typic Haplusterts) may be minimized by adoption of any soil management technology (e.g. ridges or broad-bed furrow system) that helps in improving internal drainage.

Bunding

It consists of building earthen embankments across the slope of the land, following the contour as closely as possible. A series of such bunds divide the area into strips and act as barriers to the flow of water, thus reducing the amount and velocity of the runoff.

Contour bunds are recommended for mildy sloping lands (<6%) and porous soils receiving annual rainfall of less than 800mm. Graded bunds are suitable for high rainfall areas or soils with low infiltaion rate where water stagnation in the field is not desirable. (Sharda and Ojasvi, 2005). Bunding is very effective in reducing runoff to the tune of 50-60%. However, the contour bunds in deep black soils have been a failure due to formation of wider cracks during summer and swell during the monsoon. So they are not stable in black soils. Further the poor drainage properties of deep black soils gives raise to long stagnation of water against contour bunds and make the structure unstable.

Conservation ditching: It is an alternative to bunds has been recommended in black soils where bunds are not successful. Ditching with 0.6m depth, 0.6m base width and 4.5m top width have been found effective. Conservation ditching serves the dual purpose of a terrace and storage structure within the field and can retain more than 90% of the annual runoff, which can be utilized for supplemental irrigation.

Contingency Crop Planning (CCP)

Contingency Crop Planning assumes greater significance with respect to the recurrent aberrant weather patterns over different parts of the country. It refers to mitigate any unexpected, unusual, unfavourable and hence unwanted accidental weather situations occurring at any time without prior knowledge at any time before the crops are sown or even after the crops are sown. The contingency crop planning therefore is proposed to mitigate such situation through the choice of appropriate crop and varieties, cropping systems or other necessary relevant management practices

There is a broad consensus that India will have to depend on the rainfed areas to help increase supply, often quoted as second green revolution. The semi arid tracts in India cover 84 districts, spreading over 47 mha which is more than one-third net sown area which are vulnerable for the aberrant weather patterns.

Unfavourable deviated weather pattern with particular reference to monsoons from the normally expected one is referred to as aberrant weather and is a common feature of rainfed farming. The rainfall is seasonal, erratic and highly variable with space & time. Vagaries of monsoon rains and breaks of varying duration are most not uncommon in India. Aberrant weather conditions are characterized by

i) Early onset and early withdrawal of monsoon

ii) Early onset and late withdrawal of monsoon

iii) Normal onset but early withdrawal of monsoon

iv) Normal onset but prolonged dry spells during the southwest monsoon season during which most of the dryland crops are grown

v) Rains may terminate earlier than normal cessation date or may continue beyond the normal rainy season.

Under these conditions either sowing delayed or maturity coincides with heavy showers.

Three important aberrations in the rainfall behavior have been commonly observed. These are:

Based on the analysis of historical rainfall records the following main features of rainfall distribution in India is deduced.

- There is large variation in the dates of commencement of the southwest monsoon rainfall from year to year in different parts of the country.
- The monsoon rainfall is of sequential nature with dry spells interspersed with wet spells, dry spells extending from few days to several weeks.
- There is a large yearly variation in the dates of cessation of south west monsoon rainfall in different parts of country.
- There is yearly variation in the quantum of rainfall
- High intensity rainfall occurs in association with the movement of cyclones/depressions resulting in sizeable loss of rain water through run-off and deep drainage.

Weather is the most variable factor in rainfed crop production which can hardly be managed. However, contingency crop planning offer optimum crop yields for efficient weather/ water management.

To develop a contingency plan for an area, a detailed study of the rainfall data is a prerequisite. Based on this rainfall data, the planning can be done. The following table shows the contingency crop planning for selected AERs.

Late Onset of Monsoon

During some of the years the onset of south west monsoon gets delayed so that the crops/varieties which are regularly grown in the region cannot be sown in time. Delayed sowing of the crops can lead to reduced and even uneconomical crop yields. Under these circumstances, two management options are available.

1. Transplantation-This is the assured way of compensating the delay in the onset of rains, through transplantation. Although this is labour intensive, studies indicated that finger millet should be transplanted for maintaining yield levels whenever sowing is delayed.

2. Alternate Crops/Varieties: Depending on the receipt of rains, such crops and varieties can be selected. Under Madhya Pradesh conditions if the conserved moisture is too less select short duration/low water demanding crops such as mustard in place of gram.

Mid Season Correction

After sowing of crops if there is a long dry spell the following measures can be resorted to.

1. Alteration of plant density-The rate of soil moisture depletion increases with increasing leaf area. If drought occurs at 40- 50 DAS, reduction in leaf area by thinning plant population to half can mitigate the adverse effects of drought. This is commonly recommended for sorghum and pearl millet.

   If drought occurs during pre-flowering, every third plant or alternate row may be removed to reduce the soil moisture loss and preserve it. Moisture stress of 35-50 days can be lessened by such thinning

2. Residue mulching-If the break in monsoon is very brief, soil mulching was found to be tool in extending the period of storage of water in the soil profile due to reduced heat load on the soil and hence the evaporation losses which in turn leads to extended

periods of water availability. Deep cracks formed due to drought increases the loss of stored water by evaporation

3. Rain water harvesting and recycling-In situ water harvesting and / or runoff recycling are the potent measures for crop life saving during the periods of moisture stress. There is no control over the receipt of rainwater but it can be managed in a way better what is being done at present after it is received. Supplementary irrigation provided to the crop through rainwater harvesting and recycling particularly during the grain formation stage will be very much rewarding. Run-off recycling holds immense prospects in deep black soil where the seepage losses are very less. High intensity rains are quite common under Madhya Pradesh conditions during the rainy season. The inevitable runoff could be stored in dugout ponds of appropriate sizes and recycled to the donor area, in the event of severe moisture stress.

4. Intercropping and risk distribution: Intercropping is an insurance against crop falilure in dry land farming. The choice of intercrop depends upon the pattern of rainfall. There are 3 possibilities as: A few examples on contingency crop planning is listed below.

1. Contingency crop plan and nutrient recommendations for crops in rainfed regions (Table 2)

Table 2 : CCP for selected agro-ecological regions (rainfed) of India

S. No.	Agro-Eco Region	Crop plan 1-15 August	Crop plan 16-31 August	Crop management
1	AER 4.1 Agra, Mathura, Aligarh, Bulandsahar, Meerut, Etah, Mainpuri & Western parts of Muzaffarnagar	Transplanted pearl millet, cluster bean, green gram cowpea	Cluster bean Cowpea	--
2	AER 4.4 Jhansi, Banda, Hamirpur, Lalitpur, Morena	Bajra, Guar, Cowpea (grain, fodder), Pigeon pea, Black gram	Bajra, Gaur, Cowpea as fodder	Pulses : Soybean: 20: 60:20 kg NPK & 20 kg S through inorganic fertilizers/ha General Recommendation: Apply 30-40 kg S/ha (as Gypsum) or 60 kg S/ha (as iron

				pyrite) in order to exploit full yield potential of crop and maximize NPK use efficiency. Pearl millet: 1) 40 kg N - 30 kg P_2O_5/ha to pearl millet 2) Application of 20 kg N + Azotobacter culture in pearl millet.
3	AER 6.1 Solapur, Bidar, Osmanabad, Ahmednagar, parts of Satara, Latur and Sangli	Sunflower, pigeon pea, castor, sunflower+pigeon pea (2:1)	Sunflower, pigeon pea, castor sunflower + pigeon pea (2:1), sorghum for fodder	Sunflower ; 40-60: 30-90: 20-60 kg NPK/ha Castor : 20: 40 kg NP/ha and 75% RDF+25% N (FYM)+ Azospirillum seed treatment of PSB.
4	AER 5.2 Indore, Ratlam, Ujjain, Dewas, Dhar, Khandwa, parts of Sehore	Sunflower, Sesame, Cowpea, Castor, Sorghum, Maize and Bajra as fodder	Safflower, Sunflower, Sesame, Maize	Safflower: Application of 6 t FYM/ha + 20:30 kg/ha of N & P_2O_5, respectively for soybean-safflower system. *45-40-25 kg NPK/ha in soybean-pigeon pea system. Castor: 40-50 : 30-40 kg NP/ha in soybean-pigeon pea system. Sorghum: 70-90 : 30 kg N & P_2O_5/ha *9-10 t/ha subabul loppings can substitute 25 kg N/ha. *Apply $ZnSO_4$ @ 10-25 kg/ha to Zn deficient soils. Maize: 50:30 kg N & P_2O_5/ha.

2. Contingency crop planning (Solapur region) (http://kvksolapur.org/Inn-Technologies-1.htm)

Contingent Crop Planning

On-set of Monsoon	Crops to be taken
Second fortnight of June	All kharif crops and recommended intercrops.
First fortnight of July	Peralmillet, Sunflower, Groundnut, Pigeon pea, Setaria, Horse gram, Peralmillet + Horse gram (2:1), Peralmillet + Pigeon pea (2:1), Cluster bean + Pigeon pea (2:1), Cluster bean + Castor (2:1), Castor + Ridge gourd (mixed cropping).

Second fortnight of July	Sunflower, Pigeon pea, Horse gram, Setaria, Castor, Peralmillet (ergot resistant varieties), Peralmillet + Horse gram (2:1), Pearlmillet + Pigeonpea (2:1), Sunflower + Pigeon pea (2:1), Cluster bean + Pigeonpea (2:1), Castor + Ridge gourd (mixed cropping).
First fortnight of August	Sunflower, Pigeon pea, Castor, Horsegram, Sunflower + Pigeonpea (2:1), Castor + Ridge gourd (mixed cropping).
Second fortnight of August	Sunflower, Pigeonpea, Castor, Sunflower + Pigeonpea (2:1).
First fortnight of September	Rabi Sorghum for fodder.

3. Contingency crop planning (Rewa region) (http://www.crida.ernet.in/AICRPDA/ImpAgro.pdf)

a) Normal season

- Rice

  Very early group (less than 95 days): Prasanna, Vandana, JR 3-45

  Early group (95 to 115 days): IR-64, IR 50, Basmati

  Medium duration (125 to 145 days): IR-36, Jaya, Kranti
- Sorghum: CSH 5, CSH 6
- Maize: Ganga-1, Ganga 5
- Pigeonpea: NP (WR) 15
- Soybean: JS 335
- Sesame: JT 7
- Groundnut: Jawahar Jyothi-1
- Blackgram: JU 2, DU 4
- Greengram: K851, JM 45

Intercropping

- Sorghum + pigeonpea (2:1)
- Soybean + pigeonpea (2:1)

b) Late season

- Rice (late variety): IR 50, JR 3-45
- Kodo: JK 155

- Sorghum: CSH 5
- Pigeonpea: JA 4, Asha
- Groundnut: Exotic 1-1
- Blackgram: DU 4
- Sesame: JT-1
- Safflower: JSF 1
- Sunflower: Morden

c) Late season drought

- Harvested rainwater recycled as life saving irrigation

Low land – Direct seeded

- Re-sowing of rice is needed if plant population is less than 50%
- Weeds are uprooted by manual weeding practice
- Spray of insecticide make sure if attack of insect pest observed
- Spraying of micronutrient if deficiency is noticed

Transplanted rice

- If puddling and transplanting is not possible seedlings should not be uprooted. Weeds are removed to keep the nursery beds clean. Adequate plant protection measures are taken to protect the seedling from disease and pest attack. When rainfall occurs puddling is done by tractor drawn power tiller, 30-40 days old seedlings should be transplanted, 3 to 4 seedlings/ hills be planted. Adequate fertilizer should be applied as per requirements.

Conclusion

Water harvesting has the potential to improve rainfed crop yields significantly, apart from many environmental benefits such as reduced soil erosion and overall improvement in the natural resources. However, despite localized success, broader acceptance of water harvesting techniques has been limited mainly due to high cost of implementation and higher short-term risks involved for additional inputs, cash, and labours. Water harvesting initiatives frequently suffer from lack of hydrologic data, insufficient attention during planning stages to important social and economic considerations and the absence of long term government strategy for ensuring sustainability of interventions. Greater emphasis on the involvement of the farmers right from the planning stage and a provision for appropriate extension support could help in expanding the benefits of water harvesting to large areas.

The potential WHSs in these regions is not yet fully utilized and implemented. Some of the traditional WHSs like *Haveli, Kunds,* checkdams and *Johads* offer potential even under water scarce area. Several village success stories have demonstrated that water harvesting based development paradigms were able to mitigate drought positively resulted in household economy.

With increased access to ground water reserves through intensive installation of bore wells and to surface water through canal networking these traditional methods have been grossly neglected and ecological balance has been dangerously disturbed. Communities are more vulnerable to frequently occurring droughts in the region than ever before. Traditional small WHSs and *in-situ* moisture conservation techniques have very important role to play in empowering local communities, promoting integrated village ecosystems management and creating economic wealth in rural villages. It is strongly suggested that the rain WHSs are technically, economically, environmentally and socially viable. Appropriate technology in terms of catchment area, volume of storage and capital investment for the region need to be refined/modified, incorporating indigenous experiences/technologies to suit the regional needs and requirements.

However, There is urgent need and indications to renovate the existing water harvesting/retaining structures and further adopted on a large scale would helpful in minimize the risk of water scarcity even under severe drought years. It is very important on the part of government agencies, panchayats, NGO to renovated these systems regularly and keep them workable as a vital source of subsistence.

References

Agarwal, A. and Sunita, N. (2001). *Dying Wisdom*: Rise, fall and potential of India's traditional water-harvesting systems, Centre for Science and Environment, pp:166-175.

Geddes,H.J.(1963).*Water harvesting.Proc.Nat.Symp. Water Resources.* Australian Acadamy of Science, Canberra.

Gurjar, D.S., Yadav, B.R., Nathan, K.K. (2005). Drought mitigation through traditional water harvesting systems in Rajasthan, *Intensive Agriculture,* March-April: 16-18.

http://www.crida.ernet.in/AICRPDA/ImpAgro.pdf

http://kvksolapur.org/Inn-Technologies-1.htm

Kenoyer, J.M. (2003). Uncovering the keys to the lost Indus cities. Scientific American, **289**: 58-67.

Mandal, D.K., Mandal, C, and Venugopalan, M.V. (2005). Suitability of Cotton Cultivation in Shrink-swell Soils in Central India. *Agricultural System,* **84** (1): 55-75

Navalawala, B.N. (2000). Ground water over exploitation. *Yojana.* 49-56.

Nema, R.K., M.K. Awasthi. and R.N. Srivatava. (2003). Haveli fields: An excellent rainwater harvesting system. *In*: Proc. of National Symposium on Rainwater management for water Scarcity held at JNKVV, Jabalpur 6-7th Jan 2003, pp: 360-363.

Pramanik, S.C. (2009). Rainwater management techniques for successful production of pulses in rainfed areas. *Indian Farming,* March: 15-18.

Lal, R. (2008). Managing soil water to improve rainfed agriculture in India. *Journal of Sustainable Agriculture,* **32**(1):51-75.

Samra, J.S., Sharda, V.N. and Sikka, A.K. (2002). *Water harvesting and recycling: Indian Experiences.* Central Soil & Water Conservation &training Institute, Dehradun, India.

Sethy, B. K, Shakir Ali and J. Somasundaram. (2008). Potential of rainwater harvesting for sustainable agriculture in south-eastern Rajasthan . *In: (Prasad et al.):* "Natural resources management for sustainable Development in Western India (NRMSD 2006), Allied Publishers Private Ltd , New Delhi: 57-64.

Subba Rao, A., Biswas, A.K., Sammi Reddy, K., Hati, K.M and Ramana, S. (eds). (2009). IISS Two Decases of Soil Research , Indian Institute of Soil Science, Bhopal, pp 1-132.

Sharda, V.N and Ojasvi, P.R. (2005). Water harvesting through watershed management in different agro-ecologiccal regions of India. *Indian Journal of Agricultural Sciences,* **75** (12):771-780.

Somasundaram, J. (2011). Potential rainwater harvesting systems for sustainable Production In: (eds Singh *et al.*, 2011) Efficient Utilization of Farm wastes for sustainable agriculture, Agrotech Publishing Academy, S.S.S..Prineters pp: 266-277.

Wasson, W.R. (2006). Exploitation and conservation of soil in the 3000-year agricultural and forest history of South Asia. In Soils and Societies: Perspectives from Environmental History (Eds. J.R. McNeill and V. Winiwarter) 13-50. Isle of Harris, UK, While Horse Press.

□□□

Chapter 42

Information Receiving Behaviour of Farmers of Watershed and Non-watershed Areas in Bundelkhand Region

OM PRAKASH AND V.S. KATIYAR
Central Soil and Water Conservation Research & Training Institute, Research Centre, Datia-475 661 (M.P.), India

In India, the degraded land is estimated to be about 178 million hectares. Thus, the total arable dry land/rain fed land is under constant degradation process, the degree of variation may be there from moderate level of erosion to severe level of erosion. The reasons for soil erosion in dry lands is due to many, varied and complex factors, which include not only biophysical factors but also socio-economic factors. The socio-economic factors are totally neglected to improve degraded natural factors such as land, water and vegetation (Sastry *et. al.*, 2004). Hence, Government of India initiated watershed development programme during 1983-84 to improve natural resources so as to improve the economic conditions of farmers and this programme has been continuing till date with sufficient allotment of funds in the budget. Hence, a research project was undertaken to assess the impact of watershed on natural resources for the benefit of ultimate users.

Most of the people living in rural areas depend on agriculture. It has however, been recognized that only through a holistic development like the watershed Development Programme, economic conditions of rural people can be improved. Watershed development is a risk reduction management approach, which aims at protecting the inhabitants of the poorly endowed fragile ecosystems from acute distress caused by recurring droughts, land and environmental degradation etc. It aims at providing resources that are desired by and suitable to the community. Primary objective is integration of many scattered programmes on natural resource management systems like soil conservation, afforestation and minor irrigation with production systems. Properly formulated micro-watershed project based on a study of climate, soil, water and plant resources on one hand and the human, animal resources on the other hand, offers ample scope for evolving sustained livelihood support system in these hinterlands.

Methodology

The study was conducted in ORP watershed Sheetalpur at Hamirpur, Tejpura at Jhansi & Bajni at Datia districts of Bundelkhand region. Then, from each watershed 50 farmers of different categories viz., land less, marginal, small, medium and large were selected randomly for making a total sample size 150. The data was collected by personally interviewing the respondents with the help of pre-structured and modified interview schedule. The collected data was tabulated and analysed by applying suitable statistical tools for drawing the meaningful conclusions.

Results and Discussion

1. Communication sources utilized by farmers

The major sources of communication utilized by the farmers in watershed areas were agricultural scientist (76per cent) followed by soil conservation officer (68per cent), agricultural officers Bank (42per cent), animal husbandry officers (39per cent), progressive farmers (32per cent) and sarpanch (29per cent), whereas in non-watershed areas major sources of communication were sarpanch (53per cent), progressive farmers (46per cent), soil conservation officer (19per cent), animal husbandry officer (17per cent) and agricultural officer-Bank (16per cent) (Table 1).

Table 1 : Per cent score on communication sources utilized by farmers of watershed and non-watershed areas.

S. No.	Source	Percentage of farmers*			
		Watershed		Non-Watershed	
		%	Rank	%	Rank
1.	Neighbours	26	VII	42	III
2.	Progressive farmers	32	V	46	II
3.	Surpanch	29	VI	53	I
4.	Soil Conservation officer	68	II	19	IV
5.	Agril. Officer (Bank)	42	III	16	VI
6.	Agril. Scientist	76	I	11	VII
7.	A.H. officer	39	IV	17	V
8.	Other development officer	17	VIII	09	VIII

* Indicates percentage worked out on the basis of total sample.

2. Communication channels utilized by farmers

The results indicated that the major channels of communication in watershed areas were training (82per cent), radio (68per cent), demonstration (62per cent), T.V. (59per cent), meeting (36per cent), panchayat (34per cent) and Gossip groups (19per cent) whereas in non-watershed areas major channels of communication were gossip groups (65per cent), panchayat (63per cent), radio (49per cent), meeting (32per cent), demonstrations (26per cent) and training (16per cent) (Table 2).

Table 2 : Per cent score on communication channels utilized by farmers of watershed and non-watershed areas.

S. No.	Channel	Percentage of farmers*			
		Watershed		Non-Watershed	
		%	Rank	%	Rank
1.	Panchayat	34	VI	63	II
2.	Meeting	36	V	32	IV
3.	Gossip group	19	VII	65	I
4.	Radio	68	II	49	III
5.	Training	82	I	16	VIII
6.	T.V.	59	IV	31	V
7.	Demonstration	62	III	26	VI
8.	Others	15	VIII	21	VII

* Indicates percentage worked out on the basis of total sample.

3. Impact of information receiving behaviour of farmers

The results showed that the mean scores (376.12) of communication behaviour of overall farmers of watersheds were worked out whereas in case of non-watershed areas the overall mean scores (276.39) were worked out in view of the recommended soil and water conservation practices. The mean scores of sources of communication of watershed farmers were 148.56, 133.72, 23.58, 37.40 and 32.86, respectively. However, in case of the farmers of non-watershed areas these mean scores were 119.35, 105.61, 14.68, 19.83 and 16.92 respectively. Further, the mean scores of information receiving behaviour (282.28) of farmers of watershed areas were higher than the corresponding mean scores of non-watershed areas (Table 3). This indicates that the farmers of watershed areas had better communication behaviour towards recommended soil and water conservation practices than non-watershed areas. Similar findings were also reported by Lal (1972) that 43.33 per cent of the farmers had medium, 36 per cent low and 20 per cent had high communication behaviour in respect of farm technology. The majority of the farmers, i.e. 79.33 per cent had low to medium communication behaviour with the various information sources both the receiving as well as passing of information concerning high yielding varieties. Earnest (1973) studied communication utilization behaviour variables, viz; number of sources utilized, extent of information received from sources, source credibility, frequency use of sources, number of channels utilized, extent of information received from the channels and characteristics of innovation showed significant difference in their means between small and big farmers. Akhouri (1973) revealed that there was

Table 3 : Impact of information receiving behaviour of farmers of watershed and non-watershed areas.

S. No.	Components of communication behaviour	Mean scores	
		Watershed (N = 75)	Non-Watershed (N=75)
1.	Sources of communication (SC)	148.56	119.35
2.	Channels of communication (CC)	133.72	105.61
3.	Information receiving behaviour (IRB=SC+CC)	282.28	224.96

* Indicates percentage worked out on the basis of total sample.

significant relationship between these behaviours (information input, processing and output). Similar results were obtained by Shete (1974).

Conclusion

The major sources of communication utilized by the farmers in watershed areas were agricultural scientist (76per cent) followed by soil conservation officer (68per cent), agricultural officers Bank (42per cent) animal husbandry officers (39per cent), progressive farmers (32per cent) and surpanch (29per cent), whereas in non-watershed areas major sources of communication were surpanch (53per cent), progressive farmers (46per cent), soil conservation officer (19per cent), animal husbandry officer (17per cent) and agricultural officer-Bank (16per cent) (Table 1). Similarly, major channels of communication in watershed areas were training (82per cent), radio (68per cent), demonstration (62per cent), T.V. (59per cent), meeting (36per cent), panchayat (34per cent) and Gossip groups (19per cent) whereas in non-watershed areas major channels of communication were gossip groups (65per cent), panchayat (63per cent), radio (49per cent), meeting (32per cent), demonstrations (26per cent) and training (16per cent). Further, mean scores of communication behaviour of overall farmers of watershed and non-watershed areas were worked out in view of the recommended soil and water conservation practices. The mean scores of information receiving behaviour, information use behaviour and communication behaviour of farmers of watershed areas were higher then the corresponding mean scores of non-watershed areas. This indicates that the farmers of watershed areas had better communication behaviour towards recommended soil and water conservation practices than non-watershed areas. Therefore, it is recommended that the technologies and facilities available to the farmers of watershed areas should also be provided to the farmers of non-watershed areas for sustainable production and productivity. The major constraints faced by the farmers in adoption of suitable soil and water conservation technology are high cost of technology, low benefit at initial stage of adoption, training & awareness, marketing, low land holdings and lack of inputs in watershed areas.

References

Akhouri, M.P.P. (1973). Communication behaviour of extension personnel: An analysis of Haryana Agril. Extn. System. *Ph.D. thesis*, IARI, New Delhi.

Ernest, R.S. (1973). A study of differential utilization behaviour of small and big farmers and the communication strategy. *Ph.D. thesis*, IARI, New Delhi.

Lal, D. (1972). Some correlates of communication behaviour of the Punjab farmers. *M.Sc. thesis*, P.A.U., Ludhiana.

Sastry, G., Reddy, Y.V.R. and Prakash, O.(2004). Soil and water conservation interventions in watershed development programmes - economic viability, acceptability and emerging issues, Indian *J. Soil Cons.*, **32** (2) : 81-90.

Shete, N. B. (1974). A study of communication behaviour of extension personnel of Maharashtra Agril. Extension system. *Ph.D. thesis*, IARI, New Delhi.

❑❑❑

Chapter 43

Effect of Fertilizer Doses, Methods of Nitrogen Application and Moisture Conservation Practices on Pearl Millet Under Rainfed Condition

S. KUMAR, H.P. CHAUDHARY, P.N. YADAV, U.D. AWASTHI AND H. KUMAR
Department of Soil Conservation & Water Management, C.S. Azad University of Agricultural &Technology, Kanpur

Those areas, where annual rainfall is <800 mm, is much more shortage of moisture availability to crop so in these areas, there is great need for *in-situ*, rain water conservation and having a vital scope to grow the coarse grain crops very profitably and easily. Bajra is a coarse grain crop and very suitable for these low rainfall receiving areas. Bajra is a crop grown mostly in tropical climate. It is widely grown in Africa and Asia. In Asia it is an important cereal crop of India, Pakistan, China, Nigeria, Sudan, Egypt, Russia and South eastern Asia.

India is the second largest producer of bajra in the world and production of bajra in India is 7.13 million tones and its grown in 9.41 million hectares with productivity of 7.85 q $ha^{-1.}$ Application of NPK effect the yield by 3-6 times in the upper horizons of eroded soils and by 6-10 times in lower horizons. The yield potentially of rainfed pearl millet has been very low due to very low fertilizer use low yielding cultivars and lack of good management practices. Soil moisture conservation is the main problem in arid and semi arid conditions. The moisture conservation practices have been well recognized means conservation of moisture through furrowing in sowing method under rainfed eroded condition.

Materials and Methods

The field experiment was conducted and laid out during the *kharif* season of 2007-2008 at the Soil Conservation and Water Management Farm of Chandra Shekhar Azad University of Agriculture and Technology, Kanpur. Which is situated in the sub-tropical and alluvial tract of Central Uttar Pradesh in lower doab of river Ganga and Yamuna. It lies between 25^0, 26^0 to 26^0, 28^0 north latitude and 79^0, 81^0 to 80^0 84^0 east longitude at an elevation of about 125.9 meters from mean sea level and mean annuals rainfall is 800mm a major portion of which is received during the monsoon season from the last week of June to first week of October. Sowing was don on 08.07 2007 by *desi* plough in furrows at 45 cm apart using 5 kg seeds ha^{-1} and fertilizer 80:40:40 and 40:20:20 kg Nitrogen phosphorus and potash ha^{-1} as per treatment. As per treatment weeding followed by ridge & furrow done with the help of hoe and pre-emergence application of atrazine @ 1 kg a.i. ha^{-1} was done with the help of Knapshak sprayer.

Result and Discussion

Grain and stover yield

Fertilizer doses significantly increased the grain & stover yield of maize. Maximum grain & stover yield (24.50 & 68.65 q/ha^{-1}) was recorded at the fertilizer dose 80:40:40: kg NPK ha^{-1} (F_2) and lowest yield (20.89 & 58.88 q/ha^{-1}) was recorded at fertilizer doses 40:20:20 kg NPK^1 (F_1) ha^{-1} respectively. This may be attributed to better growth and yield attributing characters due to application of high dose of fertilizer which may be helps proper supply of nutrients requirement for plant growth. Similar observations also reported by Singh and Agarwal (2004) and Sumeriya *et al.* (2005). Grain & stover yield (q/ha^{-1}) significantly influenced by method of nitrogen application. Maximum grain & stover yield (24.21 & 67.86 q/ha^{-}) was recorded under treatment N_3 (¼ basal and ¾ top dressed N) and lowest grain & stover yield (20-92 & 58.95 q ha^{-1}) was recorded under treatment N_1 (½ basal and ½ top dressed N) where as grain & stover yield (22.96 &

64.49 q/ha^{-1}) was recorded under treatment N_2 (1/3 basal and 2/3 top dressed nitrogen). These findings are agreement with those Sanjeev and Yadav (2001) and Chaudhary *et. al.,* (2002). The Grain & stover yield of bajra crop is significantly influenced by moisture conservation practices. Maximum grain & stover yield (24.59 & 69.02 q ha^{-1}) was recorded under ridge and furrow plot followed by atrazine used as pre-emergence. Similar results also reported by Yadav *et. al.,* (1995).

Table 1 : Effect of fertilizer doses, methods of nitrogen application and moisture conservation practices under rainfed conditions on grain and stover yield (q/ha^{-1}) of pearl millet.

Treatments	Yield (qha^{-1})	
	Grain	Stover
A- Fertilizer doses -		
(i) 40:20:20 (F_1)	20.89	58.88
(ii) 80 : 40 : 40 (F_2)	24.50	68.65
S.E. (diff.)	0.66	1.30
C.D. at 5per cent	1.37	2.70
B – Nitrogen application methods		
(i) ½ basal and ½ top dressed (N_1)	20.92	58.95
(ii) 1/3 basal and 2/3 top dressed (N_2)	22.96	64.49
(iii) ¼ basal and ¾ top dressed (N_3)	24.21	67.86
S.E. (diff.)	0.80	1.80
C.D. at 5per cent	1.67	3.31
C- Moisture conservation practices		
(i) Weedicide (M_1)	20.80	58.51
(ii) Ridge and furrow (M_2)	24.59	69.02
S.E. (diff.)	0.88	1.30
C.D. at 5per cent	1.37	2.70

Water use and water use efficiency

Different fertilizer doses affect the water use and water use efficiency. Highest water and water use efficiency (355 mm, 6.9 kg/ ha/mm) recorded under F_2 treatment (80:40:40 kg NPK ha^{-1}) and lowest (321 mm, 6.5 kg/ha/ mm) under F_1 treatment (40:20:20 kg NPK ha^{-1}). Different methods of nitrogen application affect the water use and water use efficiency in bajra crop, application of ¼ dose of nitrogen as basal and ¾ dose in two split as top dressed showed highest water use and water use efficiency (351 mm, 6.9 kg / ha/ mm) followed by treatment (N_1) ½ dose basal and ½ dose as top dressed and water use and water use efficiency 342 mm, 6.1 kg/ha mm and 348 mm, 6.6 kg/ha/mm respectively. Among the different moisture conservation practices, the maximum water and water use efficiency 340 cm, 7.2 kg/ha/ mm was recorded in ridge and furrow (m_2) treatment and the lowest (330 mm, 6.3 kg/ha/mm) weedicide (atrazine as pre-mergence). Higher dose of fertilizer (80:40:40 kg NPK ha^{-1}) was most efficient in water used followed by lower dose of fertilizer (40:20:20 kg NPK ha^{-1}). It is attributed to more development of root system. More developed root system extract more water and which results increasing grain yield and stover yield by increasing grain yield helps to increase water use efficiency. These results are in agreement with those Bhan *et. al.*, (1995), Jat & Gautam (2000) and Kumar *et. al.*, (2003).

Table 2 : Water use (mm) and water use efficiency kg. grain / ha / mm as influenced due to fertilizer doses methods, nitrogen application and moisture conservation practices.

Treatments	Grain yield (kg/ha)	Water use (mm)	Water use efficiency kg/ha/mm
F_1	2089	321	6.5
F_2	2450	355	6.9
N_1	2092	342	6.1
N_2	2296	348	6.6
N_3	2421	351	6.9
M_1	2080	330	6.3
M_2	2459	340	7.2

Conclusion

A field experiment was conducted during the *rainy season* of 2007-08 to study the effect of 2 fertilizer doses (80:40:40 kg NPK/ha and 40:20:20 kg NPK/ha), three methods of nitrogen application (½ basal + ½ top dressed, 1/3 basal + 2/3 top dressed and ¼ basal + ¾ top dressed) and two moisture conservation practices (atrazine @ 1 kg a.i./ha at pre-emergence and weeding followed by ridge and furrow) on growth, water use efficiency, root development, yield and gross income of rainfed pearl millet. Fertilizer dose 80:40:40 N, P, K, kg/ha produced significantly highest grain yield (24.50 q ha^{-1}), maximum figures of roots (number of primary roots 22.33 & weight of dry root

22.63 g/plot), water use (355 mm), water use efficiency (6.9 kg/ha/mm) and earned maximum net return of Rs. 6886 ha^{-1}. Application of a ¼ basal + ¾ top dressed attained maximum growth characters and gave significantly highest grain yield (24.21 q ha^{-1}), maximum values of roots (23.25 number of primary roots and 23.22 weight of dry root g/plot) water use (351 mm) water use efficiency (6.9 kg/ha/mm) and net return (Rs. 7143). Ridge and furrow treatment recorded higher growth characters and produced significantly highest grain yield (24.59 q ha^{-1}), water use (340 mm), water use efficiency (7.2 kg/ha/mm), and net return (Rs. 6977). Treatment combination $F_2N_3M_2$ has produced maximum grain yield (24.43 q ha^{-1}) and earned maximum net return of (Rs. 7002 ha^{-1}).

References

Bhan, S.; Uttam, S.K. and Shyam, R. (1995). Effect of conservation practices and nitrogen level on yield, water use and root development of rainfed sorghum 'Varsha'. *Indian Journal Soil Conservation* **23** (1) : 24-29.

Chaudhari, A.C.; Meena, N.L. and Jat, R.L. (2002). Effect of nitrogen and moisture conservation practices on growth and yield of rain fed pearl millet (*Pennisetum glauceum* L.) R. Br. *Annals of Agricultural Research* **23** (2) : 223-225

Jat, M.L.; and Gautam, R.C. (2000). Effect of summer ploughing and in situ-moisture conservation practices on growth and yield of rainfed pearl millet. *Indian Journal of Agronomy,* **45** (4) : 722-727.

Kumar, M.; Singh, H. Hooda, R.S. Kripal, A. and Singh, T. (2003). Grain yield, water use and water use efficiency of pearlmillet (*Pennisetum glauum*) hybrid, under variable nitrogen application. *Indian Journal Agronomy.,* **48** (1) : 53-55

Sanjiv, Kumar, and Yadav, M.P. (2001). Effect of tillage practices and levels of nitrogen on growth and yield of fingermillet (*Pemiseture glaucem (L.) Br.*) *Progressive Agriculture* **1** (1) : 42-46.

Singh, D.K.; and Agrawal, R.L. (2004). Nitrogen and phosphorus nutration of pearl millet (*Pennisetum glaucum*) grown in sole cropping system under rainfed condition. *Indian Journal of Agronomy* **49** (3) : 151-153.

Sumeriya, H.K.; Singh, P. and Mali, L. (2005). Studies on the effect of fertility levels on growth and productivity of (*Sorghum bicolor (L.) Moench*). *Crop Research,* **30** (1) : 6-9.

Yadav, J.P.; Singh, G.D. and Keshwa, G.L. (1995). Growth yield of pearlmillet as effected by different forms of nitrogen fertilizers and levels of zinc Haryana. *Indian Journal of Agronomy* **7** (1) : 91-93.

□□□

About the Editors

DR. S. K. DUBEY, presently Head of CSWCRTI, Research Centre, Chhalesar, Agra has a vast experience of more than 26 years of research, teaching, training and extension in the fields of Management of Salt Affected lands and Use of saline water in Agriculture, Soil and Water Conservation, Watershed Management and Utilization of urban and industrial effluents and Biosolids in Agriculture. Dr. Dubey worked on development of a new irrigation method "Pitcher Irrigation" which saves more than 70 percent water and allow high saline/ alkaline waters to grow salt sensitive crops economically. He is recipient of prestigious Netherland fellowship in the area of salinity management. In recognition of his outstanding contributions in the fields of Water Conservation and Salinity Management he has been bestowed with prestigious Dr Rajendra Prasad Puruskar of ICAR and Dr. K.G. Tejwani Award of IASWC. He has visited various countries of the world such as USA, The Netherlands, U.K., France, Belgium, Brussels, Vietnam etc. for the cause of Management of salinity and Environment related studies. As a head of centre he promoted research, training and consultancies in the field of Soil and water conservation in Yamuna Ravines. He has more than 145 publications to his credit which include research papers of journals of national and international repute, books, book chapters, bulletins, reports, excerpts in national and international conferences and also guided students for Ph.D. and M.Sc. in the field of Soil Science, Water Management and Environmental Sciences.

R.K. DUBEY, Scientist (Senior Scale), Agronomy is working in the Central Soil and Water Conservation Research and Training Institute since August, 1995. He possesses an excellent record. He is a post graduate from Chandra Shekhar Azad University of Agriculture and Technology (1990 batch). He joined the Indian Council of Agricultural Research in July, 1994. He has worked in a number of nationally and internationally funded and collaborative research and extension projects as a leader or associate. Shri Dubey has specialization in conservation agronomy, agro-forestry, soil and water conservation, watershed management, technology transfer and human resource development. He has published 24 research papers in reputed national and international journals apart from being author of bulletins, books, book chapters etc.

DR. A.K. SINGH is working as Sr. Scientist (SWCE) at Central Soil & Water Conservation Research & Training Institute, Research Centre, Chhalesar, Agra (U.P.). He has joined ICAR as a Scientist in October, 1989. He did M.E. in irrigation water management from Rajasthan Agricultural University, Bikaner (Rajasthan) in 1987. Later on he did his Ph.D. in Soil and Water Conservation Engineering from Maharana Pratap University of Agriculture and Technology, Udaipur (Rajasthan) in 2009. He has constructed multi objective tilting hydraulic flume for his Ph.D. for estimation of runoff, soil loss, percolation studies in red and black soils of South-India. Dr. A.K. Singh is having very good experience in the field of Watershed Management for red & black soils of south India as well as ravenous region of Central India. He was also associated DFID Funded Kawad (Karnataka) and APPLE (A.P.) projects as consultant. Dr. A.K. Singh has published nearly 55 research papers in national and International journals of repute. He is still working on Soil and Water Conservation related consultancy projects of U.P. State Govt.

DR. PRAMOD KUMAR PANDA is presently working as Senior Scientist (Agronomy) at Central Soil and Water Conservation Research and Training Institute, Research Centre, Agra. He did his graduation in Agriculture from Orissa University of Agriculture and Technology, Bhubaneswar, M.Sc. (Ag) and Ph.D. in Agronomy from Banaras Hindu University, Varanasi. He was a recipient of National Scholarships, Junior Research Fellowship and Senior Research Fellowship from ICAR and Governor's award. He has research experience on Dry land Agriculture, Hill Agriculture, Organic Farming, Watershed Management and Integrated Farming System. To his credit, Dr. Panda has more than 44 publications as research papers, book chapters, bulletins and popular articles.

DR. KALA, S. is presently serving as a scientist (Forestry) at CSWCRTI- Research Centre, Chhalesar, Agra (U.P). She pursued her M.Sc., and Ph.D., (Forestry) with specialization in Tree Improvement at Tamil Nadu Agricultural University in Coimbatore (Tamil Nadu) with brilliant academic record. She has also participated in Indo-U.S Science & Technology Forum sponsored training programme on Tools for conserving biodiversity in 2007. She joined Indian Council of Agricultural Research as a scientist in November, 2009. She has been actively provide her contribution in the research and development activities involved for restoration and resource conservation in degraded

Yamuna ravines. She has published more than 10 research papers in both national and international journals of repute, book chapters and bulletin exclusively, from her research findings and experiences in the field of Forestry Science.

DR V. N. SHARDA, presently Member, Agricultural Scientist Recruitment Board has a brilliant career and vast experience of more than 30 years of research, training and extension in the fields of Hydrology, Soil and Water Conservation and Watershed Management. Dr. Sharda critically evaluated the impact of planting Eucalyptus in natural grasslands, evolved the concept of conservation bench terrace system and studied hydrological behavior of mountains and ravenous watersheds. He is recipient of the prestigious Fulbright Senior Post-Doctoral Research Fellowship of United States Information Agency in the area of environmental studies. In recognition of his outstanding contributions in the fields of Watershed Management, Water Conservation and erosion modeling, he has been bestowed with several awards which include the prestigious Rafi Ahmed Kidwai Award, Dr. Rajendra Prasad Puruskar, Jawahar Lal Nehru Award for the best Ph.D. Thesis, Vasantrao Naik Award of Indian Council of Agricultural Research, Intellectual's Honour - 2003 "Uttaranchal Ratan", Bharat Singh Award of Ministry of Water Resources, Eminent Agricultural Engineer Award of IEI(I), Team Award of ICAR, a Citation by USDA-ARS, Fellowship and Recognition Award of National Academy of Agricultural Sciences, New Delhi and Fellowship of Institution of Engineers (India), Kolkata. He is a member of several national and state level committees and has more than 170 publications to his credit which include research papers in journals of national and international repute, books, book chapters, reports, etc.